To Don

With great affection
and with thanks for
all of your mentoring
over the years

Kelly Roz

November 16, 2016

The Tropical Oil Crop Revolution

The Tropical Oil Crop Revolution

Food, Feed, Fuel, and Forests

Derek Byerlee
Walter P. Falcon
Rosamond L. Naylor

OXFORD
UNIVERSITY PRESS

OXFORD
UNIVERSITY PRESS

Oxford University Press is a department of the University of Oxford. It furthers the University's objective of excellence in research, scholarship, and education by publishing worldwide. Oxford is a registered trade mark of Oxford University Press in the UK and certain other countries.

Published in the United States of America by Oxford University Press
198 Madison Avenue, New York, NY 10016, United States of America.

Library of Congress Cataloging-in-Publication Data
Names: Byerlee, Derek, author. | Falcon, Walter P., 1936– author. |
Naylor, Rosamond, author.
Title: The tropical oil crop revolution : food, feed, fuel, and forests /
Derek Byerlee, Walter P. Falcon, and Rosamond L. Naylor.
Description: New York, NY : Oxford University Press, [2017]
Identifiers: LCCN 2016017265 | ISBN 9780190222987 (hardcover) |
ISBN 9780190223007 (epub)
Subjects: LCSH: Oil industries—Tropics. | Oilseed plants—Economic
aspects—Tropics. | Tropical plants—Economic aspects.
Classification: LCC HD9490.A2 B94 2017 | DDC 338.1/73850913—dc23
LC record available at https://lccn.loc.gov/2016017265

1 3 5 7 9 8 6 4 2

Printed by Sheridan Books, Inc., United States of America

CONTENTS

PREFACE

History often has curious ways of repeating itself—even with respect to studies of food and agricultural commodities! When Stanford's Food Research Institute was founded at Stanford University in 1921, the first series of studies focused on wheat, then the major traded food grain that was also of special importance to the food security of the Western world. But soon thereafter, a second series was started on fats and oils. These commodities were growing in importance in world trade, especially vegetable oils for the rapidly expanding margarine industry, which was an emerging threat to producers of traditional fats—mainly butter. The series included books on the margarine industry, the shortening industry, whale oil, coconut oil, inedible animal fats, and the German "fat plan." In the preface of the first book, *Fats and Oils: A General View*, published in 1928, Alsberg and Taylor noted "the literature dealing with fats and oils is notably deficient in its economic and statistical aspects, largely because these materials are so diverse in origin and use, and yet to a high degree interchangeable" (p. v).

The importance of different fats and oils has changed greatly since then. What has not changed is the industry's remarkable complexity and the lack of in-depth analysis of global vegetable oil markets. For the past 25 years, oil crops have been, by far, the most dynamic crops in world agricultural growth, and their impact on land use and global trade has been profound, yet they have received little attention from agricultural economists. The vast literature on the social and environmental impacts associated with the expansion of tropical oil crops rarely relates those impacts to the broader market drivers. We embarked on this book to fill that gap—not just for agricultural economists, but for the wider community of development and environmental professionals.

Our first objective in this volume is to outline the major supply and demand drivers for the sector to gain a better understanding of why the industry began growing so rapidly around 1990, especially in the tropics, and to provide a basis for assessing future prospects. This is no easy task, because the industry is even more complex today than it was 90 years ago when our colleagues embarked on their studies. A major new biofuel sector based on vegetable oils has emerged, and the oil meals derived from oil crops are highly valued in another dynamic sector: intensive livestock. Our second objective in this book is to link our commodity analysis with the sector's sustainability record, especially social outcomes related to smallholder participation and job creation, and environmental outcomes related to tropical deforestation and conversion of savannah lands.

We find this in-depth review of the sector to be a fascinating story that is still unfolding as we write. We fill the gap in the literature in part, but recognize much more work is needed if analysts are to develop a better understanding of the critical economic, social, and environmental tradeoffs in this large and growing sector, and use that knowledge to design sensible policies for the future.

Derek Byerlee, Wally Falcon, and Roz Naylor
Center on Food Security and the Environment
Stanford University

ACKNOWLEDGMENTS

Many people have contributed their generous assistance in writing this book by providing valuable information, interviews, and reviews of draft chapters. We are extremely grateful to the following persons who are also exonerated from any remaining errors and omissions: Abdul Halim Ahmad, Nick Alexandratos, Robert Bailis, Elinor Benami, Joaquim Bento Filho, Bill Burke, Fabio Chaddad, Hereward Corley, Rob Cramb, Henry Daris, David Dawe, Cees De Haan, Chris Delgado, Dusan Drabik, Kathleen Flaherty, Rachel Garrett, Joanne Gaskell, Ken Giller, Jeremy Goldhart-Fiebert, Cheng Hai Teoh, Joseph Hanlon, John Hartmann, Paul Heytens, Mariangela Hungria, Sri Ison, Tim Johnson, Badrul Ikmal Muhamed Kamil, Valerie Kelly, Eric Lambin, Neus Escobar Lanzuela, Jim Leape, Kai Lee, Marshall Martin, Chandramohan Nair, Dimbab Ngidang, Haji Wahid Omar, Mauro Osaki, Suresh Pal, Mark Rosegrant, Paula Savanti, Jeff Sayer, Don Scott, Frances Seymour, Bhavani Shankar, Mohd Arif Simeh, Patrick Sujang, Peter Timmer, Jan van Driel, Peter White, David Wilcock, Anthony Yeow, and Liangzhi You.

When preparing this book we very much appreciated the research assistance provided by Matt Higgins, the endless work of checking citations and formatting chapters by Elissa Winters, the comprehensive and thoughtful editing provided by Kelly Cassady, and efficient financial management by Lori McVay.

ACRONYMS AND ABBREVIATIONS

ASEAN	Association of Southeast Asian Nations
CME	Chicago Mercantile Exchange
CPO	crude palm oil
DUAT	Direito de Uso e Aproveitamento dos Terras (usufructuary rights to land)
EMBRAPA	Empresa Brasileira de Pesquisa Agropecuária (Brazilian Agricultural Research Corporation)
EPA	Environmental Protection Agency
EU	European Union
FAO	Food and Agriculture Organization of the United Nations
FAS	Foreign Agricultural Service
FCE	feed conversion efficiency
FEDEPALMA	Federación Nacional de Cultivadores de Palma de Aceite (National Association of Oil Palm Producers)
FELDA	Federal Land Development Authority
FFA	free fatty acid
FFB	fresh fruit bunch
FGV	Felda Global Ventures
GDP	gross domestic product
GHG	greenhouse gases
GM	genetically modified
GMO	genetically modified organism
GOPDC	Ghana Oil Palm Development Company Ltd.
HCB	Huileries du Congo Belge
HVO	hydro-treated vegetable oil
IFPRI	International Food Policy Research Institute
IPO	initial public offering
ISPO	Indonesian Sustainable Palm Oil
KKPA	Koperasi Kredit Primer Anggota (Members' Primary Credit Cooperative)
Mha	million hectares
MPOA	Malaysian Palm Oil Association
MPOB	Malaysian Palm Oil Board
MPOC	Malaysian Palm Oil Council
Mt	million metric tons

NBPOL	New Britain Palm Oil Limited
NCRs	native customary rights
NES	nucleus estate schemes
NGO	nongovernmental organization
OECD	Organisation for Economic Co-operation and Development
P&C	principles and criteria
PNB	Permodalan Nasional Berhad (Sovereign Wealth Fund)
PSD	Production, Supply, and Distribution (data set)
R&D	research and development
RCR	resource/cost ratio
RED	Renewable Energy Directive
REDD+	Reducing Emissions from Deforestation and Forest Degradation
RFS	renewable fuel standard
Rs	rupees
RSPO	Roundtable on Sustainable Palm Oil
RTRS	roundtable on responsible soy
SALCRA	Sarawak Land Consolidation and Rehabilitation Authority
TFR	total fertility rate
US	United States
US$	US dollar
USAID	US Agency for International Development
USDA	US Department of Agriculture
WHO	World Health Organization
WTO	World Trade Organization

1

THE MANY DIMENSIONS OF THE TROPICAL OIL CROP REVOLUTION

OIL CROPS: THE WORLD'S MOST RECENT AGRICULTURAL REVOLUTION

Agricultural revolutions that sharply accelerate the growth of global food production and transform agricultural systems occur infrequently. During our professional lives, we have witnessed, participated actively in, and studied the green revolution that emerged during the 1960s and transformed rice and wheat farming in the developing world. Unparalleled in its breadth and depth of change, especially in Asia, the green revolution generated a spate of research, publishing, and debate that continues today.

Less conspicuously but no less controversially, another agricultural revolution has unfolded during the past two decades. From 1990 to 2010, world production of soybean grew by 220% and production of palm oil[1] by 300%, which is more than the increase seen in wheat production during the green revolution and much faster than the increase in rice production at the time (Figure 1.1, Table 1.1). Like the green revolution for cereal crops, this more recent revolution largely involves two crops—oil palm and soybeans—that expanded their shares in their respective crop subsector dramatically (in this case, oil crops). Another trait shared by the two revolutions is that they have played out mostly in the developing world, although not in Africa.

Despite their similarities, in crucial ways the revolution in oil crops stands in direct contrast to the green revolution, which embraced tens of millions of producers across many countries, especially where irrigation was available. The oil crop revolution has been highly concentrated in a few countries (two for each crop, in fact) and almost entirely in rain-fed areas. Unlike the green revolution, which was spurred on by rapid gains in yield, the force behind the oil crop revolution was expansion of crop area. This key difference originates with the market-driven nature of the revolution in oil crops compared with the technology-driven nature of the green revolution.

The green revolution was led by small-scale farmers whereas the oil crop revolution has been led by large-scale farmers and private agribusiness, including huge vertically integrated companies that are the world's largest farmers. Although the products of the green revolution served domestic food markets, the products of oil crops have been exported largely to global markets for multiple uses for food, feed, and biofuels.

[1] We use *oil, vegetable oil,* and *edible oil* interchangeably in this book, unless *oil* is designated specifically as a petroleum product.

Table 1.1 Contrasts between the green revolution and the revolution in tropical oil crops

Characteristic	*Green revolution*	*Oil crop revolution*
Two decades of most rapid change	1965–1985	1990–2010
Crops	Rice, wheat	Soybeans, oil palm
Production increase in developing countries in two decades	Rice: 84% Wheat: 162%	Soybeans: 222% Oil palm: 300%
Share of world production growth resulting from yield increases	Rice: 0.86 Wheat: 0.84	Soybeans: 0.26 Oil palm: 0.29
Number of producing countries with rapid expansion	Many (but not Africa)	Few (but not Africa)
Production conditions	Mostly irrigated	Mostly rain-fed
Major producers	Small-scale farmers	Large-scale farmers, companies, smallholders
Sources of technology and other support	Public sector	Agribusiness
Use of genetically modified varieties	No	Yes for soybeans
Major uses	Food	Food, feed, biofuels, industrial uses
Major markets for the large producers	Domestic	Export

Although the number of producers touched by the oil crop revolution may be small on a global scale, the products of oil crops reach a high share of the world's consumers in some way.

Like the green revolution before it, the revolution in tropical oil crops provokes controversy. Nearly all the big debates on agricultural and food systems surround tropical oil crops, including debates over the use of genetically modified organisms (GMOs), production of food versus biofuels, small-scale farming versus agribusiness, the risks of foreign "land grabs," monocropping versus diversified cropping systems, the role of agriculture in promoting healthy diets, and globalization and its environmental footprint. By far the loudest debate concerns the accusation that tropical forests in South America and Southeast Asia are destroyed to make way for oil crops.

Many specific dimensions of the oil crop revolution have been studied, but no one has developed a holistic synthesis of its origins and outcomes. Our aim in this book is to step back and review the sector as a whole, considering both the supply-side drivers and the demand-side drivers (Figure 1.2). We focus on the two most dynamic crops, oil palm and soybean, and their complex links in markets for vegetable oils. Much of the literature emphasizes the negative consequences of the oil crop revolution, especially the environmental costs of the massive changes in land use that accompanied the spread of oil palm and soybeans in the tropics. Along with these aspects of the oil crop revolution, we weigh the incomes and jobs the sector provides for millions

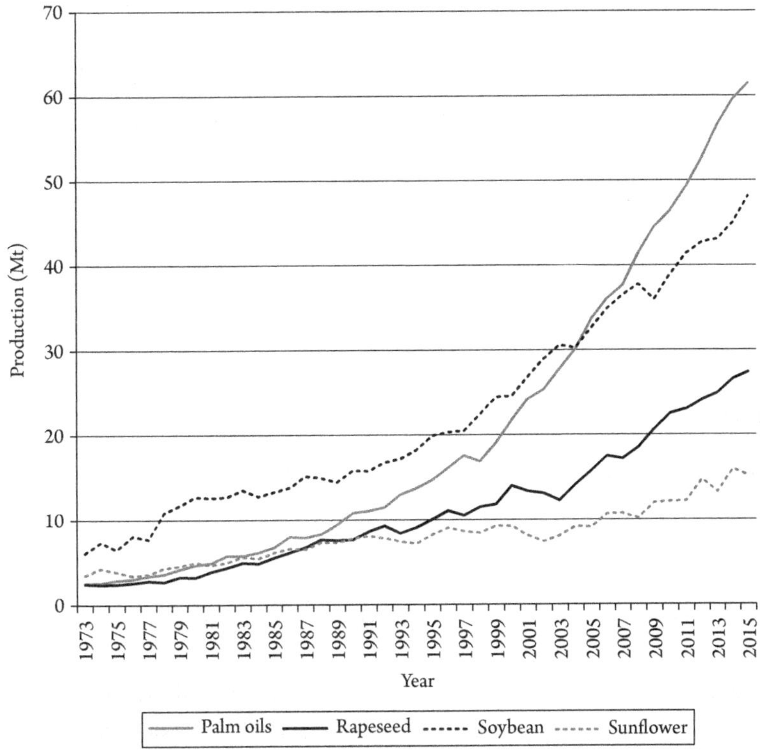

Figure 1.1 The takeoff in global vegetable oil production, led by soybean oil and palm oil (including palm kernel oil) is shown. Palm oils include both palm oil and palm kernel oil and are produced exclusively in the tropics. Nearly all rapeseed and sunflower oil is produced in temperate regions. Soybean oil is produced in both temperate and tropical regions, but since 1990 soybean production has shifted decisively toward the tropics.
Source: USDA-FAS PSD.

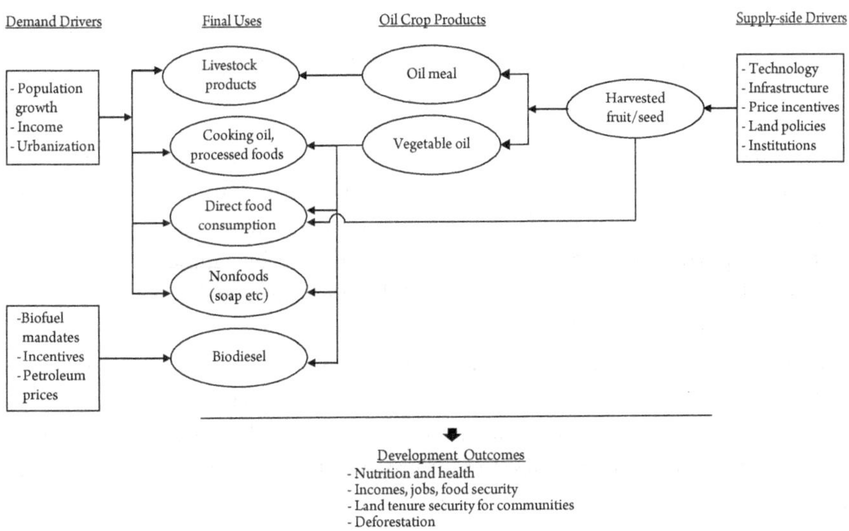

Figure 1.2 A simplified view of oil crop products and their supply and demand drivers is presented.

Table 1.2 Global statistics on the three major cereal crops and three leading oil crops

Crop	*Area, 2012 (Mha)*	*Change in area, 1991–2013 (Mha)*	*Gross value output, 2013 (US$ billion)*	*Value of exports, 2012 (US$ billion)*
Cereals				
Wheat	219	−12.3	242	54.4
Maize	185	54.1	382	35.5
Rice	165	18.2	429	23.7
Oil crops				
Soybeans	112	54.3	131	93.2
Rapeseed	37	18.9	53	24.0
Oil palm	18	12.0	37	42.3[a]

[a] The value of exports includes value added from processing and domestic shipping and handling and may exceed the value of unprocessed oil palm fruits priced at the farm gate.
Source: Calculated from FAOSTAT.

(which have lifted many out of poverty), and the critical role of vegetable oils in world food security.

Although for the most part this book analyzes and interprets the recent past, it also looks to the future. An especially pertinent issue, given that Africa is poised to join the oil crop revolution in production and consumption, is whether lessons from the recent experience with oil crops in Asia and Latin America can be applied in Africa to promote more favorable development outcomes there.

In the remainder of this chapter, we introduce facets of the oil crop revolution vital for understanding the chapters that follow. Table 1.2 presents some contextual summary statistics on the three main oil crops and three main cereal crops. Appendix A1.1 provides notes on the data sources used throughout this book.

OIL CROPS HAVE MULTIPLE USES AND SUBSTITUTIONS

The world oil crop market is not easy to understand, given the multiple crops and products involved and the numerous pathways to their final use. Two subgroups of oil crops are generally recognized. Most oil crops are annuals known as *oilseeds*; the most important oilseed crop is soybeans, followed by rapeseed, sunflower, and groundnuts. Oilseed may be consumed directly, but more than 80% of production is processed (usually referred to as *crushing*) into a high-protein meal or cake and a vegetable oil. The importance of protein meal in relation to oil varies widely among oilseed crops (Table 1.3). For example, protein meal derived from soybeans comprises 79% of processed output and two-thirds of the value; the production of soybean oil comprises 19% of processed output and about one third of value.

Table 1.3 Global production of oil crops, oil, and protein meal, 2013/14

Crop or product	*Meal content (%)*	*Oil content (%)*	*Production (Mt)*	*Processed (Mt)*	*Meal (Mt)*	*Oil (Mt)*
Perennial oil crops						
Copra[a]	33.3	62.5	6.0	6.0	2.0	3.7
Olives[b]	—	—	—	—	—	3.3
Palm oil[b]	—	—	—	—	—	58.1
Palm kernels[b]	52.9	44.4	15.4	15.4	8.1	6.8
Oilseeds (annuals)						
Groundnuts	39.5	32.1	39.2	17.4	6.9	5.6
Rapeseed	58.9	39.8	66.4	62.5	36.8	24.9
Soybeans	78.9	18.6	281.7	237.9	187.8	44.3
Sunflower	44.1	41.6	40.3	36.1	15.9	15.0
Cottonseed	45.8	15.2	44.0	34.2	15.7	5.2
Total	—	—	449.1	375.3	257.6	161.7

[a] Copra is the dried kernel of the coconut, which has a high oil content.

[b] Palm oil and palm kernels make up about 20% and 5%, respectively, of the weight of fresh fruit bunches of oil palm. Oil palm and olive fruit residues have no market value after the oil and kernels have been extracted.

Source: Computed from USDA-FAS PSD.

The second group of oil crops consists of *perennial oil crops*; oil palm is the most important crop in this group, which also includes the olive and coconut crops. These tree crops are grown primarily for their oil, although they may have important by-products. Oil palm, for instance, produces both palm oil from the fruit and palm kernel oil from the nut, and the remaining palm kernel cake is used as protein meal.

Vegetable oils find a wide variety of uses. Most oils are consumed directly as food, sometimes after further refining (margarine is one example) or are used for cooking oil. Important shares are used as an ingredient in processed foods, in industrial products other than food (such as soap), as industrial inputs (such as oleochemicals), and in biofuels. By and large, the main vegetable oils can substitute for each other, with some exceptions. Olive oil, for example, is assigned much higher value than other common vegetable oils and is highly preferred in rich countries for food uses. Jatropha oil, which has received much attention in recent years, is an inedible oil grown exclusively for biofuel. The common vegetable oils also have different fat compositions, which influence their selection as a cooking oil, their use in processed foods, and their health effects—a theme to which we return in Chapter 4.

The high protein content of meal from oilseeds (44%–48% in the case of soybeans) means the meal is valued for livestock feed. Demand for protein meal is driven consequently by the consumption of livestock products and the availability of alternative sources of protein for animal feed, such as fish meal. However, soybean is the only oil crop for which the value of the meal (two thirds its value) exceeds that of the oil. Less than 5% of the value of oil palm products comes from oil meal (palm kernel meal).

The complexity of the world oil crop sector is tempered to some extent by the dominance of only two crops in the markets for oilseeds and vegetable oils. Among oilseed crops, soybeans represent 57% of production and 85% of exports. Among vegetable oils, palm oil and palm kernel oil together constitute 40% of production and more than two thirds of exports. Soybean oil is the second most widely traded vegetable oil, accounting for 13% of vegetable oil exports—a figure that increases to 30% if we include the oil content of exports in the form of unprocessed soybeans. Among oil meals, soybean meal accounts for two thirds of global production and 76% of exports (aggregating exports of soybeans and soybean meal). During the past two decades, the dominance of oil palm and soybeans in the oil crops sector has increased significantly.

PRODUCTION OF OIL CROPS IS SHIFTING TO THE TROPICS

The increasing dominance of oil palm and soybeans in markets for vegetable oils has been accompanied by an extraordinary shift in the geography of agricultural production, with much of it toward the tropics—defined here as lying between 23° N and 23° S latitudes. Since 1970, the area planted for oil crops expanded by a staggering 150 million hectares (Mha) compared with 55 Mha for *all* cereal crops (FAOSTAT). Soybeans accounted for half that expansion, and by far the most spectacular growth was in Brazil and Argentina (Figure 1.3). This trend became apparent after 1970 and

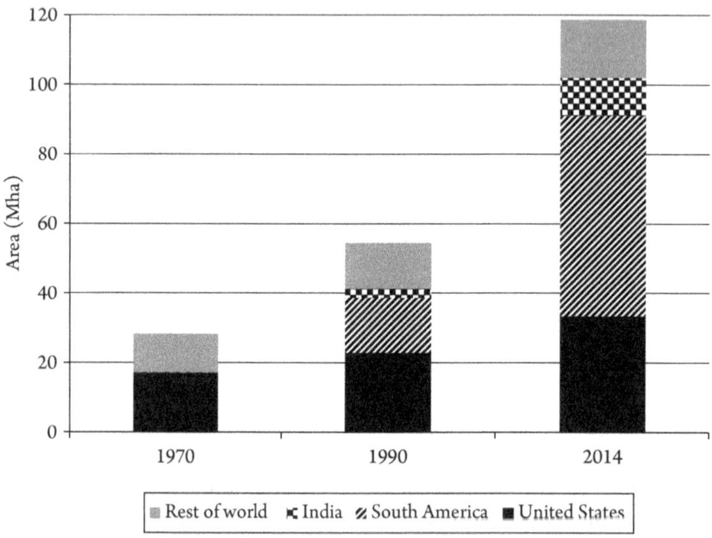

Figure 1.3 Soybean area has risen dramatically in South America.
Source: USDA-FAS PSD.

accelerated around 1990 as Chinese imports sparked a booming global market for soybeans. In 2014, Brazil surpassed the United States as the world's largest soybean exporter.

In Brazil, the soybean crop expanded initially in the more temperate south, but since 1990 the crop has moved steadily northward into the tropical region called the *Cerrado*, replacing natural grasses, scrubland, and pasture. As production in the Cerrado has expanded northward, it has encroached on the Amazon biome. Today, the state of Mato Grosso is the largest producer of soybeans in Brazil, and tropical soybeans grown north of 23° S latitude account for about three quarters of production. Soybeans have made additional incursions into tropical areas of northern Argentina, Paraguay, and Bolivia. A less well-known expansion of soybeans in the tropics occurred in central India, where area has increased from virtually zero in 1970 to around 10 Mha today, although yields remain low (Figure 1.4).

Oil palm is confined entirely to the tropics, within a narrow band from 10° N to 10° S—the zone occupied by a substantial share of the world's humid tropical forests. Although oil palm covers a much smaller area than soybeans (16 Mha vs. more than 100 Mha for soybeans), oil palm area has expanded even faster than soybean area, growing at an average rate of 4.6% annually from 1990 to 2010. Oil palm also made a major geographic shift from its original production center in Africa, where oil palm area has changed very little, to Malaysia and Indonesia (Figure 1.5).

In Chapter 2 (on oil palm) and Chapter 3 (on soybeans), we lay out the major drivers of these shifts with respect to agroclimatic suitability, investment in research and development (R&D), policy incentives, and institutions. We also discuss the evolution and efficiency of the large agribusiness concerns operating in the new producing areas for both crops. Eight of the world's largest agricultural production companies,

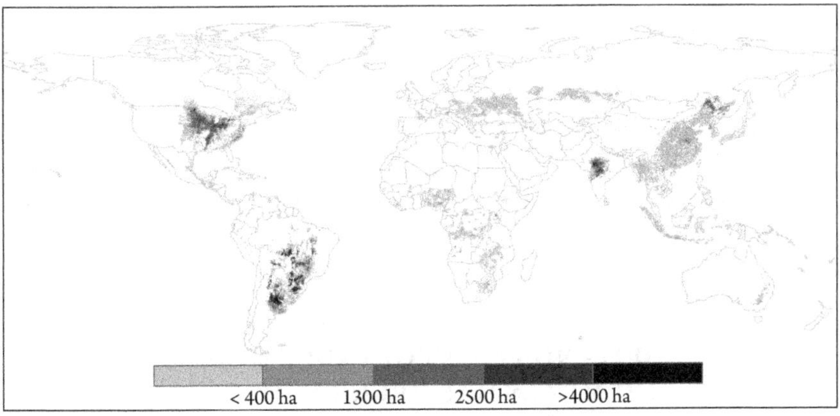

Figure 1.4 The distribution of the soybean area in 2005 is shown. Values represent the total harvested area in hectares (ha) within each 5-minute cell (about 10 × 10 km, or 10,000 ha at the equator). Note soybean area in central–west Brazil has intensified sharply since 2005.
Source: You, L., U. Wood-Sichra, S. Fritz, Z. Guo, L. See, and J. Koo. 2014. MAPSPAM. Spatial Production Allocation Model (SPAM) 2005. Version 2.0. http://mapspam.info (accessed February 2, 2016).

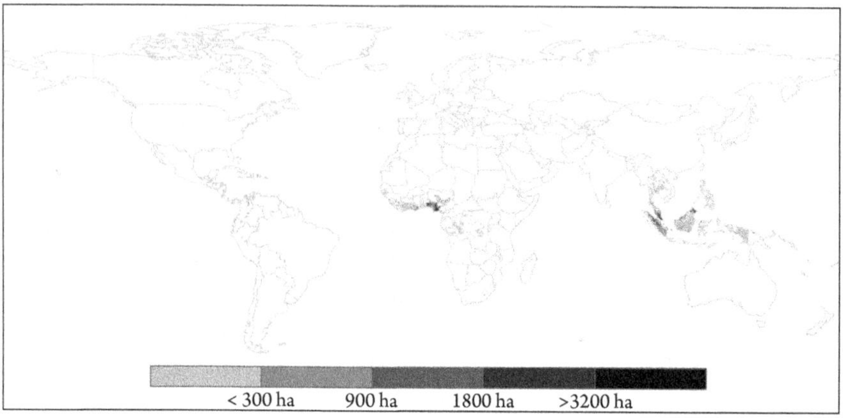

Figure 1.5 The distribution of oil palm area in 2005 is shown. Values represent the total harvested area in hectares within each 5-minute cell (about 10 × 10 km, or 10,000 ha at the equator). Note that oil palm area has intensified substantially since 2005 in Indonesia.
Source: You, L., U. Wood-Sichra, S. Fritz, Z. Guo, L. See, and J. Koo. 2014. MAPSPAM. Spatial Production Allocation Model (SPAM) 2005. Version 2.0. http://mapspam.info (accessed February 2, 2016).

based mostly in Malaysia, Indonesia, and Singapore, are involved in oil palm (United Nations 2009). Sime Darby, a company based in Malaysia, is listed as the largest producer, maintaining more than 600,000 ha of plantations and associated large investments downstream in processing, manufacturing, and marketing. These companies are now moving beyond Southeast Asia, to Africa in particular. Major energy companies such as Petrobras and Vale are also entering the industry in Brazil.

In soybeans, the large multinational companies Cargill, Bunge, ADM, Dreyfus, and the Brazilian-owned Maggi Group have invested heavily downstream in processing and shipping logistics. At the same time, the big seed companies, especially Monsanto, have invested in breeding genetically modified (GM) soybean varieties for the tropics. Maggi and a host of companies also operate very large farming operations, many surpassing 100,000 ha, with soybeans as their principal crop.

Successful smallholder production systems also exist for both crops. We examine these systems in country case studies in Chapters 2 and 3, and in Chapter 8 we present business models for further integrating smallholders into oil crop production and processing.

DEMAND FOR VEGETABLE OILS AND MEALS HAS BEEN BOOMING

The revolution in tropical oil crops is the result of the booming demand for food, feed, and biofuels, mostly from emerging middle-income countries. During the past 50 years, world soybean consumption increased 10 times over—from 26 million Mt to 260 Mt—whereas consumption of palm oil and palm kernel oil increased 25 times over, starting from about 2 Mt to surpass 50 Mt.

From 1993 to 2012, food uses of vegetable oils in developing countries have expanded at 5.1% annually—more than three times the rate of cereals (OECD-FAO). Increased consumption of vegetable oils has contributed significantly to food security, in the sense that it accounts for at least one quarter of the increase in total food calories in developing countries since 1970.[2] A positive income elasticity for vegetable oils indicates that room remains for further growth in food uses. Per-capita consumption of vegetable oils in developed countries was 25 kg versus 16.7 kg in developing countries and only 9.7 kg in sub-Saharan Africa in 2010 through 2012 (OECD-FAO). Many other drivers and substitutions must also be considered to understand the dynamics at work here. In Chapter 4, we take up this challenge, particularly the important roles of domestic pricing and tax policies, and we also discuss the controversies surrounding the impact of increased consumption of vegetable oils on health. The amount of fat consumed is the most important factor for health, but the type of fat present in vegetable oils must also be taken into account. Hydrogenated soybean oil is the most important source of trans fats in many diets, and palm oil has the highest level of saturated fats among vegetable oils. When palm oil is consumed in its unrefined form, however, as in Africa, it is an excellent source of vitamin A.

As noted, oil crops produce joint products: oil and meal. Feed use has been the major driver of demand for oil meals, especially soybean meal. Consumption of livestock products, particularly of poultry and pigs in rapidly growing middle-income countries, led the demand for oil meals to more than double from 1991 to 2014. This trend was most evident in China, which is now by far the world's largest consumer of soybeans. In Chapter 5, we review trends in soybean meal consumption and outline the determinants of future demand, looking toward 2050.

Vegetable oils are also the principal feedstock for biodiesel. Although data are incomplete, the OECD estimated that as prices of petroleum-based oil rose, consumption of vegetable oils for biofuels jumped from only 2.8 Mt in 2003 to 20 Mt in 2012 (OECD-FAO). Biofuels may make up only 13% of vegetable oil consumption, but they account for nearly half the increase in vegetable oil consumption from 2003 to 2012 (OECD-FAO) (Figure 1.6). To date, most of this increased consumption has been in the European Union. The large vegetable oil producers—Argentina, Brazil, Malaysia, and Indonesia—have started to implement mandates and incentives to produce biodiesel as well. The world's largest biodiesel plant, in Singapore, is fueled in part by palm oil from Malaysia and Indonesia.

In Chapter 6, we look at the emerging markets for vegetable oil feedstocks for biofuels, which have added to the controversy and scrutiny surrounding tropical oil crops. Many question whether it makes sense to destroy tropical forests that sequester large quantities of carbon in the name of producing so-called renewable energy to reduce

[2] These estimates, based on FAOSTAT (n.d.), appear to underestimate substantially the food calories from vegetable oils (Chapter 4).

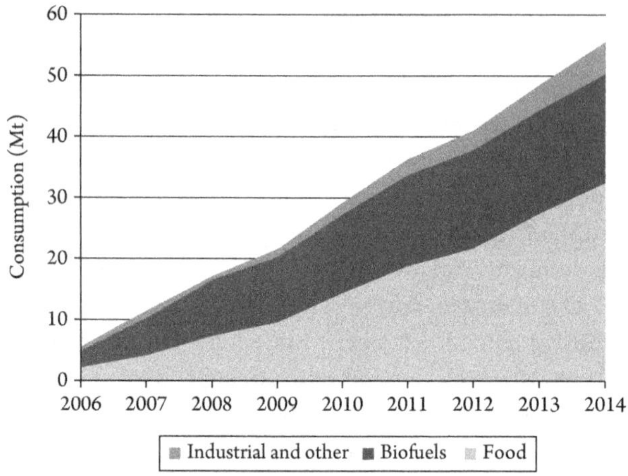

Figure 1.6 The increase in global consumption of vegetable oils from a 2005 base level. Biofuels accounted for nearly half of the increase to 2010 and one third of the overall increase to 2014. *Source:* OECD-FAO.

carbon emissions. Others question the wisdom of a food–biofuel tradeoff, especially given the critical role of vegetable oils in food security among the poor.

Finally, vegetable oils are ingredients in a host of other nonfood products, such as soap, cosmetics, oleochemicals for industrial use, and lubricants. These uses account for about 7% of vegetable oil consumption, although this share is probably greater for palm oil and palm kernel oil.

ASIAN MARKETS ARE DRIVING TRADE

In Chapter 7, we unite the supply and demand sides of the market for vegetable oils, beginning with an analysis of the extraordinary growth in international trade in oilseeds and oils, which has even surpassed growth in consumption. The analysis of trade flows is complicated by the joint production of oil and protein meal. Soybeans may be exported unprocessed and then crushed in the importing country, or they may be processed in the producing country and the oil and protein meal exported in various proportions. Argentina, for example, sets incentives to process soybeans domestically and then exports most of the oil and meal. Brazil, on the other hand, has a large domestic market for vegetable oils and exports much more meal than oil. China sets incentives to import soybeans and process them locally into oil and meal. Indonesia is the world's major exporter of vegetable oils (from oil palm), but also a large importer of soybean meal.

The analysis of trade can be taken to another level by considering the export of value-added products. For instance, in seeking to avoid the high domestic transport costs of exporting soybeans, Brazil has become the world's largest exporter of poultry meat produced from soy meal. Argentina similarly pursues a strategy of adding value by processing soy oil into biodiesel for export.

Given these complexities, and considering only the first-stage processed products (oil and meal), we can discern several major trends. First, soybeans and palm oil are among the most valuable agricultural commodities traded worldwide (Table 1.2). For more than a century, wheat was the most valuable traded commodity; but, in 2002, soybean exports overtook wheat and are now about one third more valuable than wheat exports. The export value of palm oil and palm kernel oil has increased even more rapidly to reach third place, and it also appears likely to exceed the value of wheat exports in the near future.

Second, a few developing countries drive the trade increasingly in oil crops on both the export and the import sides. During the past 50 years, first Malaysia and then Indonesia captured dramatically the market for palm oil from West and Central Africa, and now these two countries account for about 90% of the world market (Figure 1.7). The United States dominated soybean exports historically, but by 2013, Brazil and Argentina—which entered the market only during the 1970s—exported almost double the US exports of soybean products, and Brazil's became the world's largest exporter (Figure 1.8).

On the import side, the European Union has given way to Asia, led by India and China, as the major driver of world trade in oil crops. India produces barely any palm oil, yet it has become the world's largest palm oil consumer. Soybean exports have shifted dramatically to China and other emerging economies of Asia, especially since the mid 1990s (Figure 1.9). Increasing demand from China has been a major driver of the Brazilian soybean revolution (Figure 1.10).

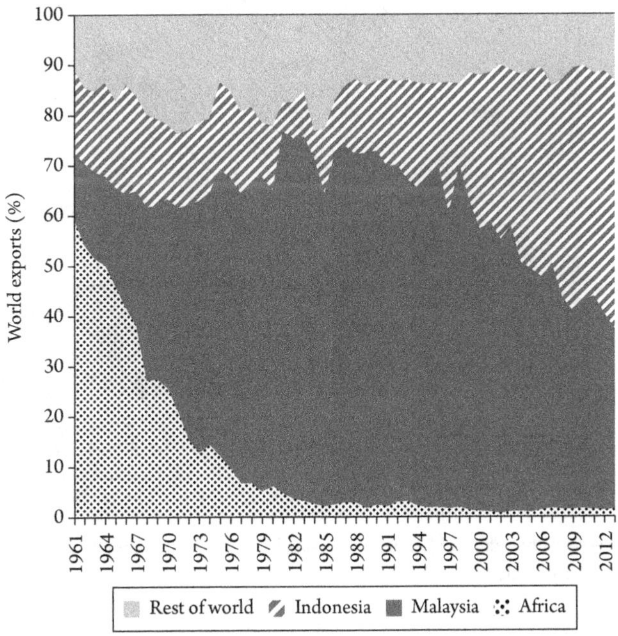

Figure 1.7 Malaysia and, more recently, Indonesia have captured the market share overwhelmingly for palm oil exports. Note that palm kernel oil and cake are included.
Source: FAOSTAT.

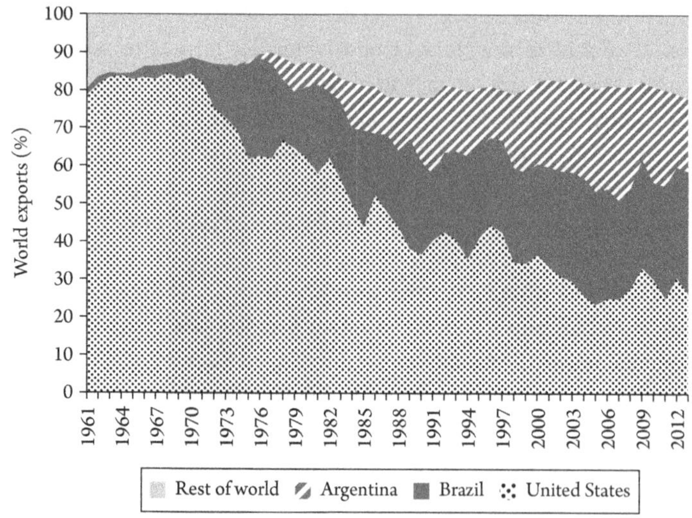

Figure 1.8 Brazil and Argentina have captured the market share decisively for soybean exports. Note that soybean meal and soybean oil are included.
Source: FAOSTAT.

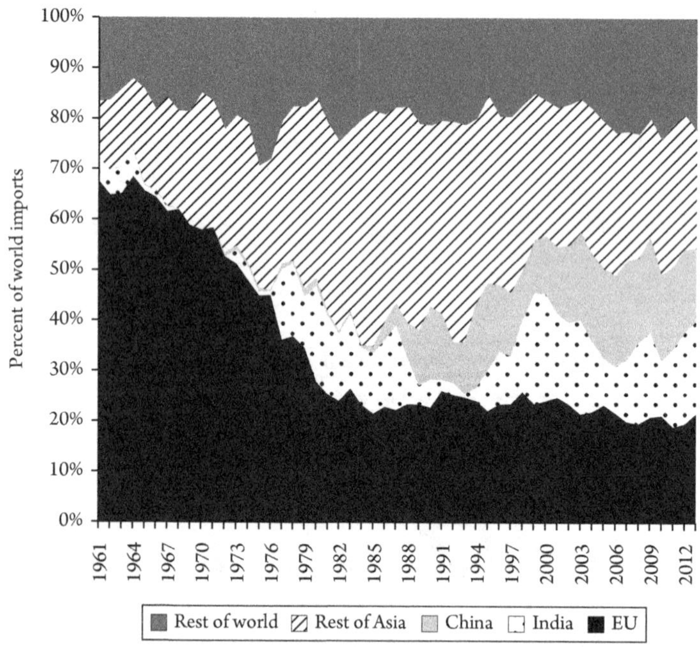

Figure 1.9 Asia, especially India and China, is the world's major palm oil importer. Note that palm kernel oil and cake are included.
Source: FAOSTAT.

One of our tasks in Chapter 7 is to untangle further the origins of these large shifts in trade flows. Obviously, the changing demand patterns detailed in Chapters 4 through 6 play a large role, along with policy incentives, in determining the direction and type

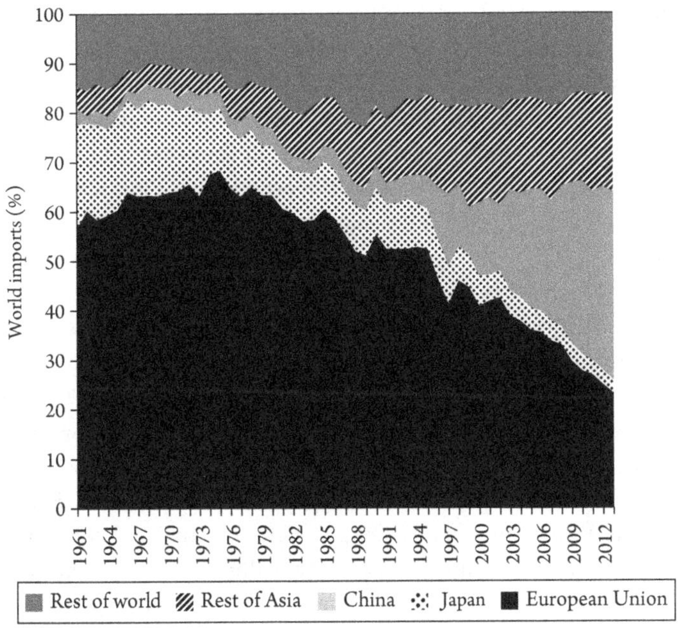

Figure 1.10 Chinese imports have been the major driver of soybean markets since 1995. Note that soybean meal and oil are included.
Source: FAOSTAT.

of product shipped. Given the instability in world markets for tropical oil crops, a major debate addressed in Chapter 7 is whether it is wise for a country to depend heavily on imports to meet demand for this major food staple. The dominance of palm oil in world markets may accentuate price instability, because oil palm is a perennial crop and little flexibility exists in the short run to adjust supply to market conditions.

An even bigger challenge of Chapter 7 is to apply our understanding of price discovery in world vegetable oil markets, especially our understanding of how futures markets and stock holdings have created efficient markets and integrated them globally. Our final task in Chapter 7 is to bring the elements of supply and demand together to provide an outlook for vegetable oil markets in 2050. Our projections indicate a sharp slowing of demand. Given reasonable assumptions on yield growth and area expansion, future market demand should be provisioned without an increase in real prices (quite possibly with declining prices), and with only modest area expansion.

TRADEOFFS BETWEEN ECONOMIC, SOCIAL, AND ENVIRONMENTAL OBJECTIVES CAN BE MINIMIZED

In Chapter 8, we turn our attention to the economic impacts of the tropical oil crop revolution. The rise of major new industries in relatively poor and sparsely settled regions can create new poles of economic development, generating employment along entire value chains. Other potential benefits are likely to include those arising from investments in infrastructure, the introduction of new technology to local populations, and the generation of tax revenues and foreign exchange. Even such seemingly obvious contributions are often debated. For example, the Brazilian soybean industry claims to have created 1.4 million jobs, including 250,000 producers (Brown-Lima et al. 2010), yet other observers conclude the high levels of mechanization accompanying the soybean expansion led to a loss of jobs in the Cerrado (Schlesinger 2004). In Indonesia, the fact that production of rice, the staple food, has virtually disappeared from many oil palm-producing areas raises questions about impacts on local food security and the vulnerability of producing regions to volatility in world prices of vegetable oils.

Another controversial issue in assessing impacts is the dominance of agribusiness companies and large commercial farms in producing soybeans and oil palm in the major exporting regions of the tropics. One interpretation of these developments is that they represent the "modernization" of backward areas; another is that they represent the inexorable advance of industrial monocropping. Yet, we have good countervailing examples of smallholder participation in these value chains, and there is considerable potential to deepen their participation. For instance, smallholders are at the center of oil palm production in Thailand, the world's third-largest palm oil producer. In India, millions of smallholders grow soybeans as a cash crop. Could African countries courting investment in the oil crop sector build on these experiences to spread the benefits of such investments? One of the big issues is how to manage the sparsely populated savannah of Africa, where the potential for crop expansion resembles that of the Cerrado of Brazil three decades ago. Could this African "sleeping giant" develop a dynamic commercial soybean sector based on small and medium-size farms (World Bank 2009)?

The massive changes in land use associated with tropical oil crops have far-reaching social and environmental impacts, which we discuss in Chapter 9. Without a doubt, oil crop area has often increased at the expense of natural areas; oil palm in particular has replaced tropical forests of high conservation value and peatlands that sequester large amounts of carbon. Companies in the oil crop sector often have been accused of "land grabs" that displace local people and remove their livelihoods. From a technical viewpoint, an apparently logical solution would be to raise yields to save land, yet this approach—associated so closely with the father of the green revolution, Norman Borlaug, that it is known as the Borlaug hypothesis (Borlaug 2007)—has been challenged. Some argue that increasing yields will only increase the profitability of oil crops and promote even faster expansion at the forest margin (Kaimowitz and Smith 2001). Even if the Borlaug hypothesis holds, technical solutions alone are not likely to save tropical forests. Institutions, especially those governing land and forest resources, require much attention. In Chapter 9, we review a number of global and local approaches to strengthen institutions that intersect with tropical oil crops, such as land rights, governance of forest resources, private standards and roundtables, and the emerging program, Reducing Emissions from Deforestation and Forest Degradation (REDD+). Brazil's recent success in arresting deforestation induced by soybean expansion, along with the recent commitments by large palm oil trading companies to zero deforestation, suggest that the social and environmental footprint of tropical oil crops can be minimized.

LOOKING TO THE FUTURE

Chapter 10 integrates our findings from previous chapters to provide a forward view of the oil crop sector. Although we see the "revolutionary" pace of change abating somewhat, the growth in food demand in late-developing countries, especially in Africa, and a projected 50% increase in demand for vegetable oils for biofuels for the next decade (OECD-FAO) will keep oil crops in the spotlight. A major question issue to which we return is Africa's potential to participate in a sustainable and equitable way in the future growth of the sector.

We also consider which policies are needed to reduce the tradeoffs between growth, poverty reduction, and the environment that have characterized the tropical oil crop revolution to date. Global players such as international agencies, consumer groups, multinational companies, and civil society are already attempting to chart a more sustainable course for the sector. Ultimately, however, the local players will be the ones who implement policies on the ground. *They* must be convinced the sector's sustainable development is good business for them and for the future.

We aspire to win–win outcomes for sustainable development, but as pragmatists we understand the real world entails messy tradeoffs. Our final theme in this book is how to improve the management of such tradeoffs. We end by giving a cautiously optimistic outlook that future development of the sector can provide balanced outcomes for inclusive economic development, food security, and the environment.

APPENDIX A1.1: A NOTE ON DATA SOURCES

The analysis of the oil crop sector is complicated by the availability of several data sources and major discrepancies among them. The three major data sources are as follows:

1. *FAOSTAT*: This is the main statistical data base of FAO and our preferred data source on the supply side. This database from the FAO has complete data for all oil crops and countries. One caveat is that major discrepancies exist between FAOSTAT data and national data for important producers, especially for oil palm. Another is that, on the consumption side, only food and "other utilization" are included, and the share of "other utilization" seems implausibly large for many important countries, possibly because of the difficulty with allocating cooking oil to food consumption and waste (Chapter 4). On a global basis, FAOSTAT gives a share of 48% to "other utilization" in 2011 for the nine major oils combined at a global level compared with 25% for the US Department of Agriculture (USDA) data set.
2. *USDA-FAS PSD. Production, Supply, and Distribution (PSD) data set of the USDA Foreign Agricultural Service (FAS)*: This data set covers the major oilseeds (soybeans, rapeseed, sunflower, groundnuts, and cottonseed) and oil crops (copra and oil palm) for all important producing and consuming countries. It provides a breakdown to food uses and nonfood uses, but does not separate biofuels. Nor does it provide area and yield data for oil palm. Most computations in this book rely on

this data set, especially on the consumption side, where we found the reported food and nonfood uses much more plausible than for FAOSTAT and more consistent with national statistics.

3. *OECD-FAO*: The OECD-FAO data set has coverage similar to the USDA-FAS PSD data set, but it provides a more disaggregated breakdown by food, biofuel, feed, and industrial uses. The estimates for food and nonfood uses are consistent with the USDA-FAS PSD, but are at wide variance with FAOSTAT, despite the involvement of the FAO in both data sets. We use this data set to report biofuel use. Note that it provides aggregate data on utilization for all vegetable oils combined and not for individual vegetable oils.

2

OIL PALM PRODUCTION AND SUPPLY CHAINS

INTRODUCTION

The takeoff that has occurred since 1970 in the production and export of palm oil and palm kernel oil—products that have been widely traded for more than a century—has no parallel in recent agricultural history. Beginning in 1970, when global production of palm oil was about 2 Mt, production doubled or more in every decade to 2010, representing a staggering 23-fold increase over 1970 levels. Given mid-decade trends, this doubling will be repeated again in the decade from 2011 to 2020, causing global production to exceed 70 Mt.

This chapter looks at this transformation from the supply side (Chapters 4–6 cover parallel developments on the demand side). We begin with a review of the essential characteristics of the crop and historical landmarks in the industry's evolution. We then deepen the analysis through four case studies, which feature Malaysia and Indonesia (the two major producers and exporters), West Africa (the original home of oil palm, where it is bound intimately to local culture and cuisine), and Colombia (the largest producer in Latin America and a country where the palm oil industry targets the market for biofuels). The concluding section of the chapter describes milestones in technological improvement for oil palm and prospects for future supply.

ESSENTIAL FACTS ABOUT OIL PALM

To understand the recent influence of oil palm on world agriculture, it is essential to understand six key facts about the crop. Most are related to the crop's specific production and processing requirements.[1]

A Perennial Crop

Oil palm (*Elaeis guineensis*) is a perennial tree crop originating in West and Central Africa. Unlike some other tropical tree crops such as rubber and cocoa, oil palm is cultivated mostly for commercial purposes as a monocrop rather than as part of a

[1] The technical details of the crop are described in an excellent book by Corley and Tinker (2016). Lai et al. (2012), Rival and Levang (2014), and Sheil et al. (2009) also provide good descriptions, largely focused on Asia.

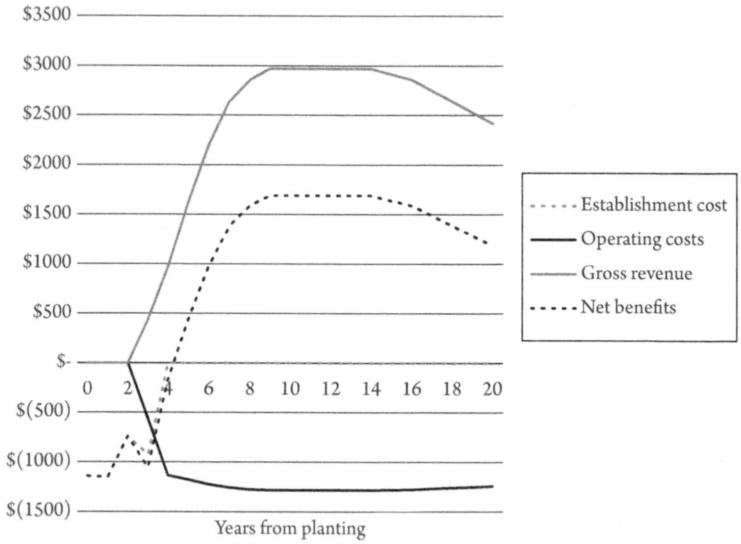

Figure 2.1 Representative cash flow for an oil palm plantation (excluding mill costs) in US dollars per hectare. Today's establishment costs are considerably higher.
Source: Data are from Fairhurst, T., and D. McLaughlin. 2009. *Sustainable oil palm development on degraded land in Kalimantan*. Washington, DC: World Wildlife, with modifications by the authors.

diverse agroforestry system.[2] Trees begin bearing fruit about 3 years after planting, reach peak production in 10 years, and have an economic life of about 25 years. This life cycle means substantial upfront investment is needed to establish an oil palm plantation—about US$10,000/ha for a greenfield investment with a mill—and a positive cash flow occurs only 5 to 6 years after initiation. Figure 2.1 illustrates a typical cash flow by year for a plantation circa 2005 (today's costs would be higher), excluding the cost of a mill.

Adapted to the Lowland Humid Tropics Only

Oil palm grows best in the lowland moist humid tropics between 10° N and 10° S latitude, with an annual rainfall of 180–2400 mm and a dry season of less than 90 days. The trees require well-drained soils, although they can be grown on relatively infertile soils, providing they are well fertilized. As noted in Chapter 1, the zone of adaptation for oil palm corresponds closely to the adaptation zone for tropical humid forests, including those in Central Africa, the Amazon, and Southeast Asia (see Figure 1.5). Consequently, the expansion of oil palm is often associated with the loss of some of the world's most biodiverse forests, which are also a critical sink for carbon (and thus a fundamental defense against global warming).

[2] Young oil palms are sometimes intercropped with food crops during their establishment phase, however.

Yields Multiple Products

The oil palm is unique in producing two distinct types of oil from fresh fruit bunches (FFBs): palm oil (high in palmitic fatty acid from the fibrous mesocarp layer) and palm kernel oil (high in lauric fatty acid from the palm kernel). The main product is palm oil, which makes up about 20% to 22% of the FFB. For international trade and in many domestic markets, although not in palm oil's traditional market in West Africa, palm oil has been highly standardized in terms of quality (chemical composition, moisture, and impurities).

About 75% of palm oil is used for food—in the form of cooking oil, as processed oils and fats (margarine is one example), and as an ingredient in a wide range of processed foods (Figure 2.2). Palm oil also has many nonfood uses for consumer goods (soap, cosmetics, and pharmaceuticals), for industrial inputs (oleochemicals), and more recently for biofuels (biodiesel).

The palm kernel accounts for about 5% of the weight of FFBs, of which about half is palm kernel oil, destined mostly for soap and industrial uses but also used in processed foods. The other half of the kernel is palm kernel meal, which is used as an animal feed.

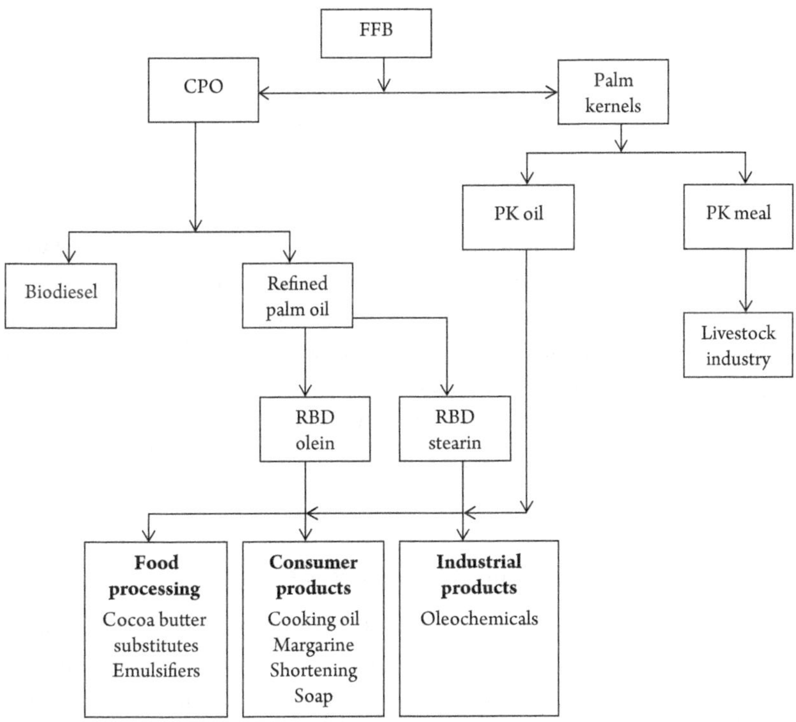

Figure 2.2 Schematic overview of the main raw products of oil palm, their downstream processing, and their final uses. RBD olein is the liquid fraction from refining; RBD stearin is the solid fraction. CPO, crude palm oil; FFB, fresh fruit bunch; PK, palm kernel; RBD, refined, bleached, and deodorized oil.
Source: Interviews with industry.

In its home setting in Africa, oil palm is valued for producing the red palm oil preferred in many traditional foods and in manufacturing local soap. The oil palm also provides palm wine and building materials.

Requires Rapid Processing

Unlike oilseeds such as soybeans, fresh palm fruit must be processed within 24 hours after harvesting to produce oil that meets the quality standards demanded in international markets. Processing of palm oil to international standards also entails substantial economies of scale. Locating production near mills not only helps to maintain quality, but also minimizes the cost of transporting bulky fruit bunches (only 25% of the fruit by weight provides economic products).

The area required to supply a profitable mill has increased from about 1000 ha in the 1960s (Hartley 1967; Tate 1996) to about 10,000 ha today (equivalent to a mill capacity of 60 t FFBs/hr). To use mill capacity efficiently and maximize quality, processing capacity has to be coordinated closely with the planting and harvesting of a large area. Coordination can be achieved through the vertical integration of milling with a large-scale production operation or through contracts or other mechanisms to coordinate milling with many suppliers. Note that the major exception to this processing setup occurs in West and Central Africa, where small-scale processors dominate palm oil production.

A Labor-Intensive Crop

Palm oil production is less labor intensive than the production of other tropical commodities such as rubber and cocoa (although not if small-scale processing is used, as in Africa), but it is still highly labor intensive compared with the commercial production of soybeans on a large scale (see Chapter 3). In economies where wages are low and labor is plentiful, oil palm can be a much-needed source of year-round employment. Given that many oil palm plantations are being established on the forest frontier in sparsely populated areas with less pressure on land, a major challenge for plantation owners is to attract and retain labor. Labor-saving innovations are used widely for some operations, but harvesting is largely manual, so the minimum labor requirement is about one worker for every 10 to 12 ha.

A Productive, Profitable Crop

Measured in terms of yields of oil per unit of land, oil palm is highly productive, even if it provides only a small amount of protein meal relative to other oil crops. Well-managed plantations in areas with good growing conditions achieve an average yield of oil (palm and palm kernel) of more than 6 t/ha, plus 0.3 t/ha of protein cake. In comparison, good rapeseed fields in Europe yield 1.8 t/ha of oil and provide about 1 t/ha of protein meal; in Brazil, soybeans grown mainly for oil meal yield about 0.6 t/ha of oil and about 2.4 t/ha of meal. Oil palm has also been highly profitable in recent years, with prices remaining well above production costs (Box 2.1).[3]

Together, these characteristics imply that oil palm is produced successfully on large plantations—often vertically integrated plantations that may also be integrated horizontally into very large regional or multinational companies. Many companies farther integrate downstream into producing palm kernel oil, refining (crude palm oil) CPO, and producing oleochemicals and biofuels. These value chains are adapted especially to supplying highly standardized palm oil and other oil palm products for the world market.

Smallholders—at least 5 million worldwide—also grow oil palm; they account for about 40% of global production of palm oil. In West and Central Africa, much of the oil palm is semiwild or cultivated on small holdings and processed through small-scale manual and semimechanized methods. These value chains serve primarily domestic markets, with the unique consumer preferences noted previously (traditional foods, soap). In between the large, vertically integrated companies and the small-scale producers and processors is a range of other business models that we explore in this chapter and in Chapter 8. They include independent small and medium-scale producers selling fruit at the farm gate or through contracts to large mills, and hybrid models combining large integrated plantations with small-scale outgrowers who sell to the plantation mill. The evolution of these different value chains is easier to understand if we first describe the industry's historical development and examine some case studies.

MILESTONES IN THE DEVELOPMENT OF THE INDUSTRY

Palm oil has a long history as an important tropical export. That history extends through three major periods of change, including the foundational period in West Africa for international trade starting during the 19th century, the rapid modernization of the industry in Malaysia starting during the middle of the 20th century, and the phenomenal globalization of the industry in recent decades.

Foundations of the Industry

Starting during the mid 19th century, a thriving trade in palm oil developed between West Africa and Britain. Wild palm fruit was harvested and processed by indigenous producers, primarily women, and exported to Europe, originally to manufacture

[3] The oil palm industry often touts oil palm's high productivity by comparing its oil yield with that of annual oilseeds. Oil palm is undoubtedly highly productive, but the gap with other oil crops narrows substantially when the value of oil meal is considered along with the ability to harvest up to three crops per year of annual crops in the tropics.

Box 2.1 Production Costs and Profits for Oil Palm

Computing the cost of producing oil palm is tricky because it is a perennial crop, and reliable survey data on costs are scarce. Many companies publish cost and profit figures for oil palm, although without details of the cost structure. Costs for Sime Darby are ranked as about average among publicly traded plantation companies (Veloo 2013). Even so, the company's plantations provided a margin of about 100% above cost of production per ton of palm products in 2012/13 (Table B2.1), admittedly a year of above-average prices.

Anggraeni and Zimmer (2014) detail costs over 5 years for representative plantations in Malaysia based on surveys by the Malaysian Palm Oil Board, excluding milling (Table B2.2). Costs are somewhat higher than for Sime Darby, but profit margins are still high and positive in all years.

Despite the paucity of good estimates, there is little doubt that oil palm has been an extremely profitable crop in recent years. The estimates of costs of production in Table B2.2 of US\$300 to US\$430/t are well below the average world price of \$650/t for 2008 to 2012 (Rotterdam price less 30% for transport and handling). Palm oil prices for 2008 to 2012 were well above trend, however, and by late 2015, they had fallen to less than US\$ 400/t (Rotterdam price less 30% for transport and handling). Some countries, such as Colombia, have much higher costs (>US\$700/t) (Veloo 2013), so caution is needed in extrapolating from these trends.

Table B2.1 Costs and returns to oil palm in Sime Darby plantations, 2012/13

	Malaysia	*Indonesia*
Yield FFBs (t/ha)	22.45	20.21
Oil extraction rate (%)	22.5	22.8
Kernel extraction rate (%)	5.08	4.70
Yield CPO (t/ha)	5.05	4.61
Yield kernels (t/ha	1.14	0.95
Selling price of CPO (US\$/t)	808	666
Selling price of kernels (US\$/t)	403	270
Weighted price (US\$/t)	733	598
Cost of production		
Production (US\$/t product)	294	250
Milling (US\$/t product)	70	50
Total	364	301
Profit (US\$/t)	369	298

CPO, crude palm oil; FFBs, fresh fruit bunches.

Source: Data are from Sime Darby Berhad. 2013. *Global reach local solutions annual report* 2013. Kuala Lumpur: Sime Darby Berhad.

Table B2.2 Production costs and returns for palm oil for representative Malaysian plantations (US$/t oil)

Year	*Fertilizer*	*Pesticide*	*Labor*	*Land*	*Establishment*	*Other*	*Total cost*	*Price oil*	*Profit*
2009	60.8	3.1	35.5	141.0	64.0	18.9	323.3	568.4	245.1
2010	85.3	3.9	62.1	165.0	81.0	29.5	426.7	774.9	348.2
2011	112.6	3.3	85.9	136.0	69.0	115.8	522.6	867.3	344.6
2012	115.7	3.5	79.3	147.0	74.0	59.2	478.7	802.4	323.7
2013	53.7	3.5	64.9	155.0	57.0	80.8	415.1	695.5	280.4
Average	85.6	3.5	65.6	148.8	69.0	60.8	433.3	741.7	308.4
% cost	19.8	0.8	15.1	34.3	15.9	14.0	100.0		71.2

The estimates assume palm oil accounts for 90% of revenues (the rest is from palm kernels) and costs are prorated accordingly. Other costs include machinery, fuel, depreciation, and general overhead. Establishment costs have been annualized.

Source: Data are from Anggraeni, D., and Y. Zimmer. 2014. Palm oil: Economics of the driver of global vegetable oil markets. Paper presented at the Agri Benchmark Global Forum, Des Moines, Iowa, August 2014.

lubricants, soap, and candles (Henderson and Osborne 2000) and not to use in food. The invention of margarine during the late 19th century and a process to hydrogenate vegetable oils around 1905 turned palm kernels into an additional major export for processing into oil for manufacturing margarine. Nigeria dominated trade in palm oil and palm kernels from the 19th century and well into the 20th century (Martin 1988; Lynn 1997).

Oil palm thus started as an African crop produced entirely by smallholders. When demand for vegetable oils in Europe expanded rapidly during the early 20th century, Lever Brothers (a predecessor of today's Unilever), the major trading and manufacturing company for palm oil, sought land concessions for harvesting wild palms and for establishing large palm plantations in British West Africa, notably Nigeria. British colonial governments denied these concessions on at least three occasions, on the grounds they would not be competitive with the existing producers and, in any event, would risk igniting conflicts with local communities (Udo 1965; Kilby 1967; Fieldhouse 1978). Local chiefs argued strongly against plantation agriculture, believing it would infringe on their land and labor rights (Udo 1965; Byerlee and Rueda 2015).

Rejected in Nigeria, Lever Brothers obtained a concession of up to 750,000 ha in the Belgian Congo and additional concessions in the Cameroon, then under German colonial rule (Wilson 1954; Fieldhouse 1978). Lever Brothers at first focused on building mechanized mills to process the wild harvest (Berger and Martin 2000). Eventually, based on early experimentation in the Cameroons before World War I, Lever's company (Huileries du Congo Belge [HCB]) cultivated oil palm in plantations successfully to supply the increasingly large, efficient mills developed through improvements in milling. HCB, headquartered in Leverville and aided by the technological innovations of colonial government scientists, was the largest oil

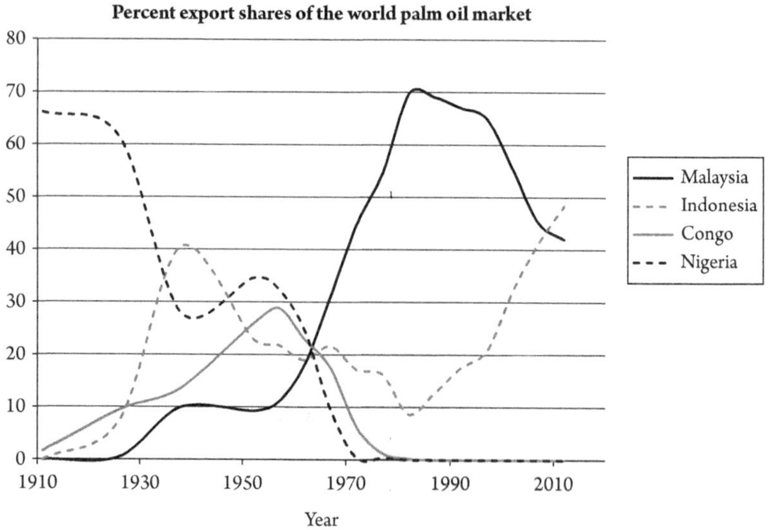

Figure 2.3 Historical percentage of major exporters in global export market for palm oil. Nigeria and the Democratic Republic of the Congo are now significant importers of palm oil, excluding palm kernel oil.

Source: Byerlee, D., and X. Rueda. 2015. From public to private standards for tropical commodities: A century of global discourse on land governance on the forest frontier. *Forests* 6 (4): 1301–1324.

palm plantation in the world by 1960, when the Democratic Republic of the Congo gained independence (Fieldhouse 1978). HCB—often associated during the colonial period with harsh labor practices and conflicts with villages of rights to harvest wild palms, in which Brussels intervened periodically (Marchal 2008)—was nationalized after independence. The industry declined and the Congo ceased to export palm oil (Figure 2.3).

Before World War I, Adrien Hallet, a Belgian national who had worked in the Congo, transferred oil palm milling technology to Sumatra (then part of the Netherlands East Indies and today in Indonesia). He was also extremely lucky in being able to adapt African oil palms grown in the Deli area of Sumatra for ornamental purposes to commercial cultivation with good fruit and oil yields (Martin 2003). With these technologies he established the forerunner to Socfin, now a major global palm oil company. As early as 1920, a Dutch scientist working in Sumatra stated presciently that "the yield of palm oil in this region well exceeds that achieved in Africa—I confidently leave to the heads of the great plantation companies the task of proving that Asia can rival Africa in this product" (Tate 1996, p. 56).

During the 1920s, the introduction of large-scale milling technologies in Sumatra and the exportation of palm oil to Europe in bulk provided standardized palm oil suitable for use in food, especially margarine, which enlarged the palm oil market. Sumatran exports of palm oil accounted for 26% of the world market by 1939, exceeding Nigerian exports (Pim 1946), although Nigeria remained the major exporter of palm kernels. The depredations of World War II and the postwar struggle for independence in Indonesia weakened the Sumatran industry severely, however.

Shortly after the first commercial plantings in Sumatra, oil palm was cultivated in Malaysia's Selangor State by a Frenchman, Henri Fauconnier (better known for his later success as a novelist), in a joint venture with Hallet's Sumatran company. Plantings by the Guthries Group, a forerunner of today's giant Sime Darby, followed. By 1941, Peninsular Malaysia had 34 oil palm plantations, although their average size (a few hundred hectares) was a fraction of the size of a modern plantation (Tate 1996).

Back in Africa, Nigeria continued to be a major exporter, but increasing competition from the Congo and Southeast Asia led to a period of considerable debate within the Nigerian colonial government on how to upgrade a local industry based on wild trees and manual processing. Efforts to foster smallholder plantations had little impact, although improved, hand-operated screw press mills were adopted fairly widely (Kilby 1967). Incentives were offered to attract foreign capital into the milling industry, but investors were unwilling to commit without an ensured mill supply from an associated plantation, which the colonial government continued to rule out. Some blame these antiplantation policies for the decline of the Nigerian industry (Meredith 1984).

With Europe facing a shortage of vegetable oils during World War II, the colonial government in Nigeria established the Oil Palm Marketing Board to stabilize supplies and foster development of the industry. In practice the Marketing Board became a significant source of taxation, further squeezing production (Helleiner 1966; Kilby 1967). The acute shortage of vegetable oils after World War II prompted a change in government policy during the 1950s to foster the establishment of large-scale plantations as well as smallholder schemes (Udo 1965). Most were sponsored by the state through grants and loans from taxes on smallholders collected by the Marketing Board, with the result that inefficient state-owned plantations and high taxation of smallholders together with the Nigerian Civil War during the 1960s essentially killed the export industry. Nigeria went from being a net exporter of palm oil around 1980 to importing more than US$1 billion worth of palm oil in 2012.

The Modern Palm Oil Industry

The present-day palm oil industry had its origins in Peninsular Malaysia. During the 1960s, as rubber prices declined, Malaysia took a strategic decision to diversify its economy and promoted oil palm as an alternative crop. After the Congo achieved independence, Unilever moved most of its operations to Malaysia, and many of the rubber companies such as Guthries and Harrisons and Crosfield (both later merged into Sime Darby) switched from producing rubber to oil palm. The Malaysian government, supported by the World Bank, also strongly encouraged oil palm production and processing, and pioneered modern industry methods through the Federal Land Development Authority (FELDA)—a parastatal organization established in 1956 to resettle and provide livelihoods to the poor and landless through tree crops, principally oil palm and rubber (Teoh 2013). Together these programs propelled the industry's rapid development. Malaysia reigned as the world's leading exporter of palm oil from 1966 to 2007.

An important technological milestone during this period was the wide adoption of the higher yielding *tenera* oil palm. The traditional fruit type in Africa (also originally grown in Asia) was the *dura,* characterized by a thick shell and low oil content (10%–12%). Research in the Belgian Congo during the 1930s had already discovered the tenera fruit type, with a thin shell and a higher oil content of around 20%, and identified it as a hybrid of dura and *pisifera,* a third fruit type without a shell that is not grown commercially because of sterility (Berger and Martin 2000). Further development of the tenera type in Malaysia led to its widespread adoption; with the use of chemical fertilizer, the tenera type provided a quantum jump in yields.

The most important phase of oil palm expansion started around 1990 in Indonesia in Sumatra and the Indonesian part of Borneo (Kalimantan), and in the East Malaysian states of Sarawak and Sabah in North Borneo. Building on the prewar legacy of oil palm production in Sumatra and on the Peninsular Malaysian experience starting during the 1960s, the expansion of the 1990s was aided by investors from Peninsular Malaysia and Singapore, who were seeking an alternative to the increasing wages and growing land scarcity in Peninsular Malaysia. Governments in the new producing regions strongly encouraged expansion by providing large land concessions (usually as part of "state-owned" forest land) and other benefits to attract investors. During this period of explosive expansion, land rights and deforestation emerged on the world stage as major issues (Sheil et al. 2009; Sayer et al. 2012), and Indonesia became the world's largest palm oil producer and exporter.

The Globalization of the Industry

In the ultimate irony of globalization, Asian firms, driven by controversy over continued deforestation in Indonesia and attractive terms for land concessions in Africa, are now moving aggressively into Africa. Large plantations—the combined investment plans amount to billions of dollars—are planned in Cameroon, Gabon, Liberia, and Sierra Leone, among other countries. The considerable potential to revitalize what was once West Africa's primary export is accompanied by the social and environmental risks inherent in large-scale plantation agriculture. The debate ignited by these plans is reminiscent of the debate that occurred a century ago, when Lever Brothers first sought concessions in Nigeria.

Oil palm is also on the move in Latin America, where a closely related species, *Elaeis oleifera,* was harvested traditionally from the wild on a small scale to extract oil for local use but was never cultivated commercially. The African species was introduced during the 1920s by the leading banana export company, United Fruit (now Chiquita), and oil palm started expanding commercially from the 1960s, in part to diversify from banana and cattle production in the lowland humid tropics (Pacheco 2012). Oil palm area expanded steadily and then accelerated during the 21st century, largely to meet biodiesel mandates. Most producers are medium scale, with up to 100 ha; others are much larger but are still owned domestically. The largest producer is Agropalma in Brazil, with 40,000 ha under plantation.

Finally, oil palm area is expanding in other Asian countries—often beyond its optimal ecological environment. Thailand is the world's third largest producer, with other countries in Southeast Asia also experiencing expansion. India has embarked on an highly ambitious program to substitute for its huge imports, largely through irrigated oil palm.

CURRENT PRODUCTION

Today, palm oil production and exports are highly concentrated in Indonesia and Malaysia, which together provided 85% of the supply of 50 Mt in 2012 (Figure 2.4). Thailand is in third place, followed by Colombia and Nigeria. Yields vary widely. They are lowest in African countries, where the area of unimproved palm is large. Globally, yields have grown by about 2% annually since 1991, mostly as a result of the increasing concentration of production in the two major producing countries, where yields are relatively high (Figure 2.5).

An additional characteristic of today's industry is that growth has been driven by exports. During the 1960s, about half of production was already exported. Today, around three quarters of production is destined for export (45 Mt in 2014). Only in its center of origin in West and Central Africa is palm oil produced mostly for domestic consumption, and even there it is highly commercialized. Indonesia is now the largest exporter, but also the second largest consumer, of palm oil in the world.

The development of the oil palm sector into a major global agribusiness industry is best understood through a series of four case studies. The studies focus on Malaysia, the pioneer of the modern industry; Indonesia, now the world's leading producer and exporter; West Africa, the home of oil palm and the recent focus of investors; and

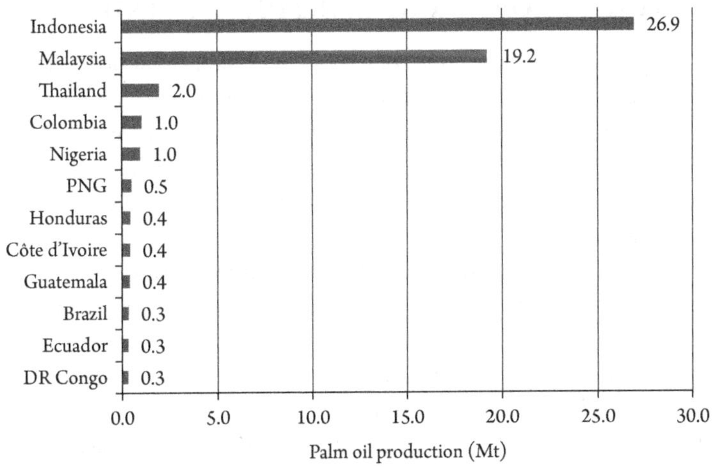

Figure 2.4 Production of palm oil by the main producing countries, 2013. Note: Major differences exist between production data in FAOSTAT and national statistical sources, especially in Malaysia and Colombia. Palm kernel oil not included. DR Congo, Democratic Republic of the Congo; PNG, Papua New Guinea.
Source: FAOSTAT.

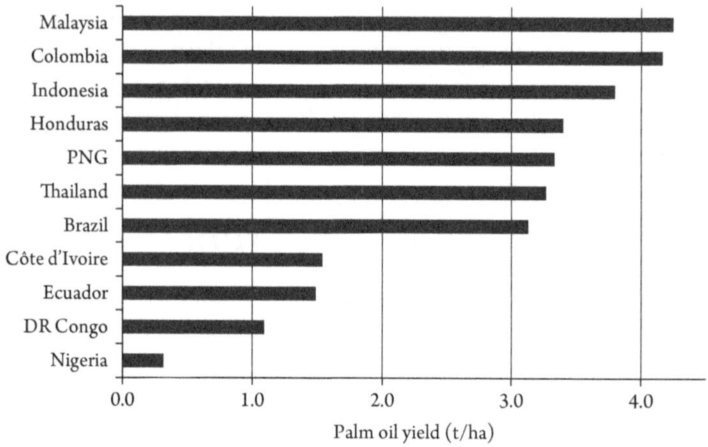

Figure 2.5 Average palm oil yields in major producing countries, 2013. Major differences exist between production data in FAOSTAT and national statistical sources, especially for Malaysia and Colombia. DR Congo, Democratic Republic of the Congo; PNG, Papua New Guinea.
Source: FAOSTAT.

Colombia, a relatively new entrant with an emphasis on oil for biofuels. These case studies have three main purposes. First, we want to provide a broad description of the supply side of the industry, including the major players. Second, we wish to analyze the roles of public policies and private actors in developing the industry. Third, and last, we are interested in understanding how some countries emerged as global leaders in the industry whereas others declined. In each case, we flag economic, social, and environmental issues, although a full analysis is left to Chapters 8 and 9.

MALAYSIA: PIONEER OF THE MODERN INDUSTRY

Malaysia emerged as the global leader of the oil palm industry shortly after achieving independence in 1957. The Malaysian oil palm industry was built on a favorable natural and human resource endowment, and the large infrastructure, R&D, and institutions for plantations established during colonial times (Pletcher 1991). In 1960, Malaysia was the world's most specialized tree crop economy; tree crops, mostly rubber, accounted for 75% of its crop area. Although oil palm covered less than 2% of the tree crop area in 1961, it provided a critical nucleus for expansion. Some companies, notably United Plantations, had been specializing in oil palm for nearly 40 years (Martin 2003).

Beginning in the late 1950s, the government of Malaysia implemented a broad-ranging program to diversify away from rubber (Teoh 2002). With an active private sector aided by government incentives and strong global market prospects for vegetable oils, the oil palm industry exploded from 55,000 ha in 1961 to reach nearly 1 Mha in 1980 and 5 Mha today, with most of the recent growth in the East Malaysian states of Sarawak and Sabah (Figure 2.6). Malaysia's share of the world export market for palm oil rose from 9% in the mid 1950s to 73% in 1983 (Pletcher 1991). At the

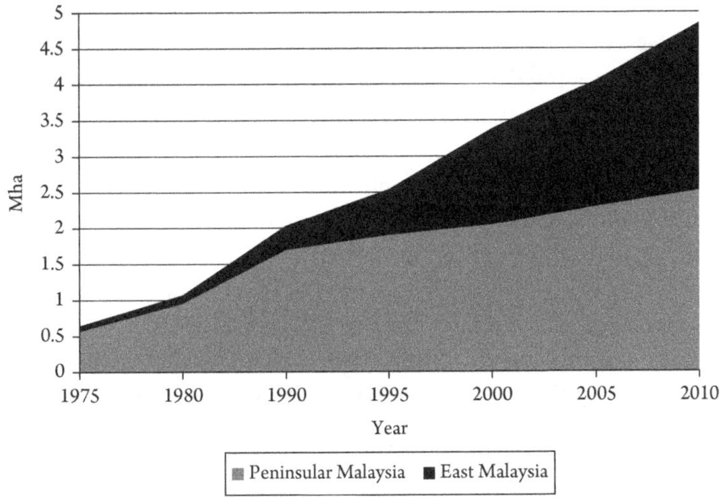

Figure 2.6 The recent growth of oil palm area has been fastest in East Malaysia.
Source: Data are from Malaysian Palm Oil Board (2013).

same time, Malaysia developed a large downstream industrial structure for processing and manufacturing based on palm oil and palm kernel oil.

Several major players led Malaysia's oil palm revolution. First, plantation-based companies inherited from the colonial period (Guthries, Sime Darby, Harrisons and Crosfield, Kuala Lumpur Kepong, United Plantations) made large investments in developing the industry. Following independence, Malaysia was one of a handful of countries where plantation companies made an orderly transition from foreign to national ownership and experienced little disruption in capital investments and productivity growth. The plantation industry was still 42% foreign owned during the 1970s when the New Economic Policy was introduced to reduce foreign ownership of plantation companies to 30% and increase ethnic Malay ownership to 30% (Pletcher 1991). The government, through Permodalan Nasional Berhad (PNB), a sovereign wealth fund, orchestrated buyouts of foreign equity in the companies from 1976 to 1982 (Ahmad and Kitchen 2008; Teoh 2013). PNB took a major stake in Sime Darby, Golden Hope, and Kumpulan Guthrie, the three largest companies, as well as 15 others (Pletcher 1991).

A second major player and pioneer in the oil palm industry was FELDA, with its effort to resettle the rural poor. Because FELDA settlers repaid only 70% of costs over 15 years through deductions from their dividends, they received a substantial element of subsidy. FELDA is described as a smallholder scheme, but in fact the plantations were often managed centrally in large, contiguous units—an approach that Pletcher (1991) called "benevolent paternalism." FELDA has supported more than 100,000 settlers, and several other federal and state schemes are pursuing similar objectives and using a variant on the FELDA model. Most settlers are now shareholders in the plantation rather than producers, and FELDA has emerged in its own right as one of the largest oil palm companies in the world (discussed later).

A third group of major players in the industry coalesced when new, home-grown, and fully private companies came into their own (Teoh 2013). For example, today's third-largest producer, IOI Corporation, was launched in 1983 and is considered one of the most innovative players in the industry, acquiring Dunlop's and Unilever's holdings in Malaysia in 1990.

The package of policy incentives provided to companies demonstrated the Malaysian government's considerable foresight in pushing for diversification and value addition, and was timed fortuitously to take advantage of growth in the market. The major policy objectives and instruments were laid out in a series of industrial master plans that aimed to promote "export-oriented industrialization." Accordingly, successive plans emphasized adding value, an objective achieved by refining CPO during the 1970s, manufacturing oleochemicals during the 1980s, and developing sectorial clusters (including biofuels) during the 1990s and 2000s (Rasiah 2006). Stimulated by these incentives, all the large companies integrated vertically into one or more downstream activities: CPO refining, kernel crushing, or manufacture of cooking oils, food products, specialty fats, oleochemicals, and, most recently, biodiesel.

The major instruments were as follows:

- *Outright subsidies to the sector*: Examples include the support to FELDA mentioned previously, as well as periodic planting grants financed by export duties on CPO to stimulate diversification from rubber and to update planting stock in aging plantations. Initially, during the late 1950s and 1960s, the grants aimed to convert plantations from rubber to oil palm when rubber prices were low. Later, replanting grants were timed to coincide with periods when prices of palm oil (and the opportunity cost of replanting) were low; at the same time, they served to withdraw supply from the market.
- *Differential export taxes and other tax incentives to stimulate downstream investments* (Figure 2.7): The tax on exports of CPO versus tax-free exports of refined palm oil was the most significant incentive. The differential was around 7 percentage points during the 1960s but widened considerably during the 1970s before declining to zero in recent years (Athukorala and Loke 2009). Palm oil refining in the early years also qualified for "pioneer status," which provided tax holidays and excluded exporting companies from national equity ownership rules (Rasiah 2006).
- *Strong support to R&D*: Public investment in R&D complemented a long history of private R&D in the sector. A specialized Palm Oil Research Institute of Malaysia was established in 1979 and is now part of the Malaysian Palm Oil Board (MPOB), supported by a levy on exports. An early success was achieved with the introduction of the pollination insect *Elaeidobius kamerunicus* from West Africa in 1982 to solve a long-standing problem of pollination (Tate 1996; Teoh 2002). The Malaysian public and private sector is estimated to invest at least US$180 million in R&D[4] (45% private) on oil palm alone, equivalent to 2% of farm-level production value

[4] In 2005 purchasing parity power dollars.

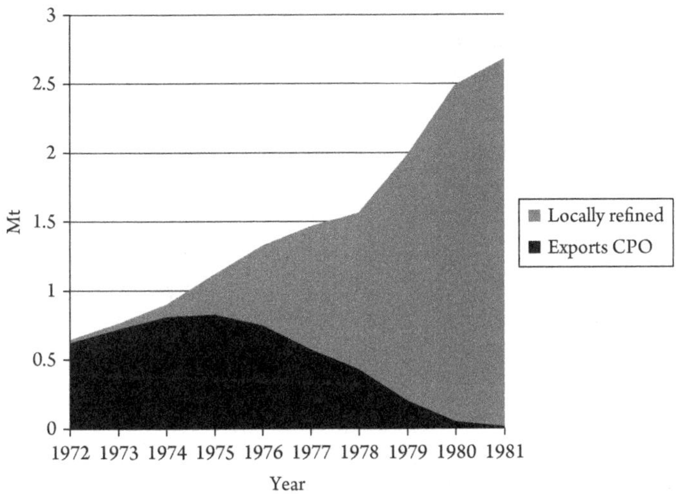

Figure 2.7 The transition from exports of crude palm oil (CPO) to local refining, Malaysia, 1972 to 1981.

Source: Data are calculated from Rasiah, R. 2006. Explaining Malaysia's export expansion in oil palm and related products. In: *Technology, adaptation, and exports: How some developing countries got it right*, ed. V. Chandra, 163–192. Washington, DC: The World Bank.

in the sector (Agricultural Science and Technology Indicators and Malaysian Agricultural Research and Development Institute 2012).

- *Development of strong marketing institutions*: The Kuala Lumpur Commodity Exchange, established in 1980, operates the world's first palm oil futures market and has become the benchmark for pricing palm oil (Pletcher 1991). In addition, the Malaysian Palm Oil Council (MPOC), also supported by an industry levy, has financed advocacy programs and trade missions to build markets, especially in light of the US soybean industry campaign against tropical oils during the 1980s and environmentalists' campaigns against oil palm in recent years. An export credit fund was also available to encourage developing countries to import Malaysian palm oil.

A hallmark of the Malaysian industry has been strong collective action by the major players and effective public–private coordination and regulation. A number of statutory bodies already mentioned engaged private players and their financing, including MPOB and MPOC. In addition, the Palm Oil Registration and Licensing Authority (PORLA), now also part of MPOB, licensed all participants in the industry and monitored the quality of exports. The Malaysian Palm Oil Association (MPOA) is the umbrella organization of private actors to improve coordination along the entire value chain (Teoh 2013). Some argue, however, that such close public–private collaboration has led to a culture of patronage and rent seeking, especially in Sarawak (Varkkey 2012a; Cramb 2013).

Currently, the palm oil industry is one of Malaysia's most important industries, accounting for 7% of foreign exchange, employing directly close to half a million people,

Table 2.1 Downstream industrial capacity in the Malaysian palm oil value chain, circa 2012

Sector	*n*	*Capacity (Mt)*
Crushing mills for CPO	408	93.2
Palm kernel crushers	41	5.2
Refineries	51	23.5
Oleochemicals	17	2.6
Biodiesel	29	2.7

CPO, crude palm oil.

Source: Data are from European Union Delegation to Malaysia. 2012. The Malaysian palm oil sector—overview. Agenzia per la Promozione all'estero e l'internazionalizzazione delle Imprese italiane. http://www.ice.gov.it/paesi/asia/malaysia/upload/173/Palm%20Oil_overview_2012.pdf (accessed 31 Jan 2016); and Sime Darby. 2009. *Palm oil industry in Malaysia: Skills and knowledge for sustained development in Africa.* Washington, DC: The World Bank.

and employing another 400,000 in downstream industries (European Union Delegation to Malaysia 2012). The depth of this industry, with nearly 100 downstream processing and manufacturing plants in addition to crushing mills, is shown in Table 2.1. The industry has maintained a strong export orientation, exporting as much as 95% up to the 1990s; since then, the increasing use of oleochemicals has accounted for 14% of the industry's exports. A recent effort to jumpstart a biodiesel industry involves the rapid establishment of 29 plants to meet a mandatory blending target of 5%. The jury is still out on this venture. Only 10 plants were operating by 2010 because of high feedstock prices (European Union Delegation to Malaysia 2012) (see Chapter 6).

The industry is also experiencing further consolidation. In 2007, Guthries and Golden Hope Plantations merged with Sime Darby to form one of the largest agricultural production companies in the world, profiled in Box 2.2. Forty-one companies are listed on the Bursa Malaysia (the largest are described in Table 2.2). Many have multinational operations in plantations and downstream industries in Indonesia and more recently in Africa. They are also investing extensively in downstream activities in importing countries, such as refining, oleochemicals, and logistics, including investments in processing other types of vegetable oils. Some evidence indicates that privately owned companies are performing better than companies with majority state equity or with mixed ownership (Ramasamy et al. 2005). The same study, conducted before the 2007 merger of the big three, also finds evidence of diseconomies of size for large companies.

At the same time that Malaysian companies have been investing outward, inward investment into the industry, mainly from Singapore, has increased sharply. Majority-owned foreign companies manage nearly 1 Mha of Malaysian oil palm plantations, led by Singapore-based Wilmar (Teoh 2002).

Box 2.2 Sime Darby Berhad: A Global Leader in Oil Palm

Since William Sime and Henry Darby started developing large rubber plantations and other operations in Malaysia in 1910, growth, mergers, and acquisitions have made Sime Darby one of the world's largest plantation companies. This sprawling conglomerate maintains interests in many sectors (automobiles, properties, industry), but the plantation division generally provides about one third of its profits. In 2013, it had 600,000 ha of planted oil palm and 61 mills, mostly in Malaysia and Indonesia, producing about 5% of the world's supply of palm oil. It owns extensive downstream operations in 14 countries to refine palm oil and produce oleochemicals, processed fats and foods, and biodiesel as well as upstream operations in Malaysia and Indonesia to produce seed and provide advisory services. The plantation division generated profits of about US$1 billion in 2012 and US$750 million in 2013, but fell to about US$300 million in 2015 as prices dropped. Also in 2015, the company invested US$1.7 billion to purchase New Britain Palm Oil Limited in Papua New Guinea.

The company is focusing on improving yields through research and replanting with high-yielding varieties and clones. It also seeks to improve labor productivity by mechanizing many operations. The company's strong commitment to sustainability includes certification of 100% of its production by the Roundtable on Sustainable Palm Oil, which has made Sime Darby the world's largest producer of certified sustainable palm oil. In 2014, the company announced its commitment to zero deforestation.

Sime Darby launched a large operation in Liberia in 2009 that projects a plantation of 220,000 ha and an investment of more than US$2 billion. This project had planted only 10,000 ha by 2014 as a result of land conflicts and an Ebola outbreak (see Chapter 9).

Source: Teoh, C. H. 2013. Malaysian corporations as strategic players in Southeast Asia's palm oil industry. In The palm oil controversy in Southeast Asia: A transnational perspective, ed. O. Pye and J. Bhattacharya, 19–47. Pasir Panjang: Institute of Southeast Asian Studies; Sime Darby Berhad. 2013. Global reach local solutions annual report 2013. Kuala Lumpur: Sime Darby Berhad; and Oxford Economics. 2014. Making FDI work for sub-Saharan Africa: Lessons from Liberia. Oxford: Oxford Economics. Company interview.

FELDA has also become a large global player. In 2012, it restructured into Felda Global Ventures (FGV) and Felda Holdings. The initial public offering (IPO) for FGV was the largest in Asia and the second largest globally in 2012, after the IPO for Facebook. FGV, in addition to 425,000 ha of plantations in Malaysia and Indonesia, operates in 10 countries, including downstream operations in canola and soybeans. It now has majority equity in Felda Holdings, bringing the total plantation area of the FELDA group to more than 800,000 ha and making it the world's largest producer of palm oil.

Today, government-linked (through PNB) and private companies manage 61% of the plantation area, and this proportion rises to 75% if FELDA is included. The remaining area is occupied by other federal and state schemes as well as independent smallholders, who expanded their share to 14% in 2012 (Figure 2.8).

Table 2.2 Largest Malaysian oil palm companies, 2014

Company	*Plantation area, Malaysia (000 ha)*	*Plantation area, Indonesia (000 ha)*	*Plantation area, other (000 ha)*	*Downstream activities, Malaysia*
Sime Darby	314	299	10 (Liberia)	KC, POR, food,[a] OC, B
Kuala Lumpur Kepong	110	140	44 (PNG)	KC, POR, food, OC
Kulim	37		79 (PNG)	B
IOI Group	169	60		KC, POR, OC
Genting Plantations	94	74		
Tabung Haji	91	82		
United Plantations	41	10		POR, food, OC
Felda Global Ventures	344	56		KC, POR, food, OC, B

B, biodiesel; KC, kernel crushing; OC, oleochemicals and specialty fats; POR, Palm oil refining; PNG, Papua New Guinea

[a] Cooking oils, margarine, and similar products.

Source: Company websites.

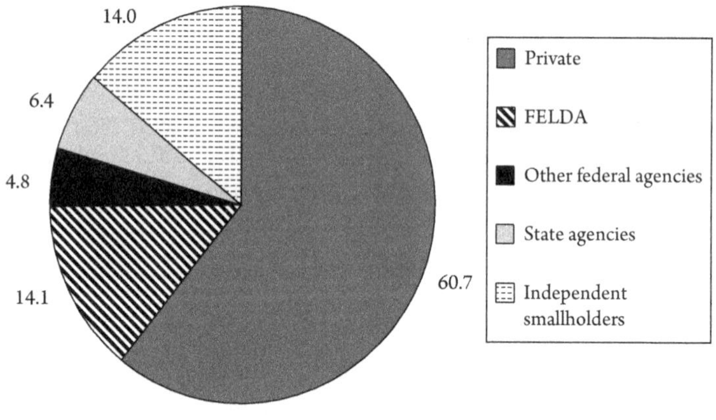

Figure 2.8 Percentage of oil palm area by ownership, Malaysia, 2012. Some of the largest "private" companies have majority state equity through Malaysia's sovereign wealth fund. FELDA, Federal Land Development Authority.

Source: Data are from Malaysian Palm Oil Board (2013).

Despite the apparent success of the Malaysian industry, it faces major challenges in ensuring competitiveness and improving sustainability. A first set of challenges is land and labor scarcity (which are growing) and yields (which are not). Based on the official policy of leaving 50% of Malaysia's land area forested, MPOB estimates that only 0.6 Mha may be available for expansion (European Union Delegation to Malaysia

2012). After yields grew rapidly during the early years to around 1980, they stagnated at around 3.5 t/ha until the early 2000s. As we discuss later in this chapter, strategies for increasing productivity include replacing the 25% to 30% of trees that are more than 25 years old (European Union Delegation to Malaysia 2012), employing best management practices in existing areas, and making long-run investments in R&D to shift out the yield frontier.

A second challenge is that rural wages are increasing quickly. Reminiscent of the colonial period, more than 80% of the labor force is foreign, a source of considerable social tension (Saravanamuttu 2013). Even with the dependence on foreign labor, wages of about US$12 per day are much higher in Malaysia than Indonesia—the main source of immigrant labor—and, with higher land prices, costs are also significantly higher (about 20%) in Malaysia.

A third challenge is to place the expansion in Sabah and Sarawak on a more sustainable footing. The social and environmental impacts of oil palm in Peninsular Malaysia have generated relatively little controversy, in part because oil palm production was built on a well-established plantation industry. The situation is different in the East Malaysian states of Sabah and Sarawak, which have most of the new oil palm area planted since 1995. That land belonged to extensive upland farming and agroforestry systems, much of it forested. In Sarawak in particular, the shift to plantation agriculture represented a sharp shift in policy from the colonial and early independence eras, when plantation agriculture was discouraged in favor of developing cash crops such as rubber as part of smallholders' existing farming systems (Cramb 2007). The shift in policy reflects close ties between state officials and their families and investors (Cramb 2013; Straumann 2014). The result has been an escalating controversy about human rights and deforestation similar to the controversy in Indonesia.

In summary, Malaysia has built a very large and dynamic palm oil industry that has contributed substantially to economic development. The industry has been not only a major source of growth and foreign exchange, but also has been associated with one of the best records of any country in reducing poverty. The poverty rate in the agricultural sector fell from 68% in 1970 to 12% in 1997 (Simeh and Ahmad 2001). The government clearly used what economists call "industrial policy" to pick a winner and has provided strategic and consistent support to the sector over time (Lin 2012).[5] Ironically, Malaysia's success in oil palm has left it with the original problem it planned to solve: the agricultural sector depends heavily on one crop, now oil palm rather than rubber.

INDONESIA: THE LARGEST AND MOST CONTROVERSIAL PRODUCER

We have already described how Indonesia's pioneering success in establishing large oil palm plantations was eclipsed during World War II and how the industry declined

[5] Similar efforts to diversify the sector through coconut and pineapple production failed, and even cocoa, which bloomed for a while, was not sustained (Tate 1996).

after independence, especially after foreign-owned plantations were reorganized into state-owned companies in 1968. Since then, the state's role in the oil palm sector has evolved steadily, reflecting changing ideological currents, donor priorities, and fiscal resources (Barlow et al. 2003; McCarthy and Cramb 2009). Starting in the 1970s, the government invested in state-owned companies with support from the World Bank to establish nucleus estate schemes (NESs), with a nucleus estate and associated smallholders. Beginning in 1986, the policy was to encourage greater involvement by the private sector, and both Indonesian and foreign (mostly Malaysian and Singaporean companies) invested heavily in the industry. The state continued to fund infrastructure and provide land concessions and cheap credit to both companies and smallholders.

After 1996, a partnership model was introduced with local communities to facilitate access to land by plantation companies and, since 2000, most land concessions have been managed by local governments in the newly decentralized administrative structure (Larson 1996; McCarthy et al. 2012). Companies still received special credit lines although they were no longer subsidized.

Today, Indonesia is the largest producer and exporter of palm oil in the world, after overtaking Malaysia in 2006 in production and 2008 in exports. The island of Sumatra, Indonesia's traditional stronghold of plantation agriculture, has about 70% of the country's oil palm area. Most of the rest is in the new frontier of the Kalimantan provinces of the island of Borneo (McCarthy and Cramb 2009). In 2011, palm oil and palm kernel oil together accounted for 12% of the value of agricultural output (second after rice) and provided US$20 billion in foreign exchange earnings, or half of agricultural exports (FAOSTAT). The downstream industries of processing and refining also contribute substantially to the economy, although Indonesia was late relative to Malaysia in emphasizing local refining capacity and downstream manufacturing of oleochemicals and other industrial products (Rifin 2011).

At least 1.7 million small- and medium-scale producers are involved in the sector; they employ as many as 3 million people, and the estate sector employs close to 2 million laborers (International Finance Corporation 2013). When employment in processing and downstream industries is included, the total number of jobs is estimated to be as high as 6 million (Varkkey 2012b; International Finance Corporation 2013).

The highly competitive position of palm oil in the Indonesian domestic market and in global markets reflects Indonesia's position as the world's lowest-cost producer, as a result of its low land and labor costs and reasonably high yields (3.5–4.0 t/ha). Much of the technology and management was a direct spillover from Malaysia, especially after the sector was opened to foreign investors.

The many similarities in the Malaysian and Indonesian experience with oil palm plantations must be set against some important differences. These differences relate to ongoing tensions over the relative emphasis on domestic versus export markets, the involvement of smallholders, and the controversies over oil palm and deforestation.

Markets, Policies, and Incentives: Food Versus Exports

Support to the Indonesian oil palm industry during the 1970s and 1980s aimed largely to maintain self-sufficiency in vegetable oils in the domestic market, and policies toward the industry reflected this objective in general. Vegetable oils were regarded as a strategic commodity for poor consumers, much like rice. The realization that palm oil could be produced at much lower prices than coconut oil, the dominant vegetable oil at the time, occasioned the remarkable switch by consumers to palm oil (described in Chapter 4).

Policies aided the transition to palm oil, including a variable export levy to reduce consumer prices, domestic price ceilings, occasional export bans, and allocation of quotas to supply the domestic market, especially for state-owned companies. Overall, the policy environment was quite unstable; Rasiah and Shahrin (2006) identify no less than 15 trade and consumer policy changes from 1978, and Rifin (2011) counts 11 changes in export tax rates in the more liberal market period from 1997 to 2007. Despite these multiple policy changes, the effective tax on the sector averaged close to zero from 1970 to 2004 (Fane and Warr 2007).

Another implication of Indonesia's early focus on the domestic market was that downstream industries to add value to exports developed more slowly than in Malaysia. Much of Indonesia's CPO was shipped to Malaysian companies to refine before being exported once again. Even in 2008, more than half of Indonesian palm oil was exported as CPO (Rifin 2011). Recent policies have clearly favored investment in local refining, however; the export tax on refined oil was reduced from 25% to 10% in 2012, and was 0% in 2014. In 2015, an export levy of US$50/t was introduced primarily to support a biofuel policy to convert 20% of diesel consumption to biofuels (see Chapter 6).

Finally, policymaking authority is more scattered in Indonesia than in Malaysia, where a strong central government—especially the Ministry of International Trade and Industry and the MPOB—provide overall policy direction to the industry. In Indonesia, the institutional environment has been dispersed across the Ministry of Agriculture (Directorate General of Estate Crops, Indonesian Oil Palm Research Institute, Plantation Crops Advisory Service), the Ministry of Forestry, and Bappenas (the Ministry of National Development Planning) without an umbrella organization such as MPOB (Rasiah and Shahrin 2006). Since 2000, this policymaking environment has become even more diffuse as decentralized provincial, district, and village authorities have been more active in setting the local development agenda.

Agrarian Structure: Estates Versus Smallholders

The tension between developing the palm oil industry through large plantations or smallholder family farms is an enduring theme in Indonesia. In fact, these alternatives may not be as stark as they appear at first glance, because both groups of producers have become increasingly heterogeneous in recent years.

Estates Companies

A handful of companies controlled the sector up to 1986 under the Suharto regime. Alongside the state-owned companies (which by definition responded to the state), the private companies that entered the industry at that time were concentrated into four groups—Sinar Mas, Astra, Salim, and Raja Garuda Mas—which maintained close ties to the government and invited patronage politics. Former government officials and others connected closely to the government acted frequently as corporate advisors or board members (Varkkey 2012a). After 1990, the company landscape gained complexity with the entry of foreign investors, mostly Malaysian and Singapore-registered companies investing directly or in partnership with local companies. Table 2.3 lists some of the largest domestic and multinational players.

With few large contiguous areas remaining in the more established areas, such as Riau Province in central Sumatra, large estate development slowed after 2000, and smaller and medium companies and individuals with up to 1000 ha entered the industry (Nagata and Arai 2013). Patronage persists in the decentralized governance environment. Rent seeking—by local officials charged with facilitating plantation development through partnerships of companies and communities—still influences outcomes strongly. McCarthy et al. (2012), for example, note that district officials often hold shares in smaller local companies.

Supported Smallholders

Since 1976, Indonesia has consciously sought to involve smallholders in oil palm development through a number of schemes, starting with the NESs. These schemes had multiple objectives. On the one side, they made it easier for companies to obtain the land and labor required to establish large-scale plantations and promote economic development in the relatively sparsely populated outer islands. On the other side, the schemes had the social objective of reducing poverty by working with programs to resettle poor people from densely populated Java to the outer islands and by providing income-generating activities for local communities dependent on subsistence farming.

The details of these schemes have varied over time. With the initial NESs, smallholders—either local people or settlers—provided labor to establish the plantation. During the subsequent production phase, smallholders, usually organized into cooperatives, managed their 2-ha plots under the supervision of a nucleus estate, the *inti*. Smallholders, or so-called *plasma,* in turn were obligated to sell FFBs to the nucleus estate mill according to a price formula based on world prices, with deductions for an establishment loan and extension services (Vermeulen and Goad 2006). The government provided financing for smallholder oil palm establishment, initial living expenses, and housing, and estates provided extension and processing services. When the loan was repaid, smallholders received title to an individual plot or the equivalent in company equity. During the early years, the ratio of inti to plasma was 20:80, but in later years the share of smallholders was smaller (Vermeulen and Goad 2006). In the

Table 2.3 Some major palm oil companies operating in Indonesia

Company/group	*Origin*	*Oil palm area in Indonesia (000 ha)*	*Other activities*
PT Perkebunan Nusantara (PTPN I-IV)	Indonesia (state owned)	650	Rubber, cacao, coffee, and others
Golden Agri Resources	Indonesia/ Singapore	470, including plasma (smallholders)	Cooking oil, margarine, and shortening
Astra Agro Lestari	Indonesia	281, including plasma and rubber	Rubber
Indofood Agri Resources	Indonesia	217	Palm oil refining, cooking oil, margarine and shortening, rubber, and sugar
Sime Darby	Malaysia	208	Research and development, seed, trading
Wilmar International	Singapore	171	Largest palm oil trader and wide-ranging food processing and manufacturing, and biodiesel
First Resources	Indonesia/ Singapore	170	Refining and biodiesel
Asian Agri	Indonesia (private)	160	Forestry and pulp
Kuala Lumpur Kepong	Malaysia	140 (land bank)	Rubber
Musim Mas	Indonesia (private)	120	Cooking oil, oleochemicals, and biodiesel
Bumitama Agri	Indonesia/ Singapore	115	NA
Genting Plantations	Malaysia	74	NA
Sampoerna Agro	Indonesia	70	Rubber, timber

Source: Company websites. NA is not available

most successful NESs, a cooperative work unit managed smaller units of about 50 ha (Jelsma et al. 2009).

Most evaluations have graded the NESs as reasonably successful (for example, see Zen et al. [2006]). Some have been highly successful (Jelsma et al. 2009). Land conflicts were common because the government allocated "state-owned" forest land to NESs, although local communities used that land under customary tenure arrangements for their extensive, long-fallow farming systems (Colchester et al. 2006). Land conflicts also erupted as settlers sponsored by the transmigration program and independent migrants flowed into the newly opened areas. Even official statistics count 3500 land conflicts related to oil palm in recent years (Jiwan 2012)—a theme we examine in Chapter 9.

In part to address these problems, the NES programs evolved into the Koperasi Kredit Primer Anggota (KKPA)[6] program during the late 1990s. That program handed many responsibilities to cooperatives, which were expected to provide land through a benefit-sharing agreement negotiated with companies. The emphasis also shifted to working with local communities as the transmigration program was phased out. During the mid 2000s, the KKPA evolved into the Partnership program facilitated by local officials under the newly decentralized governance structure. During this period, the share of participating smallholders generally declined to as little as 20%, the minimum required to receive a license to establish a plantation (McCarthy et al. 2012), and many plasma were managed centrally as part of the nucleus plantation, with smallholders receiving equity shares (Vermeulen and Goad 2006). Table 2.4 summarizes the evolution of these various schemes to develop oil palm production by smallholders.

Independent Smallholders

One of the most important developments in Indonesia has been the rise of so-called independent smallholders (defined by the Roundtable on Sustainable Palm Oil [RSPO] as having less than 50 ha) and medium-size producers (having more than 50 ha). These producers have capitalized on the increasing density of palm oil mills in established areas and the increase of standalone mills that do not have their own plantations. For example, Riau Province had 173 palm oil mills in 2009; 46 of them had no plantation and depended entirely on purchases from independent producers (World Wildlife Fund—Indonesia 2013).

The growing density of mills has provided a competitive market for FFBs within an accessible area to enable many small and medium-size farms to enter the industry—often rubber farmers or plasma participating in an established NES (Feintrenie et al. 2010). In the largest oil palm-producing province of Riau, the area of these independent producers considerably exceeds that of the plasma NESs and now rivals that of the companies (Figure 2.9).

[6] Members' Primary Credit Cooperative.

Table 2.4 Evolution of smallholder development schemes in oil palm, Indonesia

Scheme	*Period*	*Participants*	*Benefit sharing*	*State role*
NES or PIR (Proyek Inti Rakyat) schemes	1977–1985	Local farmers and transmigrants with varying emphasis	Sixty percent to 80% land to smallholders for about 2 ha, and later 1 ha for food crops and housing; 30% output to pay loans; estate charges for inputs and extension	Mainly state-owned companies, provision of land and subsidized credit
PIR *Trans*	1986–2000	Local farmers and transmigrants	Villages provide land as condition for inclusion	Government of Indonesia credit line, initially subsidized; facilitation of access to land
KKPA (Koperasi Kredit Primer Anggota)	1995–early 2000s	Local farmers and transmigrants	Village cooperatives provide land for inclusion with variable benefit sharing, with minimum 20% to smallholders in land or in company equity	Mostly facilitation
Partnership	After 2005	Mostly large scale, with local communities and farmers	Similar to KKPA but also joint ventures in which villages receive equity in exchange for land	Local government facilitation of agreements, national credit line reinstituted

Source: Constructed from information in McCarthy, J. F. 2010. Processes of inclusion and adverse incorporation: Oil palm and agrarian change in Sumatra, Indonesia. *The Journal of Peasant Studies* 37 (4): 821–850; Vermeulen, S., and N. Goad. 2006. *Towards better practice in smallholder palm oil production*. London: International Institute for Environment and Development; and industry interviews.

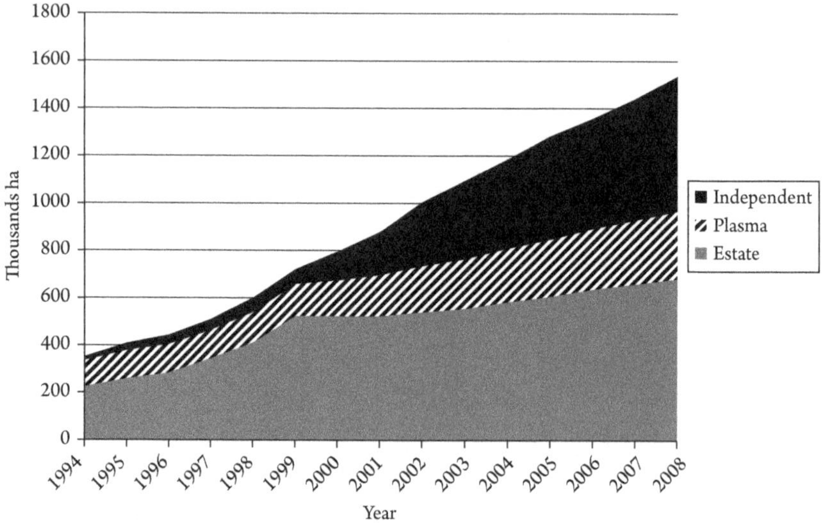

Figure 2.9 Trends in the structure of oil palm production, Riau Province, Indonesia, 1994–2008. *Source:* Data are from Nagata, J., and S. W. Arai. 2013. Evolutionary change in the oil palm plantation sector in Riau Province, Sumatra. In: *The palm oil controversy in Southeast Asia: A transnational perspective*, ed. O. Pye and J. Bhattacharya, 76–96. Pasir Panjang: Institute of Southeast Asian Studies.

Given the substantial costs of establishing oil palm (especially the cost of high-yielding seedlings and fertilizer), the difficulty of obtaining long-term financing, and many smallholders' lack of experience with oil palm, oil palm producers are becoming more differentiated. Many independent producers with 10 to 200 ha have more resources than producers with a typical NES holding of 2 ha (World Wildlife Fund—Indonesia 2013). Yet, Indonesia also has many poorly resourced independent smallholders who use inferior planting materials, apply low levels of input, lack access to extension, and obtain correspondingly low yields (Zen et al. 2006). Indeed, with land prices increasing, many smallholders cannot compete and have sold to larger farmers (McCarthy 2010).

Over time, both corporations and "smallholders" have become much more heterogeneous groups of producers. Where the NES model is still used, the ratio of inti to plasma is now much higher than in earlier years of strong state support for NESs. The growth of independent millers clearly favors the growth of independent smallholders, although a change in policy in 2007 to prohibit independent mills may slow this trend (Susanti and Budidarsono 2014). In Chapter 8 we provide a thorough review of these options.

Sustainability: Development Versus Conservation and Reduced Land Conflict

The most visible issue for oil palm in Indonesia, at least from a global viewpoint, has been its close association with deforestation and land conflict. Seventy percent of the oil palm planted from 1982 to 1999 was located on forest land owned legally by the state and managed by the Ministry of Forestry. Not all of that land was forested—much had been logged or cleared for other purposes—but much was already used by local communities under customary arrangements to which the state gave only notional recognition. At least half the expansion of oil palm area from 1990 through 2005 was achieved by converting forests to tree crop farms, mainly oil palm (Koh and Wilcove 2008).

Another environmental concern is that about 25% of oil palm was planted on peatlands, which sequester large amounts of carbon that is emitted when the land is drained and cleared (World Bank 2012). Not surprisingly, Indonesia is rated the world's third-largest emitter of greenhouse gases (GHGs), and land-use changes account for 60% of its GHG emissions (Brockhaus et al. 2012). Finally, the cost of the biodiversity lost when oil palm encroaches on biodiversity hot spots (areas of high biodiversity) adds to the tally of environmental concerns.

Indonesia's oil palm boom occurred during the 1990s, right after the Rio Earth Summit, which produced a series of international agreements strongly supporting the conservation of tropical forests, such as the International Convention on Biological Diversity and the Framework Convention on Climate Change. The threat to tropical forests became even more contentious in 1997 when huge forest fires in Indonesia seriously reduced air quality in neighboring Malaysia and Singapore. The fires have occurred repeatedly since then (and notably in 2013 and 2015), and many were traced to the clearing of land for oil palm (Sheil et al. 2009; Marlier et al. 2015). The haze from these fires has caused massive economic and health costs in the region (Chapter 9).

Although oil palm is often linked to deforestation, the role of oil palm in relation to other drivers of deforestation is complex. Official figures indicate that 28 Mha were deforested between 1990 and 2005—an area that dwarfs the roughly 5 Mha planted with oil palm. In many cases, companies receive concessions to grow oil palm, but their major motivation is to extract timber from the land or speculate on its future value rather than produce oil palm. Nagata and Arai (2013) carefully document what happened to the 2.95 Mha granted for oil palm production in the major producing province of Riau from the 1970s to 2007. Ultimately, only 1.3 Mha were planted, 0.8 Mha were held by active companies but not planted, and 0.9 Mha were allocated to companies that became inactive. Land allocated to companies that became inactive is very likely to have been used primarily for timber. For active companies holding unplanted land, an element of speculation is probably involved.

In Kalimantan, where oil palm concessions covering about 10 Mha have been awarded, only about 5% of that area was planted by 2010, although much of the unplanted area had likely been logged (Dwyer 2015). Local governments that depend on the taxation of timber for revenue have a strong incentive to allocate large forest land concessions, often disregarding national laws, in the name of oil palm plantations, but in reality for logging.

These problems reflect the generally poor governance of forest resources and weak regard for customary tenure, combined with a good dose of rent seeking by political interest groups (Brockhaus et al. 2012). The government has introduced various reforms, such as devolving land negotiations to companies and local communities and giving local rather than central government authorities responsibility for issuing plantation permits, yet the problem remains, sustained by collusion among village officials, local government, and companies. Indeed, revenue-hungry local governments appear to be engaged in a "race to the bottom" to attract investors into the plantation sector, as seen in the relationship between the rate of deforestation and the local election cycle (Burgess et al. 2012). Rent seeking aside, in general "land allocation in Indonesia is characterized by a complex landscape of institutions and legal frameworks" (Brockhaus et al. 2012, p 34) with conflicting policies and a lack of institutional clarity, especially between different levels of government (World Bank 2012).

Good estimates on the future trajectory of changes in land use are challenging to obtain. About 60 Mha are claimed to be suitable for oil palm, compared with about 9 Mha currently planted (Varkkey 2012b). The USDA estimates that companies have about 6 to 7 Mha in their land bank that has not been planted (Foreign Agricultural Service 2013). The big question is how much of that land is forest with a high conservation value or considerable carbon stocks. Officially, about 40 Mha of the state-controlled forest estate is not forested, including the *alang alang* lands infested with *Imperata* grass, but much of that land is already in use under customary tenure arrangements that are now recognized legally (Fairhurst and McLaughlin 2009). Subtracting forest land and land with a population density of less than 25 persons/km^2, Deininger and Byerlee (2011) estimate that 6.7 Mha is available for expansion—still a sizable area.

The continuing outcry over the social and environmental costs of oil palm development elicited a response. For example, the government, gearing up for incentive payments to conserve forests under REDD+, imposed a partial moratorium on further land conversion. Some powerful plantation companies and trading companies have agreed variously to certify their social and environmental sustainability, commit to zero deforestation, stop planting on peat land, and not exploit local communities. Together, these initiatives portend exciting opportunities for Indonesia as its struggles to reduce the social and environmental costs of rapid expansion in the oil palm sector. In Chapter 9 we provide a deeper analysis of the likely success of these initiatives.

WEST AFRICA: MIXED SYSTEMS AND MIXED SUCCESS IN THE ORIGINAL "OIL BASKET"

We now return to the original home of oil palm in West and Central Africa. The fact that oil palm and its products are part of the cultural fabric and the traditional farming systems in West and Central Africa has had a number of implications for the development of the industry. First and foremost, the dura oil palm was integrated into local farming systems as a wild or semiwild species by leaving palm groves as well as individual trees in long bush–fallow systems (Martin 1988). Second, on the consumption side, red palm oil processed in small mills is essential to the local cuisine (Cheyns and Rafflegeau 2005). The red oil from dura palms is processed with a very low extraction rate of about 10% of the weight of FFBs and has a high free fatty acid (FFA) content that is well above today's international standard for CPO (less than 5%) (Foundation for Partnership Initiatives in the Niger Delta 2011). Red oil is used for sauces and is also processed into soap (mostly the oil with very high FFA levels), in roughly equal amounts (Ofosu-Budu and Sarpong 2013). Third, women have had a leading role in the sector through control of processing and distribution (Martin 1988; Ibekwe 2008). Last, palm trees provide important by-products. With a ratio of palm kernel oil to palm oil of 0.4 in Africa (because of the use of dura palms and low oil extraction), palm kernel oil is relatively more important in Africa than in Asia, where the ratio is only 0.1. In Nigeria, palm kernel oil is often blended with palm oil (Gold et al. 2012). Palm trees are also prized for the production of palm wine, an important local alcoholic drink.

The traditional value chain has undergone important changes through small-scale mechanization of processing and the gradual shifting of production to smallholder plantations, yet the value chain's basic elements, especially the use of red palm oil for preferred local foods, remain unchanged. Most estimates concur that the small-scale sector still accounts for about 70% of the oil consumed in the region, although statistics on this sector are notoriously unreliable (Gold et al. 2012; Ofosu-Budu and Sarpong 2013). Table 2.5 presents a rough but dated breakdown of the production subsectors in Nigeria.

These value chains use a wide range of processing methods based on their throughput capacity, oil extraction rate, and level of mechanization, as well as the use of a wet process (fruit boiled beforehand) versus a dry process (Hyman 1990; Poku 2002) (Table 2.6). All the resulting products are characterized by a high percentage of FFAs because of the delay in processing, which makes it easier to remove individual fruits from the fermenting fruit bunches using the prevailing manual techniques (Osei-Amponsah et al. 2012; Nchanji et al. 2013). The oil with the highest FFA level, probably more than half the total, is destined to soap manufacturing.

Table 2.5 Approximate breakdown of the oil palm sector in Nigeria, circa 2008

System	*Area (000 ha)*	*Production of palm oil (000 t)*	*Yield of palm oil (t/ha)*	*Share of national production (%)*
Semiwild groves	1700	567	0.33	55
Smallholder plantations	313	204	0.65	20
Estates	118	256	2.17	25
All	2131	1027	0.48	100

Source: Calculated from Gold, I. L., C. E. Ikuenobe, O. Asemota, and D. A. Okiy. 2012. Palm and palm kernel oil production and processing in Nigeria. In *Palm oil: Production, processing, characterization, und Uses*, ed. O.- M. Lai, C.- P. Tan, and C. C. Akoh, 275–298 Urbana, IL: AOCS Press; and Foundation for Partnership Initiatives in the Niger Delta. 2011. *A report on palm oil value chain analysis in the Niger delta.* Abuja: Foundation for Partnership Initiatives in the Niger Delta.

Table 2.6 Comparison of different scales of milling technology, Ghana

	Traditional artisanal	*Small scale*	*Medium and large scale*
Mill capacity (t FFB/hr)	<0.1	0.1–1	5–60
Palm area to supply mill (ha)	Small farm	20–100	3000–10,000+
Oil extraction rate (%)	8–10	12–13	16–23
Mechanization	Manual	Semimechanized	Full
Power supply	Firewood and mill waste	Electric or diesel	Electric, diesel, or self-generated
Free fatty acid content (%)	10–20	8-12	<5, but some higher for red oil
Main product	Red oil	Red oil	CPO plus some red oil

CPO, crude palm oil; FFB, fresh fruit bunch. Some small-scale production may be sold to large mills for processing into CPO.

Source: Data are from MASDAR Ministry of Food and Agriculture. 2011. *Masterplan study on the oil palm industry in Ghana*. Eversley: MASDAR Ministry of Food and Agriculture.

Today, it is useful to think of two main value chains, with considerable overlap and crossover (Foundation for Partnership Initiatives in the Niger Delta 2011), as shown in Figure 2.10: the small-scale sector, and the medium- and large-scale sector.

Commercially oriented smallholders, cultivating either dura or tenera palm, own or sell to semimechanized processors with a capacity of up to 0.5 t/hr and with an extraction rate of about 12% to produce red palm oil containing 8% to 12% FFAs. Processing in this sector is a major source of income, estimated to employ about 100 person-days/t oil output for an investment of less than US$10,000, compared with about 0.004 person-days/t in large mills involving an investment of about US$10 million. Small-scale producers located near large mills also have the option of selling their FFBs to large mills to produce CPO to international quality standards.

The medium- and large-scale sector is oriented mainly to producing international standard CPO, although a small share may be sold in the local red palm oil market. This sector consists partly of locally owned, medium-size plantations (up to 1000 ha) that are often integrated vertically to medium-scale mills with a processing capacity of at least 5 t/hr. Another part of this sector (akin to the Asian plantation systems) is made up of large plantation companies, often multinational, with vertically integrated industrial mills capable of processing 30 to 60 t/hr. Currently, most of these plantations are owned by the long-established plantation companies from Europe: SIAT

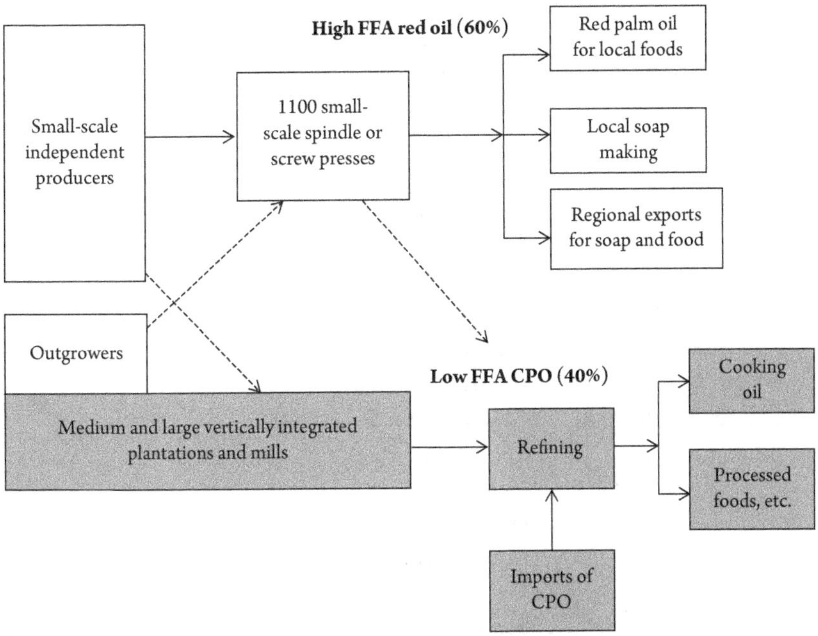

Figure 2.10 Schematic view of the Ghanaian oil palm industry. The international standard for CPO is less than 5% FFAs. Dotted lines indicate less important flows between the informal sector (blank boxes) and the formal sector (solid boxes). CPO, crude palm oil; FFA, free fatty acid.
Source: Adapted from Wilcock, D., and V. Kelly. 2014. *The economics of non-industrial oil palm production and processing in Ghana*. Utrecht: Solidaridad.

and Socfin, and SIFCA, a multinational plantation company founded in 1964 in Côte d'Ivoire (Table 2.7). In most cases, these companies took over state-owned plantations that had been established after independence and were privatized during the 1990s under structural adjustment programs. Many of the large plantations have outgrower schemes in which smallholders are obligated contractually to sell their fruit bunches to the plantation mill, although in practice they often sell in the open market to small processors.

Based on land concessions granted in recent years, these long-established companies will soon be overtaken by the very large Asian companies (Sime Darby, Golden Agri-Resources, and Olam) with long experience in oil palm and its derivatives, and by the Siva Group, a new entry from India with interests in oil palm in Asia and Latin America.[7] All these companies have interests in the region, with the largest area of new concessions in Liberia and Cameroon. Downstream in the value chain, the palm oil trading giants Wilmar, Olam, and Unilever play a key role, especially in local refining and manufacturing, often based on imports from their Asia-based operations.

[7] The Singapore-based energy company Wah Seong is another new Asian entry. It has announced plans for large oil palm investments in the Democratic Republic of the Congo.

Table 2.7 Major company actors in oil palm in west and central Africa, 2012

Company	*Origin*	*Target countries*	*Existing plantations (000 ha)*	*Associated smallholders (000 ha)*	*Total potential area (000 ha)*
SIAT	Belgium	Ghana, Nigeria, and Gabon	57	14	94
Socfinaf/ Bolloré	Luxemburg and France	Cameroon and Nigeria	52	12	64
SIFCA	Côte d'Ivoire[a]	Côte d'Ivoire and Liberia	49	89	144
Siva Group	India with Singapore registration	Sierra Leone, Liberia, Côte d'Ivoire, Cameroon, and Democratic Republic of the Congo	22	10	792
Olam	Singapore	Gabon and Ghana	11	12	224
Sime Darby	Malaysia	Liberia	10	0	220
Table total			190	125	1838
All west and central Africa			338	4115	7587

[a] Singapore-based Olam and Wilmar have a 27% stake in SIFCA.

In addition, Golden Veroleum Liberia (listed in Singapore) has obtained a 220,000-ha concession and Equatorial Palm Oil (UK listed), a 65,000-ha concession in Liberia. Felda Global Ventures (a Malaysian company) has reported to have a 100,000-ha concession in Guinea.

Source: Most data are from Hardman and Co. 2012. *Commercial viability and value creation in west African palm oil production*. London: Hardman and Co.

Palm Oil Markets

The two distinct palm oil products produced in West and Central Africa serve distinct markets: (1) red palm oil for local food and soap, and (2) international standard CPO that is refined further to produce cooking oil and to manufacture margarine and processed foods. The price of local red palm oil for domestic food use is generally below the price of international standard CPO, which is set by world prices adjusted by transport and tariff costs (although the reverse can be true in certain seasons).

In addition to domestic markets, palm oil is widely traded regionally and internationally. Traditionally, as we have seen, West Africa was the major exporter of palm oil for soap and industrial uses, and palm kernel oil for margarine. As the availability of standardized Asian palm oil low in FFAs jumped during the 1960s with the rapid expansion of oil palm in Malaysia, West African producers managed neither to improve productivity nor quality to serve international markets, nor even to supply the rapidly growing domestic market. West and Central Africa moved from being an exporter to a significant importer of vegetable oils, with palm oil making up 74% of all vegetable oil imports. In 2013, sub-Saharan Africa imported 3.5 Mt of a total supply of 6.6 Mt, or more than half the total consumed (Figure 2.11). Nigeria is a large net importer; its imports are estimated at 0.8 Mt. Only Côte d'Ivoire was a net exporter.

For many years, Nigeria's ban on importing palm oil fostered a thriving smuggling industry from neighboring countries. Even after liberalization, a 60% markup on the import price resulting from tariffs and taxes has maintained informal imports, mostly of red palm oil, estimated at 400,000 t (Foundation for Partnership Initiatives in the Niger Delta 2011). As in domestic markets, red palm oil imported informally from small-scale producers in neighboring countries is used for food and soap (Wilcock and Kelly 2014). The major part of imports is supplied as CPO or standardized refined palm oil from Asia and is destined for use as cooking oil or for the food-processing sector.

Although these market and structural characteristics are common across the traditional oil palm belt of West and Central Africa, some large differences among countries reflect variation in their colonial and postindependence policies for oil palm development, as illustrated by a brief description of the oil palm sectors of Ghana and Cameroon. These countries have similar populations and arable land area, but Cameroon, located in the Congo Basin, has less than half the population

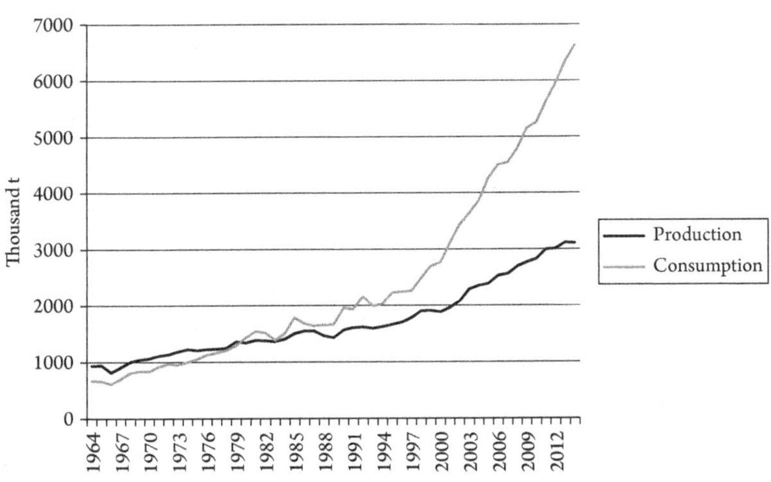

Figure 2.11 Production and consumption of palm oil and palm kernel oil in sub-Saharan Africa. *Source:* USDA-FAS PSD.

Table 2.8 Comparative statistics on the oil palm sectors of Cameroon and Ghana

	Cameroon	*Ghana*
Population density, 2011 (persons/km^2)	45	109
Arable land area, 2011 (Mha)	7.6	7.6
Area oil palm, 2008–2012 (000 ha)	108	354
Yield palm oil, 2008–2012 (t/ha)	2.7	0.4
Production of palm oil, 2008–2012 (000 t)	292	142
Growth in palm oil production, 1961–2012 (%/yr)	6.1	5.0
Palm oil supply, 2011 (kg/capita/yr)	10.2	8.3
Supply imported, 2011 (%)	6	63
Supply exported, 2011 (%)	1	40

Source: FAOSTAT.

density of Ghana (Table 2.8). The oil palm sectors of both countries have grown rapidly, at 5% to 6% annually since 1961, with important structural differences. Cameroon has the highest yields in Africa; Ghana has one of the lowest. Ghana engages much more actively in palm oil trade, both imports and exports, than Cameroon.

Cameroon: A Long History of Plantations

Cameroon has the longest experience of any country in the world with the cultivation of oil palm in plantations, dating from the first efforts by German colonial officials in 1907 (Nkongho et al. 2015). Even today, plantations remain important in Cameroon relative to other countries in the region, supplying more than half the palm oil output (Table 2.8). Hoyle and Levang (2012) estimate that 59,000 ha of oil palm are cultivated by large companies, including about 40,000 ha by three companies from the Paris-based Bolloré Group, either directly or as a major shareholder in the Luxemburg-based Socfin Group, and the remainder by two state-owned companies, Pamol and Cameroon Development Corporation. Associated with the plantations are 35,000 ha of supervised smallholders (Table 2.9).

Independent smallholders produce oil palm on another 100,000 ha, although they account for about only one third of national output (Table 2.9). A recent survey found that smallholders have relatively large oil palm plantings (20–50 ha) and that two thirds of that area is new land obtained by clearing secondary forest (Nkongho et al. 2014). Not surprisingly, farmers favor extensive management approaches employing migrant labor, but use few purchased inputs.

Smallholders either process their own produce or sell it directly or through intermediaries to small-scale mills that produce red palm oil for the local market. Nchanji et al. (2013) identified six types of processing technology in use, with capacities ranging from 0.1 t FFB/hr to 0.5 t FFB/hr. The oil extraction rate varies from 13% in the smallest mills to 18% in the largest—well above that observed in the small-scale sector

Table 2.9 Structure of the oil palm sector in Cameroon

System	*Area (000 ha)*	*Yield of oil (t/ha)*	*Production (000 t)*	*Production (%)*
Five large plantations	59	2.0	120	52
Supervised smallholders	35	0.9	30	13
Independent smallholders	100	0.8	80	35
Total	184	1.3	230	100

Source: Data are from Hoyle, D., and P. Levang. 2012. *Oil palm development in Cameroon*. World Wildlife Fund for Nature Report. Cameroon: World Wildlife Fund.

of neighboring Nigeria. Small-scale processing is also a very important source of jobs and income in Cameroon (Nkongho et al. 2014).

Average yields of FFBs in Cameroon are the highest in Africa and similar to yields in Asia, although palm oil yields are lower because of the low oil extraction rate (FAOSTAT). A World Bank study indicated that all systems are strongly competitive in import substitution, although less so for exports because of high transport costs (World Bank 2008a).

The government of Cameroon has maintained a consistent policy of support to the oil palm sector. Because a program to promote outgrowers failed—growers did not repay loans and often sold FFBs to small-scale processors who offered higher prices than the large mills—the government has focused on providing improved planting materials and extension to upgrade the independent smallholder sector. At the same time, to attract large investors into the sector, the government has created an aggressive program to provide land concessions. At least six companies are reported to be trying to secure up to 1 Mha for oil palm, the largest such program of land concessions for oil palm in Africa (Hardman and Co. 2012; Hoyle and Levang 2012). Yet, land concessions in Cameroon can pose serious risks of damaging forests of high conservation value or treading on the rights of local communities. The saga of Herakles Farms (Box 2.3) illustrates these risks and also shows how relations with the media and civil society have been mismanaged. With this experience, the government is increasingly turning to models to engage and upgrade smallholder to raise productivity.

Ghana: A Smallholder Industry

In contrast to Cameroon, in Ghana approximately 75% of oil palm is produced in the small-scale sector (MASDAR Ministry of Food and Agriculture 2011), and at least 60% is processed in the small-scale sector (as depicted in Figure 2.10). Of the roughly 300,000 ha of oil palm, only 20,000 ha is in four large-scale plantations, and another 20,000 ha is associated with outgrower schemes. Yields of small-scale producers are only about one third those of the estate sector.

The small area of plantations in Ghana is a legacy of the colonial policy favoring the small-scale sector. After independence, the focus changed somewhat to emphasize

Box 2.3 The Debacles of Herakles

The experience of New York–based Herakles Farms, through it is subsidiary SG Sustainable Oils Cameroon, illustrates the trials and tribulations of a new generation of investors in oil palm in Africa. Herakles had no apparent previous experience in oil palm production, yet in 2009 the government of Cameroon approved Herakles' project to develop an oil palm plantation of 60,000 ha. The fact that the plantation was surrounded by five conservation areas, including one national park, sparked controversy almost immediately. Although Herakles conducted an economic and social impact assessment that was accepted by the government and planned to obtain certification under the Roundtable for Sustainable Palm Oil (RSPO), the project quickly came under fire from a number of international nongovernmental organizations (NGOs), including Greenpeace, the Oakland Institute, and the Rainforest Foundation, as well as local communities and NGOs.

The most obvious issue was the threat to biodiversity, given that the project area is classified as a biodiversity hotspot and acts as a migration corridor for a number of species, including elephants. The area also includes 14,000 people whose livelihoods depend on the forest. Some of these people might be among the projected 10,000 individuals to be hired by the project, and some support it. Even so, little consultation with local communities appears to have occurred, and they have received no compensation. The contract with the government has not been released, although the annual land rental rate reportedly is US$1/ha after development, recently raised to US$6/ha.

In light of the controversy, Herakles withdrew from the RSPO, the government of Cameroon suspended the project briefly in 2013, and several civil action cases were pending in the courts in 2015. Given these high transaction costs, Herakles suspended the project and new managers have been appointed.

The Herakles case illustrates the risks and high transaction costs involved when inexperienced companies seek to invest in oil palm. In contrast, a large 100,000-ha oil palm project by Olam, in a joint venture of US$236 million with the government of neighboring Gabon, is proceeding with relatively little dissension. Olam has a track record in oil palm in Southeast Asia and has been much more transparent about assessing social and environmental impacts. The project has set aside land for local farmers to produce food, and it has assigned some 30,000 ha of plantations to outgrowers. All allocated land has been surveyed to assess its conservation value and, to date, 70% of the land concession has been set aside for conservation.

Source: Greenpeace. *2013. Herakles farms in Cameroon: A showcase in bad palm oil production.* Washington, DC: Greenpeace USA; Rainforest Foundation UK. 2013. *Seeds of destruction: Expansion of industrial oil palm in the Congo Basin: Potential impacts on forests and people.* London: Rainforest Foundation UK; Badgley, C. 2014. *When Wall Street went to Africa. Foreign Policy.* http://foreignpolicy.com/2014/07/11/when-wall-street-went-to-africa/ (accessed January 30, 2015); and Herakles Farms. 2011. Palm oil FAQ's. http://www.heraklesfarms.com/palm-oil-faqs.html (accessed January 30, 2015).

state-owned plantations, but even as late as 1970, only 2000 ha of plantation were in production (Fold and Whitefield 2012). Additional state-owned plantations with associated smallholder outgrowers were added during the 1970s and 1980s, mainly through World Bank support.

Led by the small-scale sector, the production of palm oil has increased rapidly since 1960. On the production side, a recent survey found that 93% of smallholder palms are of the tenera type, indicating that harvesting of wild dura palms has largely disappeared in the main production zone (MASDAR Ministry of Food and Agriculture 2011; Wilcock and Kelly 2014). The Oil Palm Research Institute has been instrumental in a wide distribution of improved tenera varieties, although farmers still use negligible amounts of other inputs such as fertilizer (Adjei-Nsiah et al. 2012).

Box 2.4 Economics of Small-Scale Production and Processing: Ghana

A survey of 78 producers in the main oil palm-producing areas of Ghana's central, eastern, and western regions in 2012 to 2013 indicates only about one quarter of them sell their harvest directly to an industrial mill. The remaining farmers process their entire harvest on a fee-for-service basis using semimechanized mills; 54% target the soap market and 44% target the market for unrefined red cooking oil. This group has an average farm size of 5.9 ha and obtains an average oil yield of 1 t/ha at a total cost of US$353/t (38% of this cost was for harvesting; 35% for processing; and only 24% for weeding, pruning, and other operations; the other 2% was fixed costs). The overall net return averages US$2464 per farm per year and US$21 per day of family labor, well above prevailing rural incomes and wages (although there is wide variation in returns).

Owners of mills—based on survey interviews with 44 of them—have even greater yearly incomes: US$4600 for operators of spindle presses and US$8800 for screw presses. The majority of mills use manual spindle press technology (coupled with a mechanized digester of cooked palm fruit), with an initial investment cost of US$6500 and an estimated oil extraction rate of 16.3%. Many mills have switched to the more advanced screw (expeller) press at an investment cost of US$10,500, with an estimated extraction rate of 19%, similar to a large industrial mill.

An important finding for rural development in Ghana is that the most promising approach is to upgrade small-scale processing rather than to attempt to link smallholders with large mills. Rural households invest in processing as a way to maximize the use of family labor in the off season and generate incomes well above what could be achieved through production alone. Access to finance and technical assistance to upgrade processing is the highest priority, but many mill owners also would like to increase their farm yields, given they make the greatest profits milling their own fresh fruit bunches. Yield increases will require intensification using improved genetic stock, fertilizer, and better weed control, with due attention to using scarce labor efficiently.

Source: Wilcock, D., *and V. Kelly. 2014. The economics of non-industrial oil palm production and processing in Ghana.* Utrecht: Solidaridad.

Smallholders seem to have no problem gaining access to land. As with the main cash crop (cocoa), a significant share of oil palm is cultivated under share tenancy agreements with landowners.

On the processing side, small semimechanized mills have been widely adopted and have largely replaced manual methods and hand-operated screw presses (Osei-Amponsah et al. 2012). Small-scale processing of oil is quite profitable and an important source of jobs in rural areas (Box 2.4), as in Cameroon. Because FFA levels often exceed 10%, much of the oil is used for soap, and an active export trade moves oil into neighboring countries for soap production (Wilcock and Kelly 2014).

The four major actors on the plantation side operate with major equity from Unilever, Wilmar, SIAT, and NorPalm (a Norwegian–based company with previous interests in oil palm in Ecuador) (Ofosu-Budu and Sarpong 2013). With the exception of NorPalm, all have large outgrower schemes amounting to 20,000 ha of additional palm area, although their outgrowers often choose to sell to small-scale processors (Box 2.5). To make up the shortfall, the large mills source from independent producers. Unilever, SIAT, and Wilmar also have major interests in refineries and importation of CPO. Unilever, in partnership with Nestlé, owns a large margarine factory to supply West Africa (MASDAR 2011).

The government of Ghana, on several occasions, has recognized the oil palm sector as a priority. Against the 300,000 ha currently planted to oil palm, 3 Mha are considered suited to the crop (MASDAR 2011). A President's Special Initiative on Oil Palm was launched in 2002 after a presidential visit to Malaysia. To facilitate private investment in plantations, this initiative proposed that corporate village enterprise companies could give small landowners equity shares in large plantations, but the scheme failed for lack of funding and coordination (Ofosu-Budu and Sarpong 2013). More recently, an oil palm master plan prepared for the Ministry of Food and Agriculture recognizes the challenge of finding tracts of land for large plantations and gives priority to developing smallholder production. The plan recommends establishing large outgrower programs associated with a 10,000-ha nucleus estate and upgrading the plantations of independent smallholders (MASDAR Ministry of Food and Agriculture 2011). Although smallholders are at the center of the plan, it focuses mostly on developing large mills to produce international standard CPO to substitute for imported palm oil—not on upgrading the local red palm oil value chain.

The Future: Attracting Large Foreign Investors or Upgrading Small-Scale Value Chains?

Since colonial times, oil palm strategies in West and Central Africa have wavered between emphasizing the formal sector of large plantations and associated large mills and supporting the small-scale sector, with its semimechanized processing methods and informal marketing channels. After a late start, the plantation sector was held back by state ownership and management, which undermined its efficiency. Since 1990, ample

Box 2.5 Ghana Oil Palm Development Corporation: Ghana's Largest Palm Oil Company

The Ghana Oil Palm Development Company Ltd (GOPDC) is the largest oil palm producer in Ghana and one of the most studied companies anywhere. Established in Ghana's eastern region in 1975 under state ownership (and with World Bank support), GOPDC was privatized in 1995. It is now 80% owned by SIAT, a Belgian plantation company that has interests in Côte d'Ivoire, Nigeria, and Gabon (the government of Ghana retains a 20% share). GOPDC has 6500 ha under its own plantation, employing nearly 3000 workers. Since privatization, it has invested in a palm kernel processing mill and a palm oil refinery.

The bulk of the palm fruit processed by GOPDC is provided by about 7000 outgrowers, each with around 2 ha. The outgrower scheme was the focus of a second phase of World Bank support and has expanded rapidly. A number of studies conclude contract growers have benefited from their association with GOPDC. The impacts of the scheme seem to be in doubt today, in part because farmers can sell to small processors if they offer higher prices. The resulting collapse in purchases makes it difficult for GOPDC to collect on outstanding loans. To use mill capacity more effectively, GOPDC is buying from independent producers nearby, but the mill still runs well below capacity.

GOPDC has also struggled with land issues since its inception. Ghana has a long tradition (extending even through the colonial period) of respecting customary land tenure under the control of local chiefs. Originally a state operation, GOPDC was established with land taken by the government without the consent of chiefs and without compensation. The resulting prolonged legal challenges have been inherited by the privatized company and were still underway in 2012.

Source: Huddleston, P., *and M. Tonts. 2007. Agricultural development, contract farming and Ghana's oil palm industry. Geography* 92 (3): 266–278; *Väth, S. J., and M. Kirk. 2013. Do land ownership and contract farming matter? Evidence from a large-scale investment in Ghana. Joint Discussion Paper Series in Economics no. 16-2014. Philipps-University, Marburg; and GOPDC. 2015. GOPDC. http:// http://www.gopdc-ltd.com (accessed September 10, 2015).*

evidence has emerged that when the plantation sector is put in private hands, it can compete with imported international standard palm oil. It is competitive even though production costs are believed to be significantly higher in Africa than in Asia because of lower yields and higher input costs (especially fertilizer costs) (Ofusu-Budu and Sarpong 2013). It is less clear that the plantation sector can be competitive for exports, although the low price of land concessions and low wages in some countries may tip the balance.

Deep pockets are needed to develop the plantation sector. With development costs estimated at US$8000/ha, a 10,000-ha plantation requires an investment of US$80 million. The massive 220,000-ha venture planned in Liberia by Sime Darby implies a total layout of more than US$2 billion over 20 years. Investments of this size clearly are highly visible undertakings that require access to global capital. The recent initiatives and influx of investors into the plantation sector are attracting scrutiny, generating controversy, and featuring in strongly critical reports from international civil society (Carrere 2010; Rainforest Foundation UK 2013).

The choices for governments, communities, and investors are not straightforward. The transaction costs of obtaining land for large plantations are very high in the relatively densely populated areas where oil palm is established, such as coastal West Africa. Most land is in use, and the users are defending their rights. Sime Darby's Liberia project is significantly behind schedule, stymied by the intricacies of accessing land and by protests from local communities and civil society (Evans and Griffiths 2013; Oxford Economics 2014). The alternative—investing in large mills supplied by contracted outgrowers—is also risky. Unlike smallholders in Indonesia during the early years, smallholders in West and Central Africa have an alternative market in the small-scale processing sector. Side-selling of palm fruit results in abysmal rates of loan repayment in all the African outgrower projects and causes large mills to operate at low capacity.

In the less populated areas, especially the Congo Basin, the granting of land concessions in the large forested tracts "owned" by governments is equally controversial. Concessions may overlap with forests of high conservation value, or the net GHG emission effects may be strongly negative. Undoubtedly large areas of degraded and secondary forests of lower environmental value are present, but the capacity to identify and demarcate such areas is extremely weak in the region. And even in a sparsely populated region, the customary rights of forest and land users must be respected.

Given the buoyant market for red palm oil produced in small-scale processing units, it is not clear that the best strategy for addressing the region's growing deficit of palm oil is to encourage production through the formal sector. To date, we have not identified a good study of consumer preferences and tradeoffs for local red palm oil and international standard palm oil to inform this debate, nor do reliable estimates exist on the size and competitiveness of the local soap industry, although it is certainly substantial. In any event, it seems the small-scale sector is upgrading gradually through planting improved varieties and adopting semimechanized processing methods. Programs to accelerate this improvement in a consistent manner over the long term may well be more efficient and equitable—and certainly far less controversial—than a strategy based on foreign investments in large plantations. Such a strategy would also do much more to generate employment, considering the large number of people employed in the small-scale processing sector and West Africa's growing problem of unemployed youth in rural areas.

Another side of this argument is that rising urbanization will increase the demand for cooking oil and processed foods using international standard palm oil, and there is little evidence that the small-scale sector has a comparative advantage in its production. Without large investments, the opportunity to substitute for increasing imports of CPO and even tap into the growing export market for palm oil will continue to elude Africa. It is telling that the export value of palm oil (an African product) from Malaysia and Indonesia now exceeds the total value of *all* agricultural exports from sub-Saharan Africa.

COLOMBIA: OIL PALM FOR JOBS IN A POSTCONFLICT COUNTRY

Colombia is the world's fourth-largest palm oil producer and the largest outside of Asia. The oil palm industry started expanding commercially in Colombia during the 1960s and grew steadily during the 1970s and 1980s, stimulated by domestic prices that were set above world prices (Marin-Burgos 2014). Beginning in the 1990s, the sector was liberalized and oriented toward exports by a range of incentives aimed at providing economic opportunities in the countryside, which had been racked by conflict during a long civil war. The sector was reoriented further toward biofuels in 2007 through a national oil palm policy (Consejo Nacional de Política Económica y Social 2007) supporting a biodiesel mandate of B07 (7% blend with regular diesel) introduced in 2008 and later raised to B10. The biodiesel industry has developed rapidly and today uses nearly half the palm oil produced in Colombia (Chapter 6).

Several zones are demarcated for oil palm production in Colombia; the central and eastern zones currently account for 70% of production. In 2013, Colombia passed the million-ton mark in palm oil production harvested from 300,000 ha. Another 150,000 ha of plantations will soon come on stream. Notably, this expansion came mostly from the conversion of pasture and cropland to oil palm; 51% of the trees planted in 2002 through 2008 replaced pasture, and 29% replaced cropland (Castiblanco et al. 2013). Only 16% of the expansion in oil palm area has come from converting natural vegetation, including forests, to palm production. Given its abundance of low-grade pasture, Colombia appears to have sufficient room to expand to meet targets of up to 1 Mha by 2020 (Garcia-Ulloa et al. 2012).

Medium-size producers (with 20–200 ha) made up 46% of the area in 2010, and large producers accounted for another 36%. Only 19% of the oil palm area consisted of small farms (<20 ha), but smallholders' share of oil palm area is expanding. Most small farmers are organized into *alianzas*, which are collectives that contract with larger producers and mills. In 2012, Colombia had more than 126 oil palm alianzas with a total of 6000 small farmers, each managing 10 ha on average (Martinez Pelaez 2013). Consistent with this farm structure, the average mill capacity is 25 t/hr, and half the mills have a capacity of less than 15 t/hr. Only two mills have the 60-t/hr capacity that is the standard in Indonesia (Pacheco 2012).

Oil palm producers and processors in Colombia are highly organized through Federación Nacional de Cultivadores de Palma de Aceite (National Association of Oil Palm Producers) (FEDEPALMA), which provides a number of services, including strategic planning for the industry, information, extension, and research. Cenipalma, established by Fedepalma in 1990, undertakes research with a scientific staff of about 90. Fedepalma is funded by a levy on palm oil products of 1.5% of their value at the mill gate; nearly two thirds of the levy is used for R&D, with additional cofinancing from government sources.

The government of Colombia has provided a wide range of incentives to the industry to revive the rural sector, provide jobs in postconflict areas, and achieve its biodiesel mandate (Table 2.10). Many incentives target smaller producers, although large producers that contract with alianzas may also benefit from many of these programs (Marin-Burgos 2014).

Table 2.10 Partial list of incentives offered to oil palm producers in Colombia circa 2009

Incentive	*Description*
Tax benefit for perennial crops	Exemption on taxable net income for 10 yr
Incentive for rural capital formation	A cash contribution to stimulate agricultural activities, including the planting of palm; small producers can receive as much as 40%, and medium and small producers as much as 20%
Program Venture Capital Fund	Fund to support agroindustrial initiatives that lack private investment, including biofuel projects
Price Stabilization Fund	Stabilization fund to compensate palm sector producers, sellers, or exporters to reduce the risks to producers and biodiesel refineries of price fluctuations and ensure a minimum income
Subsidized credit	Subsidized credit for oil palm crops for up to 15 yr, with a grace period of up to 4 yr
Finagro	An agricultural bank that finances working capital and investment with interest discounts depending on farm size
Agricultural Collateral Fund	Facilitation of access to agricultural credit by farmers who cannot offer the required collateral, with the guarantee depending on farm size
Technical assistance	Payment of 80% of the expenses for hiring technical assistance services
Biodiesel mandate	Requirement for 10% blend with diesel

A variety of rules set the eligibility for program support and have varied over time.

Source: Synthesized from Johnson, T. M., and J. Franco. 2009. *Economic assessment of palm oil production for biodiesel in Colombia: Case study analysis for the central and eastern zones.* Washington, DC: The World Bank Group ESMAP; and Marin-Burgos, V. 2014. Access, power and justice in commodity frontiers: The political ecology of access to land and palm oil expansion in Colombia. PhD diss., University of Twente. http://dx.doi.org/10.3990/1.9789036536851 (accessed 31 Jan 2016).

A unique incentive in Colombia is the Price Stabilization Fund, which ensures a minimum income to producers and a reliable supply of palm oil to biodiesel refineries (Johnson and Franco 2009; Selfer et al. 2013). In 2007, incentives for establishing oil palm amounted to about US$1500/ha in grants and soft loans (Selfer et al. 2013). From 1990 to 2004, the industry enjoyed a nominal rate of assistance averaging 30% of world prices (Guterman 2008). The US Agency for International Development (USAID) has played a major role in financing this support as part of its narcotics control and counterinsurgency efforts.

The industry faces at least three major challenges. The first is to ensure competitiveness, given relatively high costs and modest yields. The average cost of producing palm oil in Colombia is estimated to be 20% more than that of Malaysia and one third greater than that of Indonesia (Johnson and Franco 2009; Martinez Pelaez 2013). One of the foremost reasons for Colombia's higher production costs is the greater cost of labor—US$25/day in Colombia compared with US$11.4 in Malaysia and US$4.7 in Indonesia in 2012 (Martinez Pelaez 2013). Although less labor is employed in Colombia (about 1 person/15 ha), labor costs still represent more than 40% of costs (Johnson and Franco 2009). In addition, annual land costs are valued at US$330/ha, higher than in Indonesia, where companies receiving land concessions pay only a fraction of this amount.

Despite high costs, oil palm has been profitable financially in Colombia. The return on investment exceeds 20%, in part a result of government incentives. It would not be profitable to export palm oil at average prices prevailing before 2008, however, and the use of palm oil to produce biodiesel is not economic without the government incentives and mandates (Johnson and Franco 2009).

A second challenge is the presence, since 2010, of bud rot disease (locally known as *pudrición de cogollo*) caused by *Phytophthora palmivora Butler.* Treatment is expensive and has to be performed early during an outbreak. Producers in one area lost 30,000 ha of oil palm to the disease (Mosquera et al. 2012). The spread of bud rot adds to production costs and is difficult to halt through genetic resistance, given the long breeding cycle.

The final challenge in Colombia, as in other oil palm-producing countries, is to ensure social and environmental sustainability. Only a small share of oil palm has been planted at the expense of natural vegetation and forests, yet in a country with highly unequal land distribution and a history of land conflict, it is not surprising that oil palm has been associated with social dissension. The problem is most apparent in postconflict areas, where land was often abandoned and growers (sometimes with links to paramilitary forces) moved in to plant oil palm on land without clear rights (Marin-Burgos 2014). In other areas, plantations have encroached on public land with customary rights. Even in areas where land markets are functioning, the introduction of a profitable crop supported by public subsidies has driven up the price of land and may have displaced small farmers without the resources needed to plant oil palm.

Government programs directed at assisting smallholders and fostering alianzas were designed to move the industry toward smaller scale production models that would be more inclusive. As in Indonesia, in Colombia the tight management control companies exert over production by smallholders has sometimes been controversial. The preference of some alianzas for more diversified production systems and independent management has led to disputes with companies (Marin-Burgos 2014). Other smallholders, however, have developed strong organizations that have a powerful negotiating position with the milling companies (Rugeles 2011) (see Chapter 8).

THE CHALLENGE OF IMPROVING PRODUCTIVITY

Early Successes of R&D

Improving oil palm yields through plant breeding is slowed by the fact that oil palm is a perennial crop with a long breeding cycle of 15 to 19 years between generations, and uptake of new varieties must await replanting carried out in 25-year cycles (Soh 2012; Perera 2014). Despite these handicaps, the crop experienced impressive yield gains after World War II, in line with progress in most globally important crops. Part of this success resulted from breeding better varieties; part came from improvements in management and processing. Henson's (2012) decomposition of yield growth on a Malaysian plantation for 1951 through 1981 suggests about half the progress was the result of breeding, consistent with experience in other crops. The development and adoption of the tenera palm type was the single most important breeding success. On the management side, the biggest factor in raising yields has been the use of fertilizer (Table 2.11). Experimental data from Côte d'Ivoire support an even faster rate of gain, especially in the oil extraction rate (Henson 2012).

Much of the early success in raising yields came from investment in R&D. Important early research, including the discovery of the tenera palm, was done by the National Institute for Agronomy of the Belgian Congo[8] in collaboration with Unilever (Hartley 1967). Formation of the Palm Oil Research Institute of Malaysia in 1979 lent an impetus to oil palm research. Eventually subsumed under the MPOB, the institute has become the leading R&D organization in the industry, deploying more than 200 scientists. Significant contributions also come from the Indonesian Oil Palm Research Institute, which accounts for up to 20% of public research spending in Indonesia (Stads et al. 2007).

Even before World War II, some plantation companies in Malaysia conducted private R&D (Tate 1996). In recent years, the locus of research has shifted decisively toward the private sector. In Malaysia, private investment in R&D for oil palm now accounts for at least half the R&D investment in Malaysia (Agricultural Science and Technology Indicators and Malaysian Agricultural Research and Development Institute 2012). The private sector also leads seed production for oil palm, a US$100 million industry in itself (Soh et al. 2010).

Despite these trends, oil palm remains a relatively new crop in terms of scientific attention. Measured by the number of scientific publications per billion dollars of industry output in 2002 through 2012, oil palm—the subject of 27 publications per billion dollars—ranks well below research on cereals (100 publications per billion dollars for wheat and 60 for soybeans, for example) (Spiertz 2013).

Recent Trends in Commercial Yields

Yields of oil palm in commercial plantations appear to have grown more slowly in recent years compared with earlier periods. Colombia had the highest yields (21 t/ha of FFB) in 2012 and also experienced the highest growth in yields between 1991 and 2012 (1.6% annually). In all other cases, yields grew at 1% per annum or less. These trends are based on FAO statistics, whereas national statistics suggest even slower rates of yield growth, especially for Colombia and Malaysia. Malaysia has made most of its progress in raising yields by improving oil extraction rates rather than by improving on-farm yields (Figure 2.12). Improving yields by increasing oil extraction rates ("yield taking") is an important objective of the industry.

[8] Institut National pour l'Etude Agronomique du Congo Belge.

Table 2.11 Decomposition of yield gains on a Malaysian plantation, 1951–1981

Factor increasing yields	*Yield of palm oil (t/ha)*	*Yield gain (t/ha) resulting from*		
		Breeding	*Agronomy*	*Processing*
Initial yield, 1951	1.30	—	—	—
Fertilizer use	—	—	1.2	—
Improved *dura*	—	1.0	—	—
Use of *tenera*	—	1.14	—	—
Improved stand	—	—	0.14	—
Water management	—	—	0.24	—
Pollination weevil	—	—	0.06	—
Improved oil extraction rate	—	—	—	0.35
Total gain	4.13	2.14	1.64	0.35
Final yield, 1981	5.43			

Source: Data are from Henson, I. E. 2012. A brief history of the oil palm. In *Palm oil: Production, processing, characterization, and uses,* ed. O.- M. Lai, C.- P. Tan, and C. C. Akoh, 1–31. Urbana, IL: AOCS Press.

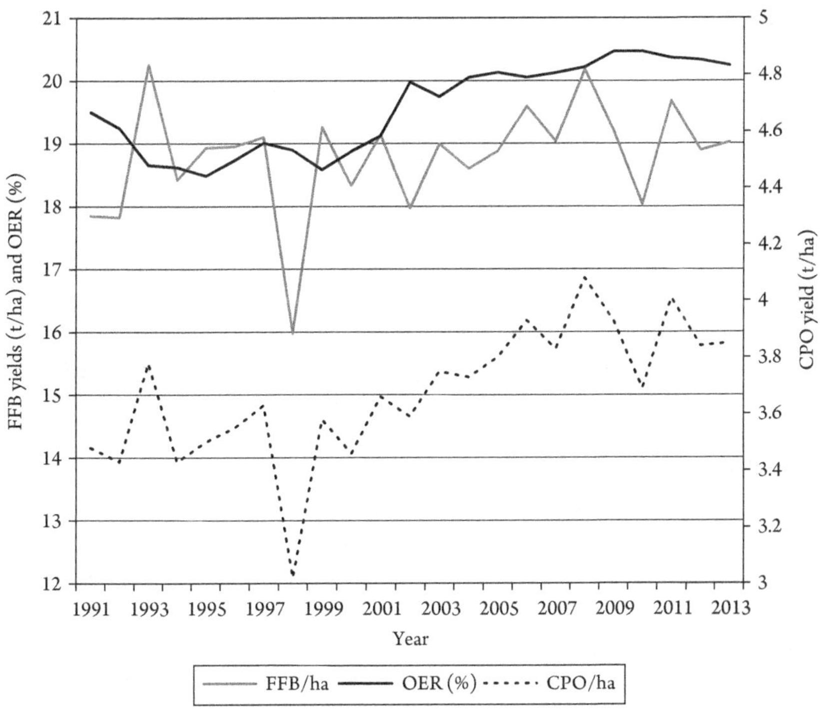

Figure 2.12 Trends in Malaysian yields of fresh fruit bunches (FFBs) and crude palm oil (CPO), and the oil extraction rate (OER). The yield of CPO is the product of the yield of FFBs and the OER. In the figure, the modest increase in CPO yield is entirely the result of improvements in the OER.
Source: Data are from Malaysian Palm Oil Board (2013).

Slow yield growth in part reflects the industry's youth. The high proportion of immature plantations in countries where oil palm area expanded very rapidly, such as Indonesia, reduces national average yields. Yields may also be growing more slowly if oil palm is expanding into areas that are less favorable for its production. The trend toward smallholder production of oil palm may also explain, in part, stagnation in national average yields. In Indonesia, for example, smallholders realize yields some 20% to 30% lower than those obtained on private large-scale plantations.

A review of all sources leads us to conclude that yield growth has been modest at best during the boom years since 1990. The rapid growth of production has been driven overwhelmingly by the expansion in oil palm area, which accounts for well more than 90% of the growth in Indonesia and 77% in Malaysia.

Progress in Raising Yields through R&D

The three paths to raise yields of palm oil are (1) to close the gap between commercial yields and potential yields through the use of technology that has already been developed, (2) to increase yield potential in oil palm through breeding and improved agronomy, and (3) to increase the oil content of the fruit or the extraction ratio. Estimates based on reviews by Fischer et al. (2014), Soh et al. (2010), and Goh et al. (2014) indicate the potential yield for existing genetic stock is around 10 t/ha of oil (range, 9–12 t/ha). Because commercial yields rarely exceed 70% of potential yields as a result of economics and producers' risk aversion, the attainable yield is considered to be 7 t/ha, which is consistent with the yields obtained on the best-managed plantations (Wahid, Abdullah, and Henson 2004; Sheil et al. 2009).

Yield gaps develop for a variety of reasons: the age of the plantation, deficiencies in nutrient management, harvesting time, canopy management, and weed and pest control are some of the more common ones. Recently, the International Plant Nutrition Institute established best management practices on a commercial scale, achieving a 13% yield increase on six plantation blocks in Sumatra over 4 years (Pauli et al. 2014). The fact that commercial yields obtained by smallholders are 3.5 to 5.0 t/ha and less suggests an opportunity to increase yields by 40% to 100% by closing yield gaps. On small holdings, an additional yield gap arises from the use of poor genetic stock and little fertilizer. Africa, where yields of palm oil are currently less than 1 t/ha, has enormous potential to raise yields on the 3.5 Mha of oil palm harvested in that region.

Over the longer term, yield potential can be increased by investing in breeding. Several estimates suggest that oil palm breeders are obtaining gains of 1.0% to 1.5% per year (Durand-Gasselin et al. 2003; Fischer et al. 2014; Perera 2014). These gains are on the same order of magnitude as those achieved in annual crops (Fischer et al. 2014). In oil palm, as in other perennial crops, cloning attracted considerable interest as a way of achieving higher yields. Cloning has proved more difficult than anticipated, although leading companies in Malaysia are replanting extensively with clones.

Development of F1 hybrids and molecular selection are also promising breeding strategies, especially now that the oil palm genome has been sequenced (Soh et al. 2010; Fischer et al. 2014).

The discovery of the SHELL gene that identifies the tenera type should accelerate selection for higher oil content (Murphy 2014). Other genes are being identified, and molecular-assisted selection is now used routinely to select for greater oil content.

For commercial producers to realize the genetic progress made through oil palm research, they must replace their trees with new, higher yielding tree varieties. This is a costly process that, at best, will take place in a 25-year cycle, when trees become too tall for harvesting. Replanting is often delayed when prices are high, as in recent years. Smallholders find it especially difficult to finance replacement and to afford a loss in income for some years.

Aside from raising potential yields, breeders are also seeking to protect yields by improving disease resistance (especially to *Ganoderma* stem rot), to facilitate harvesting by developing dwarf varieties, and to breed varieties that produce oil with high levels of oleic acid to imitate the properties of olive oil in food (Soh 2012). In West Africa and more marginal areas of Asia with a very distinct dry season (Thailand and Myanmar), oil palm requires better drought tolerance. Although GM varieties of oil palm have been developed and are being tested, they are targeted at industrial nonfood uses. Parveez et al. (2012) provide a time horizon of at least 15 years until GM varieties with consumer benefits are available.

Malaysia has set a 2020 vision of achieving 8.8 t/ha based on 35 t/ha of FFBs and 25% oil extraction—a doubling of current yields that implies a growth rate exceeding 7% per year (compared with recent yield growth of about 1% per year). In light of the steady but modest breeding progress, the slow replacement of aging plantations, and the difficulty of scaling up knowledge-intensive established best management practices, the 2020 vision seems optimistic. Fischer et al. (2014) suggest a yield gain of 2% to 3% per year can be achieved by using better management to close yield gaps and by adopting improved varieties and hybrids.

The industry is underinvesting in long-term strategic research. Private companies focus typically on applied research that has a near-term payoff. The MPOB provides some strategic research on a limited budget of less than US$ 20 million. No industry-wide R&D across the two major producers is focusing on key issues for the future of oil palm, such as broadening genetic diversity (narrow diversity is now a big risk), robotic methods for harvesting, or proactive management of new diseases and pests.

Finally, a focus on increasing yields in oil palm as a way to maintain competitiveness and reduce pressure on land and forests may be misguided for two reasons. First, some argue that higher yields, obtained by enhancing the profitability of oil palm, will only accelerate the expansion in oil palm area and heighten the risks of deforestation (Angelsen and Kaimowitz 2001). We return to a deeper analysis of this relationship in Chapter 9. Second, labor rather than land productivity may be the main issue in ensuring competitiveness of oil palm. We address that issue next.

Labor Productivity: The Key

The vaunted efficiency of oil palm as a low-cost oil crop may disappear with increasing wages in Asia. Fry (2009) suggests that labor costs per unit output are increasing by about 2% annually in Malaysia as a result of a combination of rising wages and stagnant labor productivity. Since the mid 2000s, Asia has seen a sharp jump in wages, most notably in oil palm-producing areas of Indonesia, although wages there are still notably less than in Malaysia and Thailand (Table 2.12). Brazil's even higher labor costs make ambitious plans to scale up oil palm production even more daunting. Given that a common labor-saving practice is to reduce the frequency of harvesting, which entails significant yield losses, improvements in labor productivity for harvesting are critical to enhance yields (Murphy 2014).

Gains in labor productivity have occurred, as shown by the increase in the number of hectares per worker in Malaysia from 4 ha in 1951 to 7.5 ha in 1991 (Corley and Tinker 2003), and a reported 10 to 12 ha today. These savings have been achieved through mechanical land clearing, the use of herbicides and handheld mechanical slashers for weeding, and the use of small tractors to help gather and transport harvested fruit. Just as tree crop harvesting has been mechanized in California, a concerted R&D push might lead to mechanized harvesting of oil palm (Fry 2009). Research would likely involve a combination of breeding (for dwarfness, fruit placement, and uniform ripeness) and the development of appropriate machinery, especially robotics.

SUMMING UP

Oil palm gained preeminence in world agriculture rapidly between 1971 and 2014, when production grew 28-fold at an annual rate of nearly 8%. Oil palm is now the third most valuable crop in world trade. More than 70% of palm oil is produced for world markets and reaches nearly all countries, yet 85% of production comes from just two countries: Indonesia and Malaysia.

Growth in world markets for vegetable oils for food and biofuel has been a major driver of growth in demand for oil palm, as we show in Chapters 4 and 6. On the supply

Table 2.12 Agricultural wages in Southeast Asia

Country	*Agricultural wage, 2010s (in 2010 US$/day)*	*Change in real wages, mid 2000s–2010s (%)*
Malaysia, national	11.75	9
Thailand, national	6.81	27
Indonesia, South Kalimantan	4.76	44
Indonesia, central Java (labor source)	2.67	18

Source: Data are from Wiggins, S., and S. Keats. 2014. *Rural wages in Asia*. London: Overseas Development Institute.

side, the basic technology became available during the 1960s to make oil palm a highly productive crop suited ideally to the humid tropics. With booming markets, it was also a very profitable crop that stimulated huge investments by private entrepreneurs to bring new land into production. The state also played critical roles in facilitating foreign investments and making large land concessions, often from the "forest estate," available cheaply to investors. Stable macroeconomic policy, financing from development banks, and modest but positive incentive systems have also been important.

Large, vertically integrated companies based mostly in Southeast Asia have led growth in the sector. At the same time, smallholders make up a significant share of oil palm area, and their share is growing. As the crop has expanded outside of Malaysia and Indonesia, more diverse agrarian structures have emerged based on small to medium-size producers in Thailand and parts of Latin America. Innovative business models link these producers to large palm oil mills—a link that is critical, because palm fruit requires quick processing after harvest.

With oil palm area expanding so rapidly, it is not surprising that the crop is associated with social and environmental controversies related to land-use changes—a topic we examine in depth in Chapter 9. Inevitably, the rapid expansion in area will slow as land becomes more difficult to access for many reasons, including tighter sustainability standards, enforcement of environmental regulations, and recognition of the rights of existing users. Intensification through higher yields could reduce pressure on land expansion, provided the remaining forests are protected adequately (Chapter 9). The biggest opportunities for increasing yields are with smallholders everywhere, given the gaps between smallholders' yields and those obtained in the estate sector. Nowhere is this potential more evident than in West Africa, the original home of oil palm, the main exporter of oil palm products before 1970, and the current producer of less than 5% of world output from 25% of the harvested area. Given dual markets in Africa for red palm oil and standard international grade CPO, the highest priority is to upgrade efficiency and quality in the small-scale processing sector for the red oil market. As our case studies have shown, an exclusive focus on large milling technologies is unlikely to succeed in Africa because of the importance of small-scale processing to rural incomes and employment.

Steady yield growth will also occur through breeding aided by emerging molecular tools, but the impact of this work will take time to be felt; the breeding and replanting cycles for this perennial crop are long. An equally important task is to increase labor productivity, given rising wages everywhere and the relatively labor intensive nature of current production methods. Full mechanization of harvesting operations would be an enormous breakthrough but does not seem to be on the medium-term horizon.

3

SOYBEAN PRODUCTION AND SUPPLY CHAINS IN THE TROPICS

ESSENTIAL FACTS ABOUT THE WORLD'S MOST IMPORTANT OIL CROP

During the past 50 years, soybeans (*Glycine max*) have emerged as the world's fourth most important crop in terms of area and value after rice, maize, and wheat. Soybeans are also by far the world's most important oil crop, with at least double the area and value of the two nearest rivals: oil palm and rapeseed.

Because of their high protein content (around 40%) and good balance of essential amino acids, soybeans are an important animal feed, ranking second only to maize. Nearly 90% of soybean production is processed into protein meal; the meal makes up nearly 80% of the seed weight and about two thirds of the economic value of soybeans. Soybean oil—second to palm oil as a source of vegetable oil—accounts for 18% of seed weight and about one third of the value of soybeans. Soybeans are also a nutritious and important food in some countries, especially in Asia, where they are made into a wide range of preparations such as tofu, miso, and tempeh. Direct consumption of soybeans for food, at only 1.4 kg/capita/yr, accounts for less than 5% of global use. Finally, as a legume, soybeans are valued for their nitrogen-fixing properties, which enable the crop to meet most of its own nitrogen needs, possibly with some carryover to the following crop.

Soybeans are an annual crop grown under warm, moist conditions over a wide range of latitudes—from the Equator to about 40° S and 50° N. The crop was domesticated in China around the 11th century BCE and was used mostly as food (Hymowitz 2008). Even today, domestic production in China and other Asian countries is destined largely for food.

Soybeans took off as a commercial crop during the 1920s in the United States for two reasons. First, a scarcity of cottonseed oil (provoked by high losses to boll weevil on cotton) had stimulated a search for alternatives; second, biochemists discovered that heating soybean meal greatly enhanced its suitability for animal feed. The USDA's systematic collection and release of varieties adapted to different environments, especially in the Corn Belt, also encouraged the crop's rapid uptake (Hymowitz 2008). After World War II, intensive pork, poultry, and milk production based on maize and soybean feed concentrates accelerated in the United States, Europe, and Japan; more recently, middle-income countries have also joined what has been called the "livestock revolution" (which we discuss in Chapter 5). Other countries, especially India, have promoted soybeans largely for their oil, and in a few countries such as Nigeria, soybeans quickly gained importance as a food. Since 1990, trade liberalization has provided a major boost to soybean demand by enabling land-scarce Asian countries to import soybeans to support the growth of major livestock industries (see Chapter 7).

Up until around 1990, the United States dominated the supply of soybeans, producing well more than half of global output. During the past two decades, the crop expanded extremely rapidly in South America, led by Brazil and Argentina, but appearing in Paraguay, Uruguay, and Bolivia as well. Most of the recent expansion in South America occurred in the tropical latitudes (Chapter 1). Likewise, in Central India, farmers converted some 10 Mha from other crops and fallow to soybeans during the past three decades to make India Asia's second-largest producer after China.

Soybean production expanded as well into higher latitudes, where Canada, Russia, and the Ukraine became significant producers and exporters in recent years. Even the United States has seen a definitive shift northward as the state of North Dakota, traditionally a leading producer of quality wheat, sowed more area to soybeans than wheat for the first time in 2014.

This remarkable expansion reflects the adaptability of soybeans as a relatively short-season crop (some varieties mature in less than 90 days) as well the burgeoning market demand for livestock feed and oil. Investment in R&D, initially through the public sector but more recently with strong private-sector interest, built the technological base for the soybean revolution. GM varieties are used more widely in soybeans than in any other crop. In the Americas and South Africa, more than 90% of the soybean area is sown to GM varieties, but almost none in other regions.

Today, soybeans are sown on 117 Mha with a production of 305 Mt (USDA-FAS PSD). Production of soybeans is more concentrated than for any other major crop except oil palm. The big three soybean producers—the United States, Brazil, and Argentina—account for 82% of global output (Figure 3.1). The next seven countries—China, India, Paraguay, Canada, the Ukraine, Uruguay, and Russia—account for another 16% of production, leaving only 2% for the rest of the world. Since 1991, increases in production have shifted dramatically to South America, India, Canada, Russia, and the Ukraine. With the exception of China, all the top 10 producers are net exporters of soybean products, and nearly all their domestic use is for livestock feed and oil. China is the world's dominant importer of soybeans, which are used for feed in its expanding livestock sector.

This chapter looks at the supply chain for soybeans in the tropics, where the crop is expanding most rapidly. We illustrate this expansion with three case studies, beginning with Brazil, the major player that surpassed the United States in total exports in 2014. The second (and less-well-known) case is India, where soybeans meets partly domestic oil needs in addition to supplying soybean meal for the country's growing livestock industry and for export. The third case is the region of southern Africa (excluding South Africa), which has a nascent industry and great potential to expand soybean production to meet growing regional demand. We build on these case studies in the final section of this chapter to review the outlook for further expansion of the crop. We take up the demand side in Chapters 4 and 6 (for soybean oil) and Chapter 5 (for soybean meal).

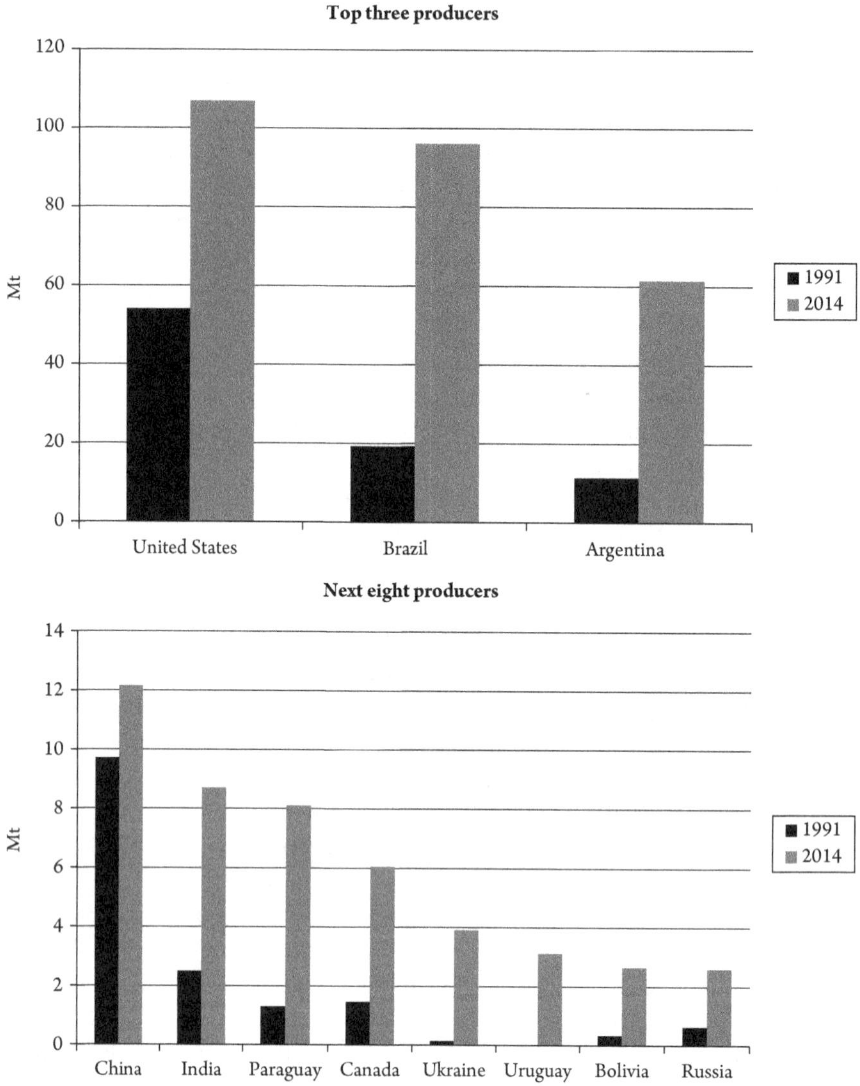

Figure 3.1 Soybean production for the top 10 producers, 1991 and 2014 (the bottom panel is scaled at one tenth of the top panel). Soybean production expanded rapidly in all the top 10 producers except China. In the United States, production has doubled since 1991, and the other major producers have increased soybean output at least four times since 1991.
Source: USDA-FAS PSD.

THE SOYBEAN TRANSFORMATION OF THE BRAZILIAN CERRADO

The most important global extension of the agricultural frontier during the past century started with the rapid expansion of soybeans in the 1970s in the Brazilian Cerrado. The Cerrado is a relatively flat region of savannah and woodland in Central

Brazil, mostly between latitudes 10° S and 20° S, covering some 200 Mha—about the size of Indonesia or four times that of France.[1] Despite receiving reasonably high and reliable rainfall of 1500 to 1800 mm, the region had highly acidic and not very fertile soils that constrained crop agriculture until the 1970s, when soil amendment technologies became available (Cassman 2005).

Once used mostly for extensive cattle ranching, the Cerrado is now the world's "soy basket." Populated with large commercial family farms (averaging more than 1000 ha) and agribusiness farms (some exceeding 100,000 ha), the Cerrado is not only a major producer of maize, cotton, and beef but it is also the most important soybean-producing region outside the US Midwest. Soybeans alone provide more than US$15 billion to the local economy, not including value added in processing. How did this remarkable transition come about?

Geopolitics and Booming Markets

Driven by geopolitical considerations, Brazil's military government of the 1960s and 1970s put in place a number of measures to open up the Cerrado, including building a new capital in the region, constructing a highway to link to the Atlantic ports, and a host of credit and incentive programs to attract settlers. During the mid 1970s, Brazilian soybean exports were stimulated by a market opportunity provided when the United States placed an embargo on its own soybean exports to protect US meat consumers from high world prices. The 1980s ushered in a period of rapid expansion characterized by better macroeconomic performance along with trade liberalization, which facilitated Brazilian access to global markets. During the 1990s, the devaluation of the exchange rate and greater macroeconomic stability were major factors in making Brazilian soybeans competitive. Richards et al. (2012) estimated that exchange rate depreciation added 6.3 Mha of soybean area in Brazil from 1991 to 2009, most of it in the tropical regions. The most important recent driver of Brazil's expanding soybean production was the rapid surge of imports by China, starting in the 1990s. Soybeans produced in the Cerrado were the major supplier of this market, which closely tracked Chinese imports (Figure 3.2). World trade in soybeans almost doubled from 1991 to 2010 (Chapter 6).

Finally, induced by rising incomes, livestock consumption has increased rapidly in Brazil and stimulated a demand for livestock feed. More than 40% of the soybeans produced are processed into meal and oil, with about half going to domestic consumption. About half the meal is exported and about one quarter of poultry and pig meat is also exported, making Brazil one of the world's largest meat exporters (Figure 3.3). A small percentage (<3%) of soybean oil is used in the chemical industry; 30% is used for biodiesel (O'Kray 2014).

[1] The Cerrado biome includes much of the states of Mato Grosso, Mato Grosso do Sul, Goiás, Minas Gerais, and parts of Maranhão, Piauí, Tocantins, and Bahia—the so-called Mapitoba Region after the first two letters of each state.

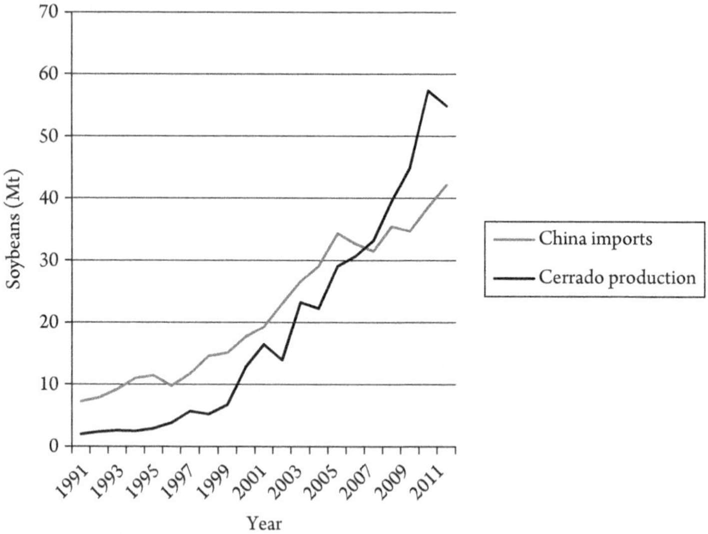

Figure 3.2 Soybean production in the Cerrado and soybean imports by China, 1991 to 2011.
Source: FAO (2015) and CONAB. 2014. Custo de Producao–Resumen. Agricultura Empresarial—Soja—Plantio Directo OGM–Alta Tecnologia. Safra de Verao –2013/14 – Primavera do Leste – MT. http://www.conab.gov.br/OlalaCMS/uploads/arquivos/13_04_05_11_12_14_soja-mt-p.pdf (accessed 31 January 2016).

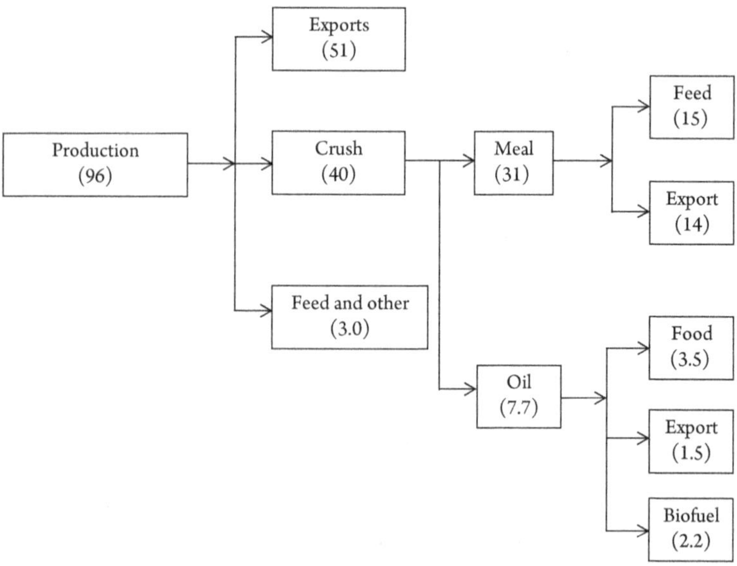

Figure 3.3 Destinations for Brazilian soybeans and soybean products, 2012. Numbers in parentheses are quantities in million tons. More than half of soybean production is exported directly, mostly to China. Most of the rest is crushed, but nearly half of the meal and one third of the oil are exported to diverse destinations.
Source: Calculated from USDA-FAS PSD. Numbers do not add exactly due to rounding errors, changes in stocks, waste, and other uses.

An Active State Role

Some claim that Brazilian agriculture and soybeans in particular succeeded because the state got out of the way of the private sector (The Economist 2013). In fact, and especially during the early years, the state had a heavy hand in the settlement and promotion of agriculture in the Cerrado through investments in infrastructure and technology and through direct support to producers. This role has been much reduced in the 21st century, but the state remains an important player.

Public Investment in Technology and Infrastructure

When the market was guaranteed, a critical factor in the soybean takeoff was the availability of technology that improved productivity sharply in the poor Cerrado soils. The state-financed Brazilian Agricultural Research Corporation (Empresa Brasileira de Pesquisa Agropecuária [EMBRAPA]) was a leader in providing this technology. EMBRAPA, founded in 1973, today has a cadre of 2000 scientists and a budget of about US$1 billion. It is the world's largest and most successful research institute for tropical agriculture.

The soil amendment technologies that enabled the highly acid Cerrado soils to be cultivated with success included heavy initial applications of lime (about 5 t/ha) and substantial initial doses of phosphorus and potassium, often bolstered with micronutrients. Later, sustainable soil management was enhanced by widespread and rapid adoption of zero tillage methods introduced from southern Brazil through networks of farmers, private dealers, and individual researchers (Ekboir 2003). Zero-tillage technology received a big push when Monsanto reduced the price of glyphosate herbicide after its patent expired in the 1990s; later, it received another with the approval of varieties engineered genetically to tolerate glyphosate.

Another critical element of the technology package was the development of soybean varieties adapted to tropical latitudes. These varieties were released by EMBRAPA based on research originally done in the southern United States and introduced into Brazil under a University of Illinois–USAID program (Kaimowitz and Smith 2001). EMBRAPA has continued to release new varieties, including GM varieties starting in 2003, when the Brazilian government approved their release. More than 90% of soybean area is now sown to GMOs in one of the most rapid and widespread adoptions of this technology anywhere (James 2014). An innovative EMBRAPA program to breed varieties for biological nitrogen fixation through suitable strains of rhizobia led to the widespread adoption of the rhizobia inoculum and an estimated savings in nitrogen fertilizer of US$150/ha (Alves et al. 2003; Hungria et al. 2005).

Although the state played a lead role in providing technology, private R&D has become increasingly active. Even during the earlier period, from 1976 to 1988, half the soybean varieties released in Brazil were from the private sector, although they were intended for southern Brazil, where imported varieties performed well (Pardey et al. 2006). Today, the big seed companies such as Monsanto, Dupont, and Syngenta are a major influence in the seed market in the Cerrado frontier. EMBRAPA has also entered a number of public–private partnerships (e.g., with BASF Chemical) to develop varieties based on genetic technology licensed by the private sector.

The state also provided the basic infrastructure to open the Cerrado. The relocation of the national capital to the new city of Brasilia in the Cerrado in 1961 was the impetus for building a number of important highways in the region, especially the BR-364, linking Mato Grosso to the southern port of Santos, and the BR-163, linking it to Santarém, a port on the Amazon River system in northern Brazil. Nonetheless, transport costs for exporting soybeans were and still are high (as discussed later in this chapter).

State Credit Programs

Land is relatively cheap in the Cerrado, but establishing soybeans requires a substantial upfront investment of US$2000/ha to US$2500/ha to convert pasture into productive cropland, and even more to convert natural vegetation (O'Kray 2014). Because most farmers migrate from southern Brazil, they require additional capital to purchase farm equipment and to settle and live until the first harvest. Various government programs financed much of this investment through the National Development Bank (Banco Nacional de Desenvolvimento Econômico e Social) and a range of development programs designed specifically for the Cerrado, such as the Development Program for the Cerrado (Programa de Desenvolvimento dos Cerrados) and the Japanese–Brazilian Cooperative Program for Development of the Cerrado (Programa de Cooperação Nipo-Brasileira para o Desenvolvimento dos Cerrados) (Garrett 2013).

During the early years when inflation was high, the effective real rate of interest on state financing was strongly negative, so as much as half of the loans provided could, in reality, be considered a grant (Bento Filho and Vian 2013). Access to subsidized credit was biased toward larger farmers and companies, who were provided strong incentives to invest in labor-saving mechanization. At its peak, subsidized credit accounted for 15% of the value of Brazilian agricultural production (Helfand 1999). Even today, farm credit continues to be subsidized, although with a positive real rate of interest (about 8% below market rates) (Garrett 2013). McFarland et al. (2015) estimated the credit subsidy was worth nearly US$5 billion in 2008; this value may be lower now, because many farmers obtain financing through contractual arrangements with buyers and processors.

Private Entrepreneurship and Collective Action

The major investments in processing and logistics infrastructure in the Cerrado came from private sources. The big multinationals ADM, Bunge, Cargill, and Dreyfus—often dubbed the ABCD group—control more than half the processing capacity and have also invested substantially in storage and port logistics (Sauer and Leite 2012; Oliveira 2015). Cargill, for example, opened a river port terminal to export soybeans from Santarém through the Amazon River system in 2003. The André Maggi Group, a Brazilian agribusiness conglomerate, is another major player in soybean production, processing, and exports from the Cerrado. In 2014, the Chinese food trading giant, COFCO, entered the Brazilian market through acquisitions by investing in export logistics. These companies largely control the export business. Smaller companies and cooperatives are important for supplying the domestic market.

Private colonization companies attracted settlers to the Cerrado; in Mato Grosso, some 35 private colonization companies organized 104 settlement programs on 3.9 Mha (Jepson 2006). These companies made their money from buying, developing, and then selling the land, and in return provided much of the initial investment to convert the land into productive cropland, and many initial services, including access to credit, securing of land titles, extension advice, and investment in soil amendments and some infrastructure.

Cooperatives also helped to build processing and logistics capacity and to facilitate settlement (Garrett 2013). Cooperatives of migrant farmers from southern Brazil, often linked to established cooperatives there, aided the settlement programs by reducing the transaction costs of acquiring suitable land, obtaining a secure land title, and gaining access to government credit programs. In a not atypical example, 100 small farmers (originally operating less than 50 ha) from a cooperative in Rio Grande do Sul obtained 60,000 ha in Piauí State through a subsidized credit of US$12 million and established the town of Nova Santa Rosa in 1998 (Bickel and Dros 2003). At least one cooperative ran its own research station and seed production company—again, using collective action to reduce the pioneering risks of growing a new crop in a new area (Jepson 2006).

Although the private colonization companies and cooperatives helped many family-type farmers, the high pioneering costs and risks also gave an advantage to large agribusiness farms. Many of these companies got their start in the settlement of Mato Grosso and, as the Brazilian public sector reduced its role, they came to lead the development of new frontiers. The most important frontier in recent years has been in the so-called Mapitoba Region, which consists of parts of the states of Maranhão, Piauí, Tocantins, and Bahia. This region is a bit drier and riskier, but is closer to Atlantic ports than Mato Grosso. Starting from a very small area in the early 1990s, some 2.5 Mha of soybeans is now sown in this region, where at least 2 Mha of high-quality land is thought to remain (much more could be productive with investment in irrigation) (Rabobank 2012). With production of more than 8 Mt in 2011/2012, this region contains some of the world's largest agricultural farms (Rabobank 2012).

Soybean farms in the Cerrado have always been large by world standards. Census data from 2006 indicate that in the central–west region, which includes much of the Cerrado, 72% of farm area consisted of farms exceeding 1000 ha; the average farm size in this category surpassed 3000 ha (Bento Filho and Vian 2013) (Table 3.1). These farms are much larger than the median-weighted average size of 200 ha for a U.S. soybean farm (MacDonald et al. 2013). Most large Brazilian farms are family farms, in the sense that families own and provide day-to-day management of the operation (Cacho 2016).

Organizationally very different from family farmers, the giant agribusiness companies operate geographically dispersed farm units, hire professional managers, and have access to global capital markets. In the Cerrado, 38 companies are estimated to

Table 3.1 Distribution of farm size in Central–West Brazil, 2006

Farm size (ha)	*Area (%)*	*Farms (%)*
<10	0	17
10–100	6	52
100–1000	24	24
>1000	70	6
Total	100	100

Source: Data are from Bento Filho, J., and C. Vian. 2013. The Brazilian experience with the occupation of the Cerrados: The dynamics of large farms and small farms. Programa Cohesion Territorial para el Desarollo, Documento de Trabajo no 5. Santiago: RIMISP.

farm more than 30,000 ha, and at least seven companies farm more than 100,000 ha (Chaddad 2014) (Table 3.2).

BrasilAgro is an example of a business model that emphasizes land improvement. The company was founded through a US$584 million IPO in 2006, in which a large Argentinean agribusiness firm, Cresud, holds 40% equity. Under its chief executive officer, Julio Piza, who has a master's degree in business administration from Columbia University, the company has grown to farming 180,000 ha in the Cerrado, achieving a positive cash flow by 2011 (Chaddad 2014). The company purchases underdeveloped properties with potential and converts degraded pasture or cropland to highly productive cropland. It is diversified geographically into nine units, each with its own professional manager.

SLC Agrícola is an even larger operation. In 2013/2014 it planted 344,000 ha spread over 16 units, and it employs around 2000 full-time workers and another 1500 temporary workers. The company is part of the SLC Group, founded in 1945 by three German immigrants to Brazil who specialized originally in farm machinery in partnership with John Deere. The company owns 340,000 ha, of which 114,000 ha is under permanent conservation, as required legally. Like BrasilAgro, it formed SLC LandCo, which partnered with an asset management company to raise US$240 million in 2013 to develop Cerrado land. Its investor prospectus states it expects to purchase land for US$2000–3000/ha and sell it after 2 years of investment for US$6000–7000.

BrasilAgro and SLC Agrícola are not unusual, as shown in the range of similar companies in Table 3.2. A big question is why very large "superfarms" have emerged in the Cerrado frontier. During the early days, access to credit strongly favored businesses that were well connected to the military government in power at the time. Before 1980, more than half the credit was captured by 3% to 4% of very large farmers (Helfand 1999). Two studies (Helfand and Levine 2004; Bento Filho and Vian 2013) found economies of size up to about 3000 to 4000 ha, but most of these economies arise from off-farm transactions for inputs, credit, and farm sales. Minimum wage laws

Table 3.2 Examples of major agribusiness farms producing soybeans in the Cerrado

Company	*Size*	*Financing*	*Other enterprises*
Bom Futuro	220,000 ha planted to soybeans (50% leased), 160,000 ha conservation area	NA	Cattle
SLC Agrícola	320,000 ha owned, 344,000 ha planted	Publicly traded with majority ownership by the Logemann family; part of the SLC Group of companies	SLC LandCo with 80,000 ha is a partnership with an asset management company for land development
Vanguarda Agro	254,000 ha (two thirds leased), 308,000 ha planted	Publicly traded	NA
El Tejar	250,000 ha planted	Argentinian company	NA
André Maggi	220,000 ha	Private	Large soybean processor and trader
BrasilAgro	180,000 ha own land	Publicly traded; IPO of US$584 million in 2006; 45% shareholding by Cresud (Argentinean)	Land development
Horita Brothers	150,000 ha	NA	Large cotton producer
Insolo AgroIndustrial	110,000 ha owned	NA	NA
Agrinvest	90,000 ha	Ridgefield Capital (US) was a major investor	NA
Ceagro	65,000 ha	Mitsubushi (Japan) has majority ownership	Major business is grain trading, logistics, and inputs
Tiba Agro	320,000 ha	NA	NA
Agrifirma	70,000 ha	British capital from Lord Rothschild family	NA

IPO, initial public offering; NA, not available.

Source: Constructed from Internet searches, based mostly on company websites.

and regulations increased the real cost of labor and transaction costs, and provided additional incentives for large-scale mechanized farms (de Rezende 2006). Yet, even with economies of scale, the large agribusiness farms are as much as 100 times the efficient size of 3000 ha suggested by these studies. The emergence of these companies may have occurred in part because of the state's withdrawal from subsidizing high upfront colonization and establishment costs, especially the cost of transforming degraded pastures and land with natural vegetation into cropland. These activities are likely to incur economies of scale in accessing the needed finance and in the specialized knowledge required for obtaining land titles and environmental permits—activities that have high transaction costs for individual farmers. It seems many of these companies also assume other pioneering risks of opening the frontier, such as the development of appropriate technologies and basic infrastructure (Rabobank 2012).

The future could well see many superfarms broken into smaller commercial units (although they would still be large by world standards), especially as the land is developed and pioneering risks are reduced. The advantage of the superfarms may also be eroded by a reinterpretation of Brazil's foreign investment law in 2010, which has curtailed foreign purchases of farm land and access to international capital severely.

How Competitive Is Brazil?

There is little doubt that soybean production in the Cerrado is competitive globally. In 2001, nonland costs per ton in Mato Grosso were similar to those in the United States (Cassman 2005). Since then, it appears Mato Grosso has become more competitive as land is fully developed and the industry matures, so nonland costs were significantly less in 2013 in Brazil than in Illinois (Table 3.3). In addition, land prices in Mato Grosso, although increasing rapidly, are still only about one quarter of land prices in the US Soy Belt, even more so in areas of the Cerrado that allow double cropping (Table 3.3).

On the other hand, the logistics costs to export soybeans from inland Brazilian locations are much higher than in the United States. In Brazil, transport, logistics, and port charges average more than US$100/t and are higher in more interior locations. Some 60% of Brazil's soybean exports reach ports at distances that sometimes exceed 2000 km by truck, whereas 60% of US soybean exports move by rail and water at a cost per ton-kilometer that is about one quarter that in inland Brazil (Fliehr 2013).[2] In general, land prices reflect the residual rent after subtracting transport costs, and are less in inland locations. Transport costs were reduced somewhat when Cargill constructed the port at Santarém on the northern end of highway BR-163, making it possible to export through the Amazon, although another effect of the port's presence was to increase the incentives to grow soybeans in the heavily forested areas nearby.

The high logistics costs of grain in Brazil have long been recognized. A large public–private investment program, the Growth Acceleration Plan, was initiated in

[2] Ocean freight rates from the port to China are similar in each case—about US$50/t.

Table 3.3 Comparative production costs for soybeans, Brazil and Illinois, 2013

	United States	*Brazil*	*United States*	*Brazil*
	Central Illinois	*Mato Grosso*	*Central Illinois*	*Mato Grosso*
Yield (t/ha)			3.9	3.0
Costs	(US$/ha)	(US$/ha)	(US$/t)	(US$/t)
Seed	180	79	46.3	26.4
Fertilizer	161	202	41.2	67.4
Pesticide	99	75	25.4	25.0
Labor	111	18	28.5	6.0
Machinery	274	103	70.4	34.2
Interest	114	17	29.2	5.8
Overhead and other	161	169	41.2	56.3
Total nonland	1099	663	282.1	221.0
Land	667	163	171.1	54.3
Total	1766	826	453.2	275.3

US costs are the actual costs for a sample of farmers. Brazil costs are anticipated costs for farmers using genetically modified organisms and high-input technology. Investment costs in land development in Brazil are assumed to be included in the land rental price.

Source: Data are from Krapf, B. M., D. D. Raab, and B. L. Zwilling. 2014. Costs to produce corn and soybeans in Illinois—2013. *Farmdoc Daily*, March 21. http://farmdocdaily.illinois.edu/2014/03/cost-to-produce-corn-and-soybean-in-illinois-2013.html (accessed 31 Jan 2016); and CONAB. 2014. Custo de Producao–Resumen. Agricultura Empresarial—Soja—Plantio Directo OGM–Alta Tecnologia. Safra de Verao –2013/14 – Primavera do Leste – MT. http://www.conab.gov.br/OlalaCMS/uploads/arquivos/13_04_05_11_12_14_soja-mt-p.pdf (accessed 31 Jan 2016).

2007 to improve road, rail, water, and port infrastructures with a planned investment of about US$100 billion. The aim is to transport 60% of exports to ports by rail and water by 2025 (Fliehr 2013), but this seems very optimistic. A logistics investment program initiated in 2012 focuses on private investment to upgrade transport and port infrastructures, including investment from China.

Environmental and Social Sustainability of the Soybean System

The major sustainability issue in the Cerrado has been the changes in land use associated with soybeans. Soybean area has expanded at the expense of natural vegetation and through the conversion of degraded pasture. During the early years, the conversion of Cerrado scrubland predominated, but as the soybean frontier moved northward, the conversion of forest land assumed a greater role. An ongoing debate (which we review in Chapter 9) surrounds the extent to which the conversion of forest land for pasture—the major source of deforestation—is driven indirectly by the expansion of soybeans.

Since 2006, forest conversion in Brazil has decreased dramatically for all uses as environmental rules have been tightened and exporters have imposed a moratorium on exports from deforested land. Although only about 10% of the Cerrado is cropped, double cropping increased from 39% in 2001 to 62% in 2011 in Mato Grosso, thus reducing pressure to expand area (VanWey et al. 2013).

Has the expansion of soybeans at the expense of natural vegetation run its course in the Cerrado? Brazilian experts project soybean production of 100 Mt by 2030, compared with a projected 2014/2015 harvest of about 94 Mt. They also believe average yields of 3.9 t/ha are attainable. If so, soybean area would stabilize around current levels (Santana et al. 2011). Even if growth in yields slows as a result of technological stagnation or the adverse effects of climate change, growth in area should not exceed 1.1% per year, which is achieved easily by the conversion of degraded pasture area, estimated at 40 Mha (Gibbs et al. 2015). Environmental rules that require 35% of farm area to be set aside for conservation also provide some protection for native vegetation (discussed further in Chapter 9).

In terms of social development, the expansion of the Cerrado was also a missed opportunity (World Bank 2009) (Chapter 8). Although the intention of the government programs was to develop the Cerrado through family farms, the result has been a growing concentration of land ownership (Bento Filho and Vian 2013). Many small-scale farmers (*posseiros*) had established only informal rights by settling and clearing land for extensive cattle grazing. A weak land administration system often conferred overlapping and duplicate rights to specific plots, which led to much ambiguity over tenure (Reydon et al. 2015). This weakness in tenure security, together with a reduction in state support for small farmers after the mid 1990s, favored larger, better capitalized farmers with legal resources to take over large tracts of pasture land and secure the title to that land for more profitable soybean production. Accordingly, land rights and tenure security have been recurring sources of tension on the Cerrado frontier. Many small farmers sold out or were displaced, sometimes violently, either to urban areas or further north (World Bank 2009). Using Brazil's restrictive definition of family farmers, only 16% of soybean farmers are classified as family farmers today (Bailis 2014).

Changes in land use and ownership are not the only sustainability issues. Large areas planted with nothing but soybeans represent a risky lack of diversity that encourages pests and diseases to proliferate. Already farmers have incurred additional pesticide costs to control an outbreak of soybean rust disease, which was made more serious by the double-cropping of soybeans in the same year, until the practice was banned (Godoy et al. 2015). Overdependence on glyphosate for weed control and for zero tillage has fostered the development of herbicide-tolerant weeds. Some associate these risks with the adoption of GMOs, but in fact the same risks are present with conventional varieties (as in the European Union), and little evidence indicates that GMOs have exacerbated the risks (Bindraban et al. 2009). In either case, however, more diversified cropping systems and integrated crop–livestock systems are needed to reduce risks and enhance sustainability. Efforts to introduce diversified crop–livestock–forestry systems have had limited success to date (Gil et al. 2015) (Chapter 9).

INDIA: A SMALLHOLDER SOYBEAN REVOLUTION

Soybeans were traditionally a niche food crop grown on a very small area in northern India. During the 1960s and 1970s, with limited success, scientists promoted direct food use of soybeans to address the shortage of protein in Indian diets (Bisaliah 1986; Agarwal et al. 2013). Soybeans took off as a cash crop only after they were introduced into Central India during the late 1970s, especially in the states of Madhya Pradesh and (later) Maharashtra. From less than 0.1 Mha in 1975, soybean area surged to more than 12 Mha in 2013, making soybeans the third most widely grown crop after rice and wheat, and one of India's major cash crops. Today, soybeans are the most important oil crop in India, accounting for 46% of oil crop area (excluding cotton). The other major oil crops produced are groundnuts and rapeseed as well as cottonseed. Soybeans account for 23% of domestic vegetable oil and 44% of oil meal production (Table 3.4). Palm oil, practically all imported, is the major vegetable oil consumed.

The spectacular rise of soybeans in India is second in scale only to Brazil, and it parallels the Brazilian experience in some ways. Even more than in Brazil, in India the recent soybean expansion has been concentrated in the tropics under rain-fed conditions, although in India growing conditions are generally characterized as semiarid—somewhat drier than in the Cerrado. In addition, growth has been propelled by area expansion (again, like Brazil), although yield growth has contributed as well.

In other ways, India's experience is quite different from Brazil's. First, the original motivation for expanding production of oilseeds, and soybeans in particular, was to substitute for escalating imports of vegetable oils to meet rapidly increasing demand from Indian consumers (see Chapter 4). It would be a mistake to characterize soybeans as an import substitution crop, however, because soybean meal has emerged as India's most valuable agricultural export in most years, in part because the high incidence of vegetarian diets limits domestic demand for livestock feed. The value of soybean meal exports far exceeds the value of soybean oil imports, although India is a large net importer of vegetable oils.

Second, unlike the large and highly capitalized farms that emerged in the Brazilian Cerrado, the rapid expansion of soybeans in India has been led by small farmers using hired machinery, bullock power, family labor, and low inputs. Nearly two thirds of the soybean area in India is planted by farmers operating less than 5 ha (Department of Agriculture and Cooperation 2012). The availability of a ready market for soybean products, especially soybean oil, along with the crop's easy fit into existing farming systems and low use of purchased inputs, contributed to its rapid uptake by smallholders.

Third, India has achieved success with soybeans through the use of conventionally bred varieties. The Indian government, operating in a noisy democracy with strong civil society voices, has adopted a conservative approach to releasing GMOs. The non-GMO soybean meal also commands a premium price in international markets.

Despite this general success, soybeans in India face considerable challenges to remain competitive with imported vegetable oils (mostly cheaper palm oil) and with the major South American players, Brazil and Argentina, in the export of soybean meal.

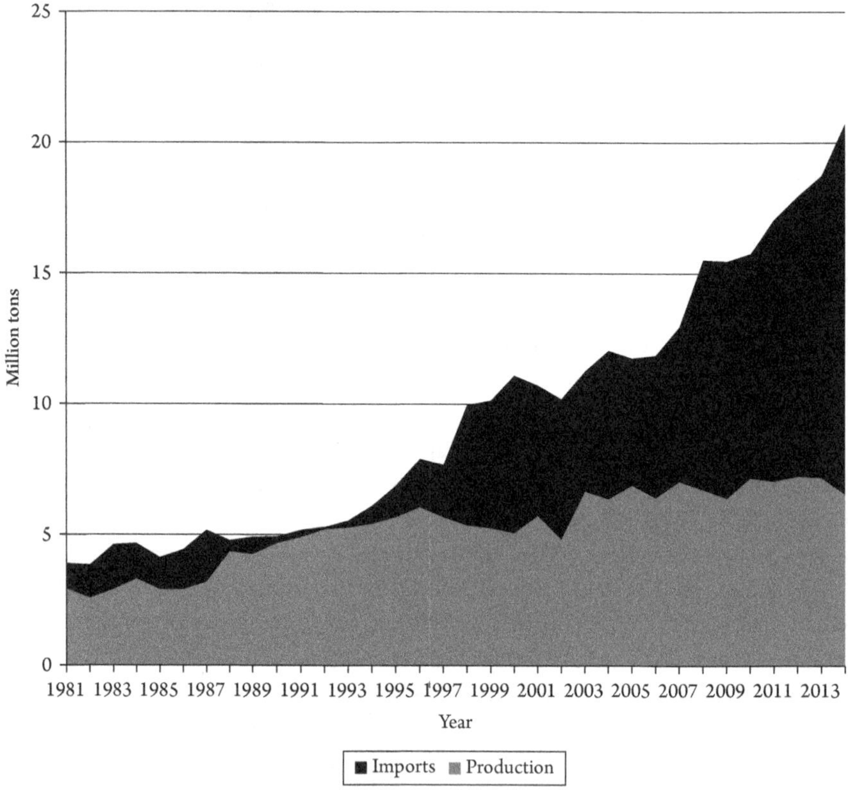

Figure 3.4 Production and imports of vegetable oils in India, 1981 to 2014.
Source: USDA-FAS PSD.

Debate continues over policies to protect domestic oil crop producers from imports and to subsidize inputs as a means of stimulating development of the Indian edible oils market. Indeed, after India was nearly self-sufficient in vegetable oils during the early 1990s, imports surged, so that today more than half the vegetable oil consumed in India is imported (Figure 3.4). The challenge of continuing the expansion of soybeans is reviewed here in relation to India's edible oils policies, the competitiveness of soybeans versus other crops and imports, and the prospects for accelerating yield growth.

The Indian Edible Oils Policy

India's soybean story has to be analyzed within the wider context of national policies on edible oils. Overall, India grows some 28 Mha of oil crops plus nearly 12 Mha of cotton (cottonseed is an important source of oil). Total oil crop production, cottonseed included, is 37 Mt, which yields 7.3 Mt of oil and 17.0 Mt of cake (Table 3.4). Because India consumes twice as much vegetable oil as it produces, the deficit is made up through imports, mostly of palm oil (Table 3.4).

As a large country that has consistently pursued a policy of self-sufficiency in food, India responded to its growing dependence on edible oils by launching a Technology

Table 3.4 Oil crop production and processing, India, average 2011/2012 to 2013/2014

Oil crop	*Crop*			*Meal*		*Oil*	
	Area (000 ha)	*Production (000 t)*	*Milled (%)*	*Production (000 t)*	*Exports (000 t)*	*Production (000 t)*	*Imports (000 t)*
Coconut	NA	673	97	231	0	605	2
Cottonseed	11,967	12,454	70	4063	27	1245	—
Groundnuts	5233	5383	67	1405	11	1187	—
Rapeseed	6827	6767	89	3541	950	2350	69
Soybeans	11,090	11,167	84	7507	3948	1673	1270
Sunflower	762	663	595	283	0	212	1118
Oil palm	NA	NA	NA	NA	NA	59	8265
Total	35,878	37,107	78	17,030	4937	7331	10723
Soybean (%)	30.9	30.1	32.5	44.1	80.0	22.8	11.8
Oil palm (%)	—	—	—	—	—	0.8	77.1

Minor oilseeds not included. Oil palm covers a small area (<10,000 ha), although India has a national program to expand the area substantially.
Source: USDA-FAS PSD.

Mission on Oilseeds in 1986, with the goal of achieving self-sufficiency by 1990. A major thrust of this policy was to generate new varieties, develop suitable inoculum, support a seed system, and promote a package of improved management practices through extension campaigns. The Technology Mission on Oilseeds was supported by protectionist policies that restricted edible oil imports to the state trading system and that tightly controlled the quantity imported. A minimum support price was also introduced (similar to the policy for rice and wheat, India's major food grains), although in general the market price for oilseeds was higher than the support price.

In 1994, as part of wider economic reforms, the government liberalized imports by the private sector and shifted to tariff-based protection consistent with World Trade Organization (WTO) guidelines (Persaud and Landes 2006). Tariffs were very high in the early years and differentiated by type of oil. For example, in 2005, soybean oil was subject to a 45% tariff and palm oil to a 90% tariff (Persaud and Landes 2006). With world prices increasing from 2007, the tariff was equalized across oils and reduced to 10% on refined oils and 2.5% on crude oils in 2014 before being raised to 20% and 12.5%, respectively, in 2015. Given that oil constitutes only about one third of the value of soybeans, the effective tariff protection to soybeans is quite small. At the same time, other types of support have been provided through a series of successor programs to the Technology Mission on Oilseeds, currently subsumed under the National Mission on Oilseeds and Oil Palm (the latter targeted at 125,000 ha by 2017).

Overall, oil crops have been heavily protected as part of India's self-sufficiency policy, but big differences exist among crops. The nominal rate of assistance, which

Table 3.5 Indicators of subsidies, protection, and economic efficiency of soybeans and other oil and competing crops in India

Crop	*Agricultural value, 1997–2004 (% all crops)*	*Irrigation and fertilizer subsidy by crop, 2004 (% all crops)*	*Nominal rate of assistance, 1990–2004*	*Resource–cost ratio, 1981–1993*
Soybeans	1.6	1.1	3.0	0.84
Groundnuts	2.1	2.7	12.8	1.41
Rapeseed	1.3	4.5	55.5	1.47
Maize	1.3	2.1	5.9	0.85
Sorghum	0.9	1.8	14.7	0.76
Cotton	2.1	5.6	13.6	0.70

Source: Data are from Pursell, G., A. Gulati, and K. Gupta. 2007. Distortions to agricultural incentives in India. Agricultural Distortions working paper 34. Washington, DC: World Bank. Resource–cost ratio data are taken from Gulati, A., and T. Kelley. 1999. *Trade liberalization and Indian agriculture: Cropping pattern changes and efficiency gains in semi-arid tropics.* New York, NY: Oxford University Press.

aggregates total tariff, subsidies, and other policy effects as a percentage of the border price, is high for groundnuts and especially rapeseed, but close to zero for soybeans (Table 3.5). In part, the low rate of assistance to soybeans reflects the fact that the meal is the major value component and that about half of it is exported, sometimes with an export tax. In addition, soybeans are a rain-fed crop that fixes its own nitrogen and does not benefit from large irrigation and nitrogen subsidies, in contrast to rapeseed (Chand 2007).

Production and Processing

Soybeans are grown mostly in the heavy black clay soils (Vertisols) of the states of Madhya Pradesh (55% of area) and Maharashtra (30% of area), where annual rainfall averages 900 mm (Agarwal et al. 2013). Soybeans were introduced initially in these states to occupy land left fallow in the rainy season—an estimated 15 Mha of Vertisols that were difficult to cultivate in wet conditions (Chand 2007). Cropping intensity has indeed increased, especially where there is irrigation for the following winter crop, mainly wheat (Bisaliah 1986), although less than 1% of the soybean area is irrigated. This practice of growing wheat after soybeans is most pronounced in Madhya Pradesh, where cropping intensity in rain-fed areas increased from 1.20 in 1990 to 1.72 in 2012. Even so, much of the expansion of soybeans and, to a lesser extent, other oilseeds has been at the expense of sorghum, millet, and pulses, which have experienced declining consumption (Figure 3.5). Many farmers also intercrop soybeans with other crops that mature after soybeans are harvested (for example, sorghum, pigeon pea, and cotton) (Shah 2014).

Soybeans are a major source of cash income, especially for poor farmers (Chand 2007). The main production costs are provided in kind in the form of

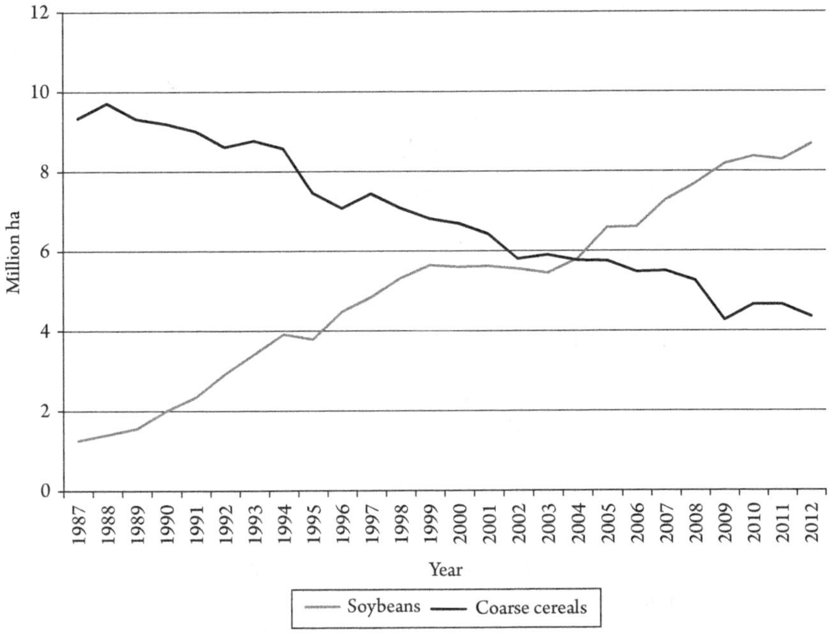

Figure 3.5 Trends in area of soybeans and coarse cereals, Madhya Pradesh, 1987 to 2012. Coarse grains include maize, sorghum, and various millets.
Source: Directorate of Economics and Statistics. 2015. Department of Agriculture and Cooperation, government of India. http://eands.dacnet.nic.in/Default.htm (accessed August 1, 2015).

bullocks and family labor, with purchased inputs making up less than 10% of all costs (Shah 2014). Total labor inputs are high—at about 50 person-days per hectare (Srivastava et al. 2015).

The harvest is processed in private and cooperative solvent extraction mills that are generally larger and more modern than the mills used for other oil crops, the maximum size of which is restricted by prevailing regulations. A solvent extraction mill may process 100,000 t of soybeans every year. This capacity is still small relative to soybean crushers in the United States and Latin America, which commonly crush more than 1 Mt annually (Persaud and Landes 2006; Meena et al. 2013). The lack of storage facilities means that crushing is quite seasonal, and capacity utilization averages only 30% to 40%.

Collective action has been important in supporting smallholders and the development of the processing industry. The Madhya Pradesh Cooperative Oilseed Growers Federation has played a central role in coordinating the value chain, establishing grower cooperatives for marketing, supporting research and extension, and investing in processing facilities (Chand 2007). Many cooperatives operate under contract arrangements with processors or have established their own cooperative processing facility. The Soybean Processors Association has strongly promoted the expansion of soybeans.

The Indian agricultural research system has also provided support to expand soybean production. Testing and breeding of soybean varieties began during the 1960s at

G.B. Pant University in Uttar Pradesh in collaboration with the University of Illinois (Bisaliah 1986). Breeding for the main growing areas in Central India was formalized with the establishment of the All India Coordinated Soybean Improvement Program, and accelerated in 1987 with the founding of the National Research Center for Soybean at Indore. Some 102 varieties have been released, and breeding is estimated to have increased yield potential by 23 kg/yr from 1969 to 2008 (Ramteke et al. 2011)—similar to the rate of breeding progress observed in Brazil and higher than progress in China (Fischer et al. 2014).

Profitability and Competitiveness

Beyond fitting easily into the cropping system, soybeans were more profitable than most other *kharif* (summer) crops. From 2007 to 2011, soybeans provided a return of 27% over total costs, and this return was positive in all years, indicating low risks (Table 3.6). Most other rain-fed *kharif* crops, such as maize and sorghum, experienced negative returns during this period. Only cotton was more profitable than soybeans, but the much higher cost of cultivation and longer growing season (which conflicts with winter crops) negate this advantage.

Soybeans also seem to be the only oil crop that is efficient economically, after policy support to oilseeds is stripped away. Using the resource–cost ratio (RCR) to measure the cost of domestic resources (land, labor, and capital) to produce savings in one unit of foreign exchange, Gulati and Kelley (1999) found that soybean production was efficient (RCR, 0.84) whereas the RCR of other oil crops was above unity (Table 3.5). The private profitability of crops such as rapeseed reflects the high protection and high levels of subsidies accorded the crop. These findings are dated, yet we see no major changes in the underlying technology that would alter them fundamentally.

Further evidence of the competitiveness of soybeans in India is provided by the comparative cost of production. From 2007 to 2011, the cost of production and price received by farmers for soybeans in Madhya Pradesh (see Table 3.6) averaged about 5% less than that in Iowa (Iowa State University 2015). Costs in India were up to 20% higher than in Brazil, which, as we have seen, is a low-cost producer. Growth in total

Table 3.6 Cost of cultivation and returns to *kharif* (summer) crops, Madhya Pradesh, average 2006/2007 to 2010/2011

Crop	*Total cost (US$/ha)*	*Net returns (US$/ha)*	*CV of returns (%)*
Soybeans	343	113	61
Cotton	653	241	92
Sorghum	295	−29	29
Maize	310	−85	85

CV, coefficient of variation.

Source: Calculated from Directorate of Economics and Statistics. 2015. Department of Agriculture and Cooperation, government of India. http://eands.dacnet.nic.in/Default.htm (accessed August 1, 2015).

factor productivity may also be negative in India, further undermining competitiveness (Kurup et al. 2015).

Prospects for Raising Yields and Improving Efficiency

Although by most counts the expansion of soybeans in India has been a success, it has been achieved primarily by the expansion of area. Yields at 1.2 t/ha are very low by world standards. (Recall that yields obtained in the three big producing countries—Brazil, the United States, and Argentina—are around 3 t/ha.) The low yields in India are explained in part by the relatively drier conditions in which the crop is produced. Bhatia et al. (2008) estimated a yield potential of 3 t/ha under irrigation and 2.2 t/ha under rain-fed conditions in Central India. Yields from on-farm demonstration trials under rain-fed conditions have averaged around 2 t/ha (Chand 2007; Jha et al. 2012), suggesting a large yield gap of 40%. Farmers operating in risky rain-fed conditions are unlikely to achieve yields exceeding two thirds of potential yields, however, and on-farm demonstrations may not be representative of farmers' conditions.

The major factors associated with the yield gap are slow adoption of improved varieties (in part because of weaknesses in the seed system), losses to pests (including weeds) and diseases (especially soybean rust), nutrient deficiencies, and harvest and postharvest losses (Jha et al. 2012; Shah 2014). Slow development and low adoption of suitable *Rhizobium* cultures limit the fixation of nitrogen for the following crop. Scope also exists to improve productivity by mechanizing key operations, especially as rural wages in India continue to increase. Finally, soybean value chains could be upgraded by adopting larger and more efficient processing mills, cutting postharvest losses, and reducing transaction costs for exporting soybean meal, given the interior location of the main soybean growing area and India's notoriously weak transport and port infrastructures.

One notable difference between India and the other major producers (except China) is that India has not released GM soybean varieties and appears unlikely to do so in the medium term. As noted, India's stance on releasing GM varieties of crops has been very cautious, given a very active civil society campaign against the technology. Cotton is the only crop for which GM technology has been approved and widely adopted. Ironically, India already consumes substantial quantities of vegetable oils derived from GM crops. The government of India approved the importation of soybean oil derived from GM varieties, and cottonseed oil—now mostly derived from GM cotton—is also an important cooking oil. Although oils from GM and non-GM crops cannot be distinguished chemically, the widespread consumption of oils from GM crops seems to have slipped under the radar of anti-GM campaigns.

The use of GM soybean varieties could have some advantages in India. For example, they may help to reduce costs and increase sustainability by facilitating the adoption of conservation tillage, or they may curb losses to insects. At the same time, these benefits could be negated by the fact that oilseed cake from India—a major export—fetches a significant premium in world markets (>10%) because it is produced from non-GM varieties.

Oilseeds are targeted to grow by 4% per year under the National Mission on Oilseeds and Oil Palm. In India, much debate surrounds the role of import protection as a way to stimulate domestic production and reduce the ever-widening deficit in vegetable oils (Gulati et al. 1996; Jha et al. 2012). To date, soybeans have been the most competitive oil crop, as reflected in their rapid growth. Soybean area is likely to continue to expand at the expense of coarse cereals and pulses, and also as a result of increased cropping intensity, but the easy gains have probably been made. Future growth will depend increasingly on raising yields, which poses a challenge in the risky rain-fed environment where the crop is grown. Major investments in irrigation (at least supplementary irrigation), R&D, and extension are needed to close the yield gap. Eventually, soybeans will gain additional impetus as the domestic demand for oilseed meal for livestock feed catches up to supply.

WILL SOUTHERN AFRICA BE THE NEXT "SOY BASKET?"

Africa has played a miniscule part in the global soybean industry to date. It accounts for less than 0.5% of production, of which more than one third comes from large commercial farms in the temperate and subtropical ecologies of South Africa. The remaining area is scattered across tropical Africa, where Nigeria is the largest producer.

All the same, Africa has the largest area suited for soybean production of anywhere in the world (Gasparri et al. 2016)—an estimated 250 Mha, concentrated in a few countries, notably Sudan (including South Sudan), Democratic Republic of the Congo, Mozambique, and Zambia (Deininger and Byerlee 2011). Indeed, some speculate that the Guinea Savannah of Africa could become the next Cerrado, provisioning not only Africa's growing demand for soybeans and other grains, but also global markets (World Bank 2009).

Although still tiny, soybean production in tropical sub-Saharan Africa (outside of South Africa) has grown at nearly 7% annually from 1991, and, unusually for crops in the region, half of this growth was from improved yields. Strong yield growth reflects the almost universal adoption of improved varieties, mostly developed by the International Institute of Tropical Agriculture (Walker et al. 2015). Viewed through these statistics, soybeans are a success relative to other crops in Africa.

Most soybeans produced in sub-Saharan Africa outside of South Africa are used directly for food. Nigeria, by far the largest producer, has based a major new industry on soy foods, which are an important source of protein for vulnerable groups, such as children (Box 3.1). In the future, however, the major impetus for the soybean market will be the rapid rise in meat consumption, which will increase the demand for soybeans to produce edible oil and animal feed; currently, this demand is mostly met through imports.

Across southern Africa, the experience with soybeans in Brazil and South Africa has reinforced the perception that soybeans can be produced efficiently for feed and oil only on large, mechanized commercial farms. Yet, India's experience indicates the development of a competitive smallholder sector could provide big dividends by

Box 3.1 Soybeans as a Food Crop in Nigeria

Nigeria is the largest producer of soybeans in Africa, with an area of 0.57 Mha and a yield of about 1 t/ha. The crop is grown mostly for food both in the home and for sale, after undergoing local processing using simple methods. It is used for a wide variety of products: fermented beans used as a sauce, steamed and fried soybean cakes, soymilk, soy cheese or tofu, and soy flour. Notably, use of soybean oil in Nigeria is negligible; the most important edible oil is palm oil. In one of the main producing regions in Benue State, subsistence food use of soybeans averages over 10 kg/capita/yr. Villagers, especially women, recognize the nutritional value of soybeans and target consumption to the nutritionally most vulnerable. For example, soybeans accounted for 34% of children's protein intake, in part because soy products are used in school feeding programs. Many soy products are also commercialized in local and national markets.

The success of soybeans in Nigeria derives from the development of more than 20 tropical varieties by the International Institute of Tropical Agriculture for various agroecologies in Africa. Use of these varieties has nearly doubled yields since 1974. The breeding program was supported by a concerted effort to introduce small-scale processing technologies accompanied by a strong outreach and nutritional education program that reached 25,000 rural people. A Japanese cooperation program was especially effective in adapting local methods to develop soy cheese or tofu.

Source: Sanginga, P. C., A. A. Adesina, V. M. Manyong, O. Otite, and K. E. Dashiell. 1999. *Social impact of soybean in Nigeria's southern Guinea savanna.* Ibadan: International Institute of Tropical Agriculture.; and Osborn, D. Z. 2008. *Soybean and soybean products in West Africa. In The world of soy (The food series),* ed. C. M. Du Bois, C.-B. Tan, and S. Wilfred Mintz, 276–298. Urbana, IL: University of Illinois Press.

introducing a cash crop with strong market prospects. Soybean production could also help to address chronic soil fertility problems in Africa's smallholder farming systems.

Zambia is a case in point. Some 85% of its soybean production comes from relatively few large commercial farmers, who grow it mainly as a second crop after wheat in irrigated systems. The remaining 15% comes from smallholder farmers who have an average of about 0.5 ha of soybeans. Neither system is very profitable; even the large commercial farmers are barely competitive (Technoserve 2011). Large farms suffer from high input costs and relatively low yields for irrigated conditions (about 2.6 t/ha). Because of the high transport costs, which raise the costs of imports, Zambia's large soybean producers can compete with imported soybeans but cannot produce the crop competitively for export. Small-scale farmers with yields of around 1 t/ha struggle to obtain improved seed and extension advice and to market their production (Lubungu et al. 2013). Even so, they can supply small local feed processors and markets for edible oils. Larger processors such as Cargill and Dunavant are now buying soybeans from smallholders, sometimes through contractual arrangements.

The underlying tensions in deciding whether to focus on commercial farmers or smallholders are even more evident next door in Mozambique. There, large areas with a relatively low population density of 10 to 50 persons/km^2 and very low productivity have been identified for the development of commercial agriculture (Ekman and Macamo 2014). For example, the government of Mozambique allocated 10,000 ha to Quifel, a Portuguese company, to produce soybeans. The project performed poorly in terms of area and yield and was eventually closed (Norfolk and Hanlon 2012; Smart and Hanlon 2014). A more ambitious effort, the ProSAVANA project, involves a tripartite agreement among the governments of Mozambique, Brazil, and Japan to build on the experience of the Cerrado. This project excited much speculation that Brazilian farmers would invade Mozambique and has been strongly criticized by civil society and others (Classen 2013). Recent analyses point to a focus on private investment in processing, and nucleus farms as a way to stimulate smallholder production (Ekman and Macamo 2014; Tawa et al. 2015). Three private companies have already developed contract farming arrangements with smallholders, with varying success as a result of problems of side-selling (Smart and Hanlon 2014).

Meanwhile, efforts by international donors and nongovernmental organizations (NGOs) for a decade have helped to develop a vibrant industry of about 27,000 smallholders to produce soybeans for the growing poultry industry. Soybean area has expanded rapidly, reflecting the crop's profitability, but the area that smallholders can manage is constrained severely by the lack of tractors for land preparation (only about one quarter of the area is plowed by tractor), by the fact that inoculum is not used, and by the generally low yields of a little more than 1 t/ha (Smart and Hanlon 2014). An added concern is that competition for land is already considerable; the Quifel project occupied some 600 ha where smallholders had grown soybeans, heightening tensions over land rights (Norfolk and Hanlon 2012).

A program in Zimbabwe has also demonstrated that smallholders can produce successfully soybeans for the feed and edible oil markets and replace, in part, production from the commercial farmers who supply those markets before land reform (Rusike et al. 2000). This program suggests a four-prong approach is needed, incorporating (1) recognition of multiple products for food, feed, and oil, including local small-scale processing options; (2) a concerted effort to provide and train input dealers and farmers to use inoculum for enhancing nitrogen fixation; (3) a strong extension system involving NGOs; and (4) collective efforts to market in volume (Giller et al. 2011). An estimated 100,000 Zimbabwean smallholders may be producing soybeans. Yields are still low at a bit more than 1 t/ha, but the value of the nitrogen provided to the following maize crop could replace one application of fertilizer. New "promiscuous" soybean varieties allow nitrogen to be fixed using the indigenous rhizobia present in the soil, and reduce the need for inoculum (Mpepereki et al. 2000).

A concerted approach (built around innovation platforms that assemble expertise in breeding, adaptive research, extension, and support to marketing) could allow smallholder farmers in southern Africa to seize a major opportunity for cash crop production in the region. In areas where population density is very low, larger commercial farmers may have a role, but the Cerrado model requires significant modification to overcome the disadvantages prevalent in the region—the weak capacity for agricultural R&D, regulatory barriers that make GM technology unavailable, lack of access to state financial resources, and high transaction costs of acquiring land. Each of these factors, negative in southern Africa, was positive in the Cerrado—strong R&D, few barriers to GM technology, and so on—and instrumental in producers' success.[3] Even if such barriers are overcome, Africa's capacity to provision global soybean markets is a long way in the future because the costs of transport and logistics are high. Recent analysis also flags the high environmental costs in terms of carbon emissions and loss of biodiversity from converting African savannah land to soybean production for global markets (Searchinger et al. 2015).

LONG-RUN SUPPLY PROSPECTS FOR SOYBEANS

The rapidly growing demand for soybean meal has been met mainly by expanding the area on which the crop is grown. Although demand growth will slow (Chapter 5), the big issue for the future remains the role area expansion will play in relation to yield growth.

Fischer et al. (2014) decompose progress in raising yields into two components: (1) progress in raising potential yields (a component that measures the maximum yield achieved in a given environment with the best available management and varieties) and (2) progress in closing the gap between potential yields and farm yields. Global progress in raising farm yields for soybeans has been similar to that for the three major cereals, but progress in raising potential yields of soybeans has been much slower (Fischer et al. 2014) (Table 3.7). As a result, the yield gap has been closing faster for soybeans, leaving a yield gap of only 31% of potential yields over current farm yields, which is much less than the gaps for other major crops. Experience indicates it is not economic to attempt to close a yield gap of less than 25% to 30%, especially in rain-fed conditions where input use is exposed to more risk. In Brazil, Sentelhas et al. (2015) also concluded the yield gap is quite small unless farmers invest in irrigation.

Part of the reason for the small yield gap in soybeans is that the global averages are determined largely by the situation in the big producers (United States, Brazil, and Argentina), which account for 82% of world production, grow the crop under high levels of management, and receive strong support from public and private research and extension systems. Other crops are distributed more widely; a greater share is grown in countries with weaker R&D systems.

Nor does GM technology offer much prospect for increasing soybean yields. The share of soybean production from GM technologies already is larger than that of any other crop. Although GM technology has reduced the cost of production in Brazil, where seed costs are low because intellectually property rights are enforced loosely (Qaim 2009), the technology does not seem to have contributed to yield growth (Xu

[3] Much of the African savannah suffers from acidic soils like the original Cerrado, but the feasibility of improving soil through liming has received little attention.

Table 3.7 Summary of progress in farm yields, potential yields, and the current yield gap around 2010 for the major cereals and soybeans

	Rice	*Maize*	*Wheat*	*Soybeans*
Growth in farm yields (%/yr)	1.0	1.5	1.0	1.3
Growth in potential yields (%/yr)	0.7	0.8	0.6	0.5
Rate of closing of yield gap (%/yr)	−0.3	−0.7	−0.4	−0.8
Current yield gap relative to farm yield (%)	72	98	50	31
Share of global production from top 4 producers (%)	65	67	46	84

Source: Data are from Fischer, T., D. Byerlee, and G. Edmeades. 2014. *Crop yields and global food security: Will yield increases continue to feed the world?* Canberra: Australian Centre for International Agricultural Research.

et al. 2013). This situation is unlikely to change, because much of the current emphasis on GM technology is to improve insect and disease resistance to replace expensive pesticides and to diversify sources of herbicide tolerance given the increase in weeds tolerant to glyphosate (Chapter 9). GM technology is also turning to quality traits such as higher oil content for biodiesel and lower linolenic acid levels that enable soybeans to substitute for hydrogenated oils in food processing (Chapter 4).

Assuming yield growth of 0.6% annually (0.5% from yield potential and 0.1% from further closing of the yield gap), at least half of future growth in soybeans will have to be met by area expansion. Although the growth in soybean area surely will slow sharply from the 3.3% annual growth of the past 20 years, growth will continue to be faster than for other major crops, except oil palm.

As in the past, potential sources for area growth are the conversion of natural areas to cropland, the replacement of low-grade pasture or fallow in current farming systems, and the substitution of soybeans for other crops. We see a role for all three sources. Brazil and other areas of Latin America still have much scope to intensify production by converting pasture to cropland. Similarly, India has potential to intensify production by using fallow land in the monsoon season, although substitution for lower value crops such as sorghum will continue to be the major source of growth. And in sub-Saharan Africa, soybean expansion, if it takes off—a big if—will likely occur through the conversion of natural areas in the savannah, and any conversion of natural land can be expected to imply environmental tradeoffs (Searchinger et al. 2015)—a subject we return to in Chapter 9.

SUMMING UP

During the past 25 years, soybeans (including oil and meal) went through a remarkable transformation to become the most valuable agricultural commodity in world trade. This transformation was accompanied by a distinct move in the center of gravity of soybean production toward the tropics. As we have seen, the major drivers of this transformation have been the demand for livestock feed in emerging economies, together with the use of soybean oil for food and biofuel, topics we return to in Chapters 4 through 6.

The three case studies examined in this chapter—the Cerrado of Brazil, Central India, and southern Africa—highlight common elements in the soybean revolution but also reveal sharp differences. In all cases, soybeans expanded rapidly through a major increase in area. Soybeans are the global leader in land-use changes. Between 1990 and 2013, soybean area expanded by 54 Mha compared with 134 Mha for all other crops. In some cases, soybean area increased through the conversion of natural vegetation or pasture to cropland, as in Brazil, whereas India expanded soybean area through the intensification of cropping systems. In the future, Africa could be an important source of growth in soybean area, largely by bringing new land under cultivation.

Even so, producers can increase soybean yields in tropical areas, in part through the use of better technology and in part through investments that make infertile lands more productive. Brazil (and to a smaller extent India) are cases in point. Yields in tropical Brazil are now similar to those in the temperate areas of the United States and Argentina.

Large off-farm investments in processing, infrastructure, and logistics have been as important—or even more important—than on-farm investments. The private sector has been the major investor, but public investments in roads and R&D have also been critical.

The major differences among the case study areas are quite important to note. In Brazil, soybean expansion is based on large-scale family farmers and agribusiness farms, in contrast with India, where poor smallholders have led the soybean revolution. If soybeans take off in Africa, a big issue is which type of farmer will lead the industry; current systems include a mix of farm types. We revisit this question in Chapter 8, where we also review the role of soybean expansion in local economic development. Given the importance of land in expanding soybean production, the sustainability implications of the soybean revolution are in the global spotlight—both the encroachment of soybeans on natural areas of high biodiversity and high carbon sequestration capacity, as well as the potential of large-scale soybean production to spark land conflicts where tenure security is weak or ill-defined. In Chapter 9, we examine how Brazil has curtailed the encroachment of soybeans into tropical forests successfully, even as it continues to convert natural woodlands of the Cerrado to soybean production.

4

FOOD DEMAND FOR VEGETABLE OILS

THE SURGE IN VEGETABLE OIL CONSUMPTION

The surge in vegetable oil consumption during the past 35 years has been nothing short of remarkable. From a global total of about 35 Mt in 1980, usage more than quadrupled to 165 Mt in 2013.[1] This growth of 4.8% annually was more than twice the rate of the population increase during the same period. Palm and palm kernel oil combined are currently the leading consumption component at 38%, having recently displaced soybean oil (now at 27%) as the major source. Rapeseed oil (15%) and sunflower oil (9%) round out the top four (Table 4.1).

Vegetable oils for food (edible oils) currently total 125 Mt, or three quarters of all oil consumption. They are the main focus of this chapter; we defer the discussion of biodiesel until Chapter 6.[2] The task at hand is to explain the causes of the recent and rapid increases in human consumption of edible oils as a prelude for assessing future patterns of their use in food. From a consumption perspective, demand is global, and the distinction between tropically and temperately produced vegetable oil is much less interesting than on the supply side. The very size of international trade in vegetable oils—about 65 Mt in 2013 (US Department of Agriculture 2014)—is one major indication of the extent to which production and consumption locations do not coincide (Chapter 7).

Slightly more than half the vegetable oil used for food in 2013 was consumed in Asia, and much of that oil originated in the tropics. Europe (at 16%) follows Asia as a distant second. In South and Southeast Asia, Africa, and the Middle East, palm oil tends to dominate consumption; in the Americas, East Asia, and North Africa, soybean oil is preferred; in the former Soviet Union, sunflower oil tops the list; and in Europe and Oceania, rapeseed has the highest percentage. Yet, because vegetable

[1] There are two major data sets for vegetable oil consumption. One series is collected by the Foreign Agricultural Service of the USDA and the other is compiled by the FAO. (In the text these sources are referenced simply as (USDA-FAS PSD) and (FAOSTAT)). At the country level the two data sets exhibit substantial differences (Box 4.1). We regard the USDA series as more reliable and use these data whenever possible in our consumption analyses. In this chapter, we refer to split years (for example, 2013/2014) as the earlier year (that is, 2013).

[2] Significant portions of this chapter draw on Gaskell (2012), whose PhD dissertation was an integral part of the ongoing palm oil project at the Center on Food Security and the Environment, Stanford University.

Table 4.1 Global consumption of vegetable oils, total use, and food use, in million metric tons, from 1980 to 2013

	1980	*2000*	*2013*
Source of oil			
Palm oil	4.8	23.6	57.0
Rapeseed oil[a]	3.9	13.4	25.0
Soybean oil	12.4	26.5	45.0
Sunflower oil	4.4	8.2	15.0
Other	10.3	16.6	23.6
Total use	35.8	88.3	165.4
Food use			
Palm oil	4.7	19.8	40.0
Rapeseed oil[a]	3.6	12.3	17.5
Soybean oil	11.9	25.4	36.8
Sunflower oil	4.4	7.8	14.3
Other	8.9	13	16.7
Subtotal	33.6	78.3	125.1

[a] Includes rapeseed oil, mustard seed oil, and canola oil.

Source: USDA-FAS PSD.

oils are very good substitutes for one another, virtually all vegetable oils are in the markets of all regions.

VEGETABLE OIL DEMAND: THE CONTEXT AND ITS COMPLEXITIES

Every food group presents its own set of difficult consumption questions. Vegetable oils are no exception; indeed, they may be the most complex of all commodity groupings. Seven sources of complexity in vegetable oil consumption help to explain this conjecture. The first difficulty arises from data sources. The discrepancies in basic consumption estimates are so large that, realistically, analysts must choose either USDA or FAO data in making comparisons across time and space. For reasons explained in Box 4.1, we have opted to use the USDA series whenever possible.

The second source of complexity is simply the sheer number of commodities involved. Each oil in this wide array differs slightly from the others in its origins, taste, fatty acid composition, and uses, yet all are highly substitutable with one another, both in direct consumption and as an ingredient in processed foods (Table 4.2).

A third source of complexity involves the form of the products being consumed. In household consumption surveys, for example, it is difficult enough for a respondent to recall how many kilos of rice or bags of flour the household has consumed. But remembering (or even knowing) the amounts of margarine, salad dressing, and oil used

Box 4.1 Data Dilemmas

Analysts undertaking global or comparative studies of vegetable oil consumption face immediate multiple data dilemmas. The first of these is whether to use FAO or USDA data sets (FAOSTAT and USDA-FAS PSD). Country-level differences between the two sources are very large and are exceedingly difficult to resolve.

Consumption data for three of the largest consumers of vegetable oil illustrate the discrepancies (Table B4.1). Analysts are thus forced to choose one source or the other, then use it consistently. Literature reviews are especially treacherous when varying conclusions are based on the different sources. We believe the onus is on the FAO to reconcile their estimates because we believe quantities such as those shown in the table for India and China are implausible and inconsistent with those countries' own estimates of food demand.

A second dilemma, related to the first, includes the accounting and conceptual problems in tallying vegetable oil consumption. Consider, for example, the following situation. Suppose a bakery buys 100 L of vegetable oil and uses the oil to deep-fry doughnuts. When the oil becomes stale, the bakery sells the used oil for use in making biodiesel. From an accounting point of view, should the 100 L be tabulated as food or industry, because the bakery is not the final consumer? What is being measured—vegetable oil purchase or vegetable oil ingestion? If placed in the food category, should the entire 100 L be allocated to food or just the oil retained in the doughnuts? And what part, if any, should be tabulated under biodiesel?

We believe these questions are at the heart of the FAO/USDA data differences. In this chapter we chose to use USDA data whenever possible because we believe them to be more reliable. This is also why our primary focus is on the demand for crude vegetable oils rather than on final demand by food category. Comparative use tables seem, to us, to have limited reliability unless grounded in clear accounting rules consistent through time and space.

Finally, for global calorie consumption, we had no alternative but to use FAOSTAT, and calories from vegetable oils may well be underestimated.

Table B4.1 Vegetable oil consumption for food, by data source, for 2013 in million tons

Country	*Commodity*	*US Department of Agriculture*	*Food and Agriculture Organization of the United Nations*
China	Palm oil	3.6	2.7
India	Palm oil	8.1	1.6
Indonesia	Palm oil	5.2	1.2
China	Soybean oil	13.7	2.9
India	Soybean oil	3.3	2.3
Indonesia	Soybean oil	0.03	0.4

Table 4.2 Summary characteristics of various edible oils

Characteristic	*Palm oil*	*Soybean oil*	*Rapeseed oil*	*Sunflower oil*	*Palm kernel oil*	*Groundnut oil*	*Coconut oil*	*Olive oil*
Saturated (%)	52 (high)	15	7 (low)	11	80 (high)	18	92 (high)	14
Smoke point (°C)	230	241	242	246	232	231	177	225
Nutritional properties	Carotene, tocopherols, sterols in unrefined oil	Gamma-tocopherol (antioxidant)	High in omega-3 fatty acids	High in omega-6 fatty acids	Highly saturated	High monosaturate content, aflatoxin risk	Highly saturated	High in monounsaturates, vitamin E, phenolics
Common end uses	Cooking oil, margarine, cosmetics	Cooking oil, margarine, shortening	Frying, baking, biodiesel	Cooking oil, margarine, shortening	Processed foods, soap	Frying, cooking oil, margarine	Baked goods, processed sweets, shortening	Salad oil, margarine, cooking oil
Global market share, 2010 (%)	34	29	16	9	4	4	3	2
Food use (%)	72	82	69	94	27	100	53	98
World price, US$/t, 2000–2010 average (SD)	624 (271)	751 (289)	914 (316)	1018 (431)	841 (419)	1332 (452)	863 (429)	4072 (950)
Top producing countries, 2010	Indonesia, Malaysia, Thailand	United States, China, Brazil	China, Germany, India	Ukraine, Russia, Argentina	Indonesia, Malaysia, Nigeria	China, India, Nigeria	Philippines, Indonesia, India	Spain, Italy, Greece
Top consuming countries, 2010	India, Indonesia, China	China, United States, Brazil	Europe, China, India	Europe, Russia, India	Malaysia, Indonesia, Europe	China, India, Nigeria	Philippines, Europe, United States	Europe, United States, Turkey

SD, standard deviation.

Source: Gaskell, J. C. 2012. The palm oil revolution in Asia. PhD diss., Stanford University.

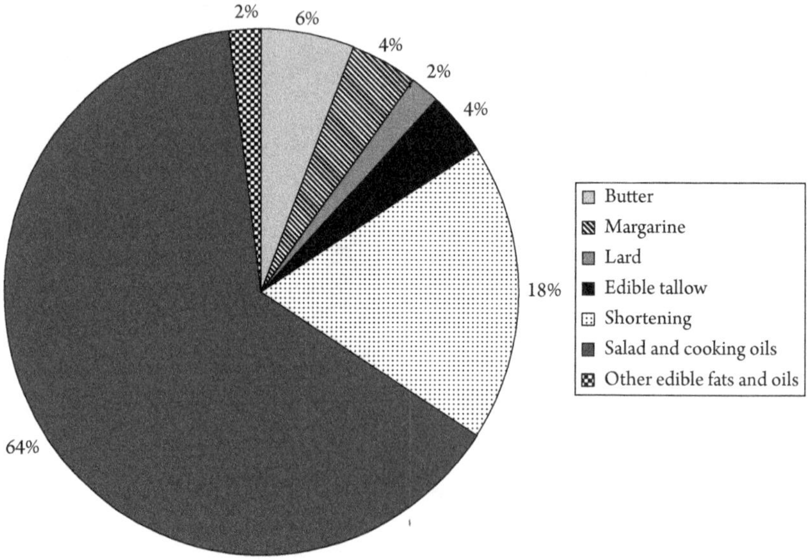

Figure 4.1 Percentage shares of edible fats and oils in US diets in 2010.
Source: US Department of Agriculture. 2011. Malaysia: Obstacles may reduce future palm oil production growth. http://www.pecad.fas.usda.gov/highlights/2011/06/malaysia/ (accessed May 2, 2016).

in baked, fried, and other prepared products is virtually impossible. It is no accident that the literature on edible oils is fragmentary, and that estimates of parameters for oil consumption tend to show high variances relative to other food groups. The United States is a partial exception on this point because the USDA has long collected data on fats and oils consumption. Unfortunately, the usage data for the United States, shown in Figure 4.1 (38 kg total in 2010), are no longer being collected.

A fourth source of complexity is the high degree of consumption substitution among the vegetable oils in processed foods. When asked about specific oils, many consumers are simply unaware of the particular oil they are consuming. Strong taste preferences prevail for some oils, such as sesame and olive oil, but the major vegetable oils are fairly indistinguishable from one another in many food products and for many cooking purposes. Moreover, this substitution occurs frequently near the end of the value chain. Labels on margarine and many other products typically list multiple oils; the exact blend changes depending on relative costs among the oils. The high elasticities of substitution among oils mean that small changes in relative prices among oils can create large swings with respect to the use of particular oils (Kojima et al. 2014). These substitutions make the estimation of market shares an exceedingly difficult task, but it is a key task if oil demand is to be linked to particular commodity markets.

A fifth complication arises because of the range of oil substitutions that occur outside the vegetable oil complex. Aside from the biodiesel and energy linkages taken up in Chapter 6, there are connections to animal fats—for example, to butter in North America and Europe, *desi* or animal-based *ghee* (clarified butter) in South Asia, and

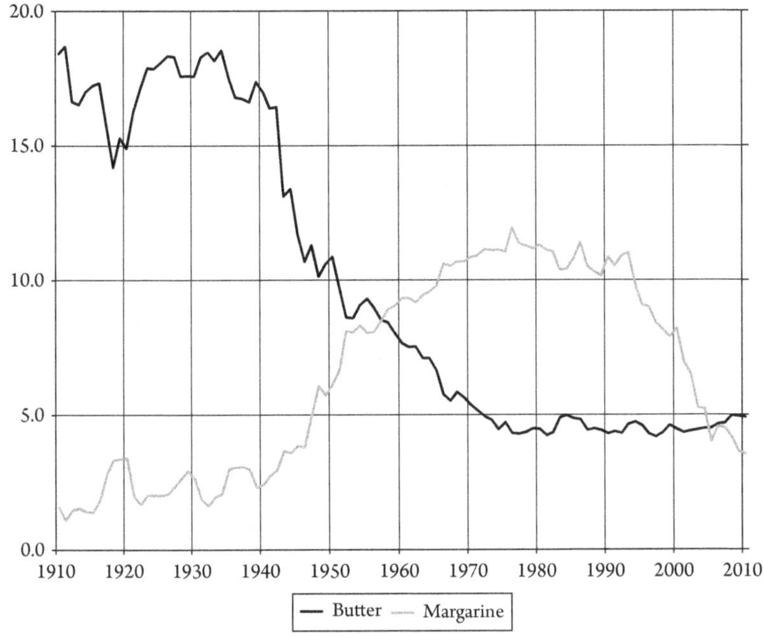

Figure 4.2 US consumption of butter and margarine (in pounds per capita), 1909 to 2010.
Source: Reproduced from Ferdman, R. 2014. The generational battle of butter versus margarine. *The Washington Post*, June 17. http://www.washingtonpost.com/blogs/wonkblog/wp/2014/06/17/the-generational-battle-of-butter-vs-margarine/ (accessed May 2, 2016).

lard in China. Dairy products are particularly subject to major policy interventions in many countries, which further confuses matters (Organisation for Economic Co-operation and Development 2013).

A single example from one country highlights many of the analytical difficulties outlined so far. Figure 4.2 provides a fascinating picture of butter and vegetable oil–margarine consumption in the United States. It illustrates the importance of substitution among fats and oils, as well as changing views of their health impacts, which are discussed later. The figure also exemplifies the conundrums facing anyone trying to project forward to 2050. Few forecasters in 1980 would have even come close to predicting what actually happened to margarine consumption in 2010.

Coproducts constitute a sixth source of complexity, as spelled out in some detail in Chapter 5. In Chapter 1 we noted that meal and oil are produced jointly by crushing whole soybeans (and other oilseeds). In most countries, meal "drives" demand for soybeans, but in a few cases (and for some time periods) it is oil. In China, for example, the importation of soybeans to produce soybean meal for the country's rapidly growing livestock industry—5 billion chickens and 500 million pigs—now seems dominant (The Economist 2011). In 2013, China imported 69 Mt of soybeans—62% of the world's total trade (US Department of Agriculture 2015d). Those imports, in turn, produced about 13 Mt of soybean oil for the domestic market. Had pig and poultry production not been surging, it seems likely that more palm oil would have been demanded/imported (instead of soybeans/soybean oil) because of palm oil's price advantage.

Finally, great complexity surrounds the recommended daily dietary allowances for fats and oils. Human health and nutrition problems emerge when too few or too many calories come from fats and oils or if the share of various fats and oils is unbalanced. The consuming public has a poor understanding of dietary recommendations, nor are they summarized easily, especially when the focus is specifically on the role of tropical vegetable oils. Usually, oil recommendations are embedded within macronutrient recommendations for combined fats and oils, which in turn have both upper and lower limits for healthy diets. Specific recommendations for vegetable oils are thus complicated, as are generalizations about the health effects of oil consumption patterns (discussed later).

The most recent expert report from the FAO and the World Health Organization (WHO) (Food and Agriculture Organization of the United Nations 2010) recommends that at least 15% of the calorie requirement for adults should come from fats[3] and that women in reproductive ages should obtain 20% of their calories from fats and oils. Children in poorer countries, especially where grain-based diets are common, also require substantial fats and oils because they are calorie-dense foods. For example, an earlier FAO/WHO expert report (Food and Agriculture Organization of the United Nations 1994) recommended that fats and oils should supply 30% to 40% of calorie requirements for children between weaning and 2 years of age. But, there can also be too much of a good thing. The experts indicate that active adults should consume less than 35% of total energy from fats and oils, and that the level of saturated fats should not exceed 10%; for sedentary individuals, the upper limit is 30% from fats and oils.

The role of vegetable oils in total calorie consumption varies enormously across countries. In 2010, diets in Bangladesh had 6% of total calories coming from vegetable oil; China, 6%; India, 9%; Indonesia, 9%; Nigeria, 12%; Ethiopia, 3%; and Brazil, 13%.[4] At the upper end of the scale are Australia, New Zealand, several Pacific Islands, the European Union, and the United States (FAOSTAT). Moreover, the average absolute number of calories per day coming from oils in the United States (676) and the European Union (485) were more than double those found in most other countries.[5] We also note that the foregoing estimates come from the FAO food balance sheets, which may understate the role of vegetable oils in diets (Box 4.1).

Data from the 1993, 2004, and 2009 national sample surveys from India are particularly interesting in that they show changes through time and by location (Gaiha et al. 2012). Figure 4.3 shows the composition of fats and oils in India. Almost half of total fats and oils in both rural and urban households now come from vegetable oils,

[3] The term *fats* in the FAO documents refers to a macronutrient, whereas *fats and oils* refers to an aggregate food category.

[4] Significant discrepancies often appear between the FAO food balance sheet data and household consumption data. For example, the China Health and Nutrition Survey of 2011, well known for its high-quality data, shows about 13% of total calories came from edible oils (Zhai et al. 2014).

[5] In the United States during the 20th century, for example, soybean oil consumption increased from 0.09 kg/capita in 1909 to 11.64 kg/capita in 1999 (Blasbalg et al. 2011).

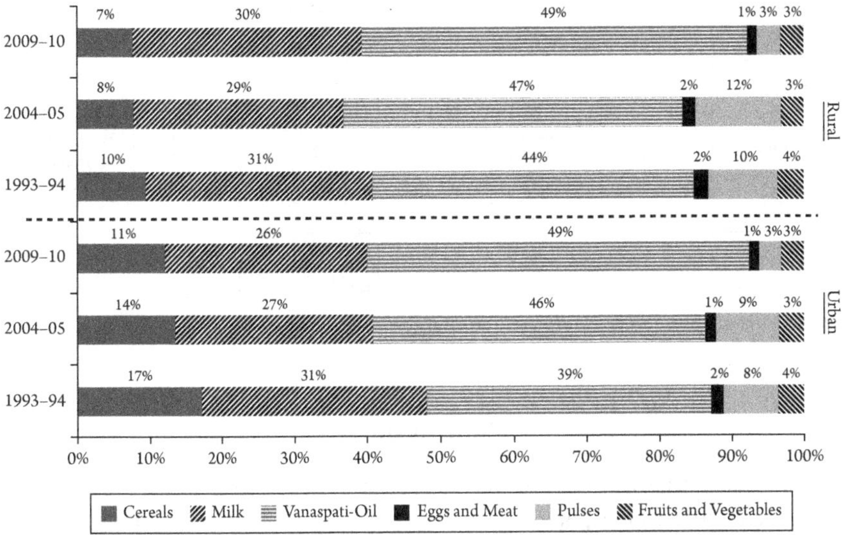

Figure 4.3 Fat intake by food commodity in rural and urban India, 1993 to 2009.
Source: Reproduced from Gaiha, R., N. Kaicker, K. S. Imai, and G. Thapa. 2012. Demand for nutrients in India: An analysis based on the 50th, 61st, and 66th rounds of the N.S.S. Discussion Paper Series, RIEB, Kobe University. http://www.rieb.kobe-u.ac.jp/academic/ra/dp/English/DP2012-14.pdf (accessed May 2, 2016).

including *vanaspati,* which is a hydrogenated (semisolid) oil product. That share has increased substantially since 1993, especially in rural areas.

PAST AND FUTURE DETERMINANTS OF DEMAND

The previous section highlighted several broad problems that complicate research on oil demand. Fortunately, considerable knowledge also exists about the regularity of oil consumption patterns. Research on household consumption behavior, and studies on the demand for agricultural commodities, have long and distinguished histories (Engel 1857; Prais and Houthakker 1955; Timmer et al. 1983; Deaton 1997; Huang and Lin 2000; Dutta and Gulati 2009; Kruse 2010). Basic estimation methods are well known, although unfortunately many of the more recent econometric developments have data requirements beyond the scope of the data currently available. For this reason, we do not seek new theoretical or estimation breakthroughs here, but instead base our analysis on previous research and on rather traditional partial equilibrium models, often using cross-sectional data.

Our approach is shown schematically in Figure 4.4. We first focus on five variables and their associated parameters that hold keys for understanding future edible uses of total vegetable oil (population, urbanization, income, health, and prices). We then present four case studies to help frame the difficult issue of market shares and how total vegetable oil consumption is divided among various sources.

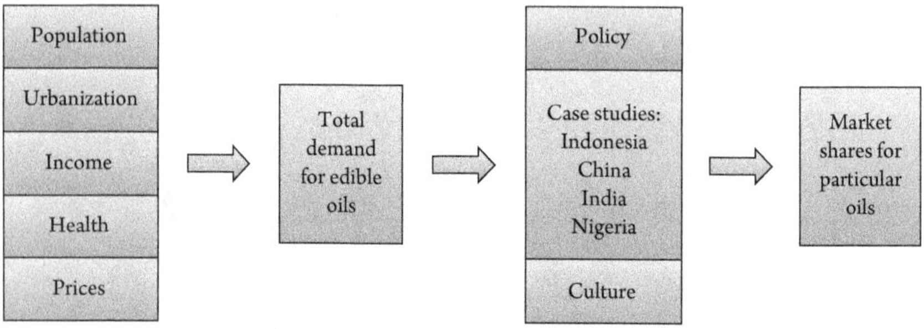

Figure 4.4 Determinants of oil consumption, total and by commodity

Population

World population growth rates peaked in 1967 at slightly more than 2.1% per year (World Bank 2015a), but population momentum, driven by the number of young women entering their childbearing years, continues to be a major determinant of population growth. When analyzing future vegetable oil demand, three population questions are critical: How many people will be added? When will they appear? And where will they be located? Decadal (net) additions, shown in the bars of Figure 4.5, are of particular interest for the "how many" and "when" questions.

Five decades—the 1960s, 1970s, 1980s, 1990s, and 2000s—saw the greatest global population increments in human history. A rapidly growing population was clearly a prime determinant of the rapidly increasing global demand for vegetable oils during the past 35 years. Looking forward to 2050, the story is likely to be quite different. Decadal population increments are falling, and they are falling fast. In a dynamic sense, population growth, although still positive, will be of much less relative importance as a determinant of global demand for vegetable oil for the next 35 years compared with the past 35 years.

The "where" part of the population question may reinforce the effects of declining decadal increments. Although Asia will still have the greatest absolute numbers, Africa will dominate population increases during the 21st century. Africa is expected to add 1.3 billion persons between 2013 and 2050; Asia, 1.0 billion; and Latin America, 0.2 billion. North America and Europe are expected to remain constant (Population Reference Bureau 2014). Of the 40 countries with the highest total fertility rates (TFRs) in 2013, 37 were within sub-Saharan Africa. The speed at which these TFRs fall will have profound implications for per-capita income growth, vegetable oil demand, and dozens of other important variables.

Africa, in particular, is a region where quite strong opposing forces will likely be at play. The region's income and oil consumption are relatively low. There is uncertainty about how fast Africa's income will grow, and whether its cultures and cuisines will result in edible oil consumption patterns similar to those in Asia. The key point is that the greatest future additions to population are likely to be in a region with low average consumption, but where rapid income growth could make for large marginal changes.

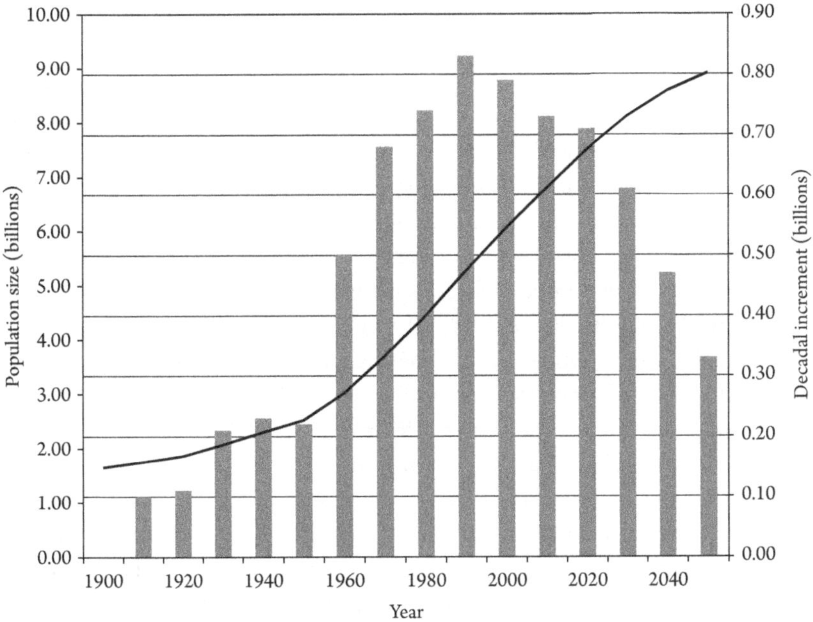

Figure 4.5 Population growth in historical perspective. The solid line is total population and the bars are the decadal increments.
Source: United Nations. 1999. *The world at six billion*. New York, NY: United Nations Population Division; and authors' calculations therefrom.

Urbanization

Virtually everywhere in the world, migration from rural areas to cities has been very rapid during the past 25 years. In 1990, the world's urban percentage was 43%; by 2014, the urban ratio had grown to 54% of a much larger global population. Urban growth rates in less developed regions have recently averaged about 1.2% annually, or about four times the rate in more developed regions (United Nations 2014).

By 2050, the United Nations (2014) expects that two thirds of the world population will live in cities, which are also growing rapidly in size. Just three countries—India, China, and Nigeria—will account for nearly 40% of urban growth between 2014 and 2050. Even by 2030, some 40 cities are expected to exceed 10 million, and a substantial number of those mega-cities have seacoast locations. It seems quite possible that such cities will secure their food increasingly via international trade rather than from their domestic hinterlands.

The housing, infrastructure, and employment implications of the migration to cities will be enormous. Our concern in this chapter, however, is limited to what urbanization will do to edible oil consumption. Virtually all studies on consumption show that rural and urban dwellers have differing consumption patterns for most food groups. The causes are, in part, a result of the range of food products available in urban areas, the proportion of meals eaten outside the home, and the changing opportunity costs of household labor, especially for urban women.

Studies on rural–urban differences in vegetable oil consumption are scattered at best, and the differentials that have been reported do not appear as large as for some other food groups.[6] The best insights have come from countries with nationally representative samples that are repeated on a regular basis. One of the clearest pictures comes from Indonesia. Gaskell (2015), using income-pairing techniques, found a statistically significant differential between rural and urban oil consumption, although urban consumption of edible oils was only about 5% greater than rural consumption at comparable household income levels. For India, there are numerous comparisons between per-capita consumption of oils in urban versus rural regions. For example, Jha et al. (2012) indicate that in 2004, urban households in India expended about Rs 37 per capita, whereas rural households expended about Rs 26 per capita—a difference at the mean of about 35%. The problem with this percentage is it embraces much more than just urbanization effects. Urban dwellers participating in the national Indian survey have higher incomes on average and face higher prices than rural participants, and thus the 35% is almost surely an overestimate of the urban difference per se. On the other hand, Huang et al. (2014) indicate that Chinese household data, at comparable income levels, also reveal large differences. Vegetable oil consumption in rural areas is only a bit more than half of urban consumption. Our sense from these large household surveys is that urbanization effects vary widely according to specific circumstances, with a range from about 5% to 40%. As countries become richer, and as their agricultural sectors become relatively smaller, the aggregate effect of rural–urban differences will also become reduced.

Income

Per-capita gross domestic product (GDP) has long proved to be a key determinant of food demand. Detailed analyses across countries, households, and time periods provide additional insight into income–oil relationships. Figure 4.6 shows a 2013 cross-sectional mapping of vegetable oil consumption per capita against GDP per capita for 50 countries with populations surpassing 20 million. The fit of the curve is significant but far from perfect, indicating that even for a given income, culture and cuisine play important roles in edible oil consumption.

As incomes rise, the slope of the income–consumption relationship levels off. The income elasticity coefficient measured across the full range of 50 countries in Figure 4.6 is 0.44,[7] with real GDP per capita explaining about half the variance—a correlation of 0.7. In some sense, this constant elasticity formulation can be thought of as the edible oil income elasticity for the world.

[6] A major problem in rural–urban comparisons is determining "equal" incomes between rural and urban groups. Urban and rural food prices are typically different, markets often are missing in rural areas, and the composition of market baskets is almost always dissimilar. Meals eaten outside the home also present difficult data collection problems, especially among surveys of urban consumers.

[7] Estimated as $\log_e$ Per capita consumption $= a + b \log_e$ per capita GDP, where b is the measured income elasticity, sometimes called the *Engel coefficient*.

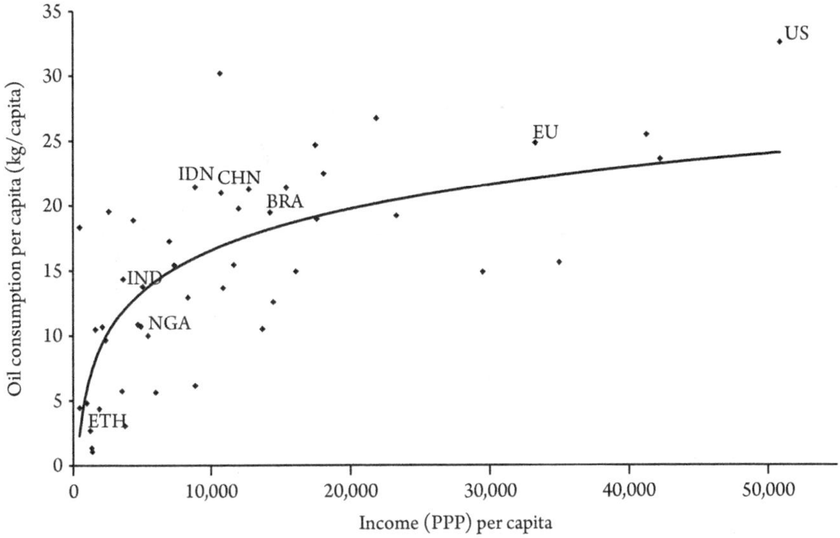

Figure 4.6 Per-capita vegetable oil consumption for food in 2013 versus 2013 per capita gross domestic product (2011 purchasing power parity base) for countries with more than 20 million people. Note: Excludes four countries lacking income data. European nations are included collectively as one nation (EU). BRA, Brazil; CHN, China; ETH, Ethiopia; EU, European Union; IDN, Indonesia; IND, India; NGA, Nigeria; US, United States.
Source: Adapted from USDA-FAS PSD and World Bank. 2015. Data. http://data.worldbank.org (accessed May 5, 2015).

Table 4.3 Representative income elasticities by food group, 1996

Food group	*Income elasticity*			
	Mean	*Median*	*Minimum*	*Maximum*
Breads/cereals	0.44	0.45	0.11	0.62
Meat	0.69	0.70	0.27	0.86
Fish	0.80	0.80	0.31	1.04
Dairy	0.76	0.76	0.30	0.96
Fats, oils	0.46	0.47	0.17	0.63
Fruits, vegetables	0.56	0.57	0.22	0.71
Other	0.69	0.70	0.27	0.86

Source: USDA-FAS PSD and; World Food Programme. 2007. *PDPE market analysis tools: Price and income elasticities*. Rome: World Food Programme.

In an earlier contribution, the World Food Programme (2007) compiled a useful handbook for practitioners that includes representative income (also price) elasticities by food group (Table 4.3). On average, edible fats and oils have income elasticity coefficients toward the lower end of the various food groups; however, oils show comparable ranges across income groups.

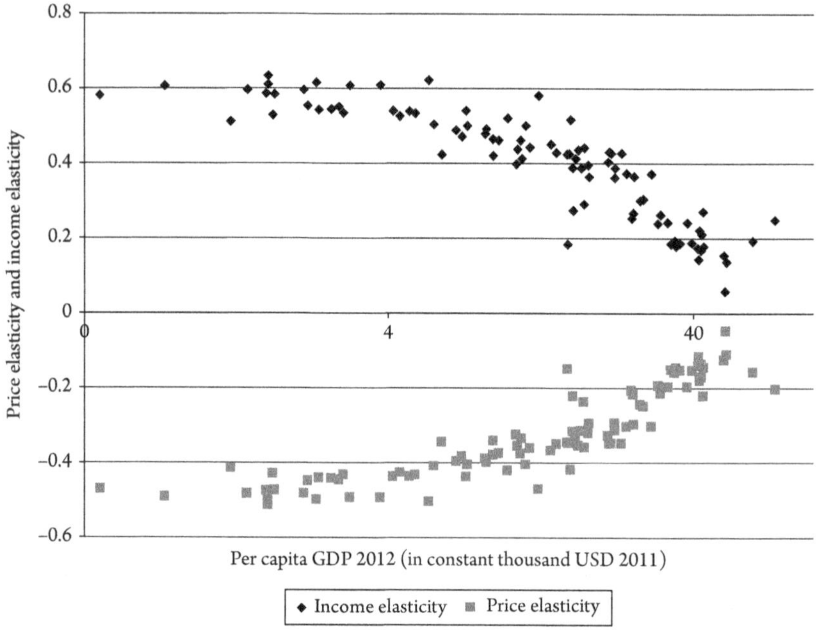

Figure 4.7 Income and price elasticity for vegetable oils by per-capita gross domestic product (GDP). USD, US dollars.

Source: Adapted from USDA-FAS PSD; US Department of Agriculture. 2012. Commodity and food elasticities. US Department of Agriculture, Economic Research Service. http://www.ers.usda.gov/data-products/commodity-and-food-elasticities/demand-elasticities-from-literature.aspx#.U724IF5X_1o (accessed May5 2015); World Bank. 2015. Data. http://data.worldbank.org (accessed May 5, 2015); and World Food Programme. 2007. *PDPE market analysis tools: Price and income elasticities.* Rome: World Food Programme.

The World Food Programme handbook also provides country-by-country elasticity estimates, drawn extensively from the USDA-ERS Elasticity Data Base (US Department of Agriculture 2012b).[8] We have combined these estimates with (purchasing power parity) per capita GDP data for 2013 to create Figure 4.7. This figure shows the $\log_e$ of per capita GDP against income elasticities for 90 countries. The data in the figure indicate that, with incomes of US$4000 per capita, a 10% increase in per-capita income would result in about a 5% increase in oil consumption, whereas at US$40,000 per capita, the response would be only about 2%. Some variation appears across countries, but the underlying regularity is impressive.

The cross-sectional income elasticity estimates presented here seem to confirm the role of per-capita GDP as a driver of per-capita consumption of vegetable oils. It turns

[8] The elasticities reported here may possibly have changed during the past 20 years. In particular, fats and oils in processed foods may have altered income consumption relationships with edible oils. To our knowledge, however, Table 4.3 still represents the most current and complete elasticity tabulation.

out there is a bit more to the story, however. Cross-sectional estimates, which compare incomes at one point in time across countries, regions, or households, present a strikingly consistent story, yet they do not always equal the income elasticities derived from looking at a single region or country through time. This dilemma, and what to do or not do about it, is discussed in Box 4.2.

Box 4.2 A Cautionary Tale about Time-Series Versus Cross-sectional Estimates

Reconciling cross-sectional and time-series income elasticity estimates is not a trivial task; indeed, it has been a hotly debated topic in economics for more than 50 years (Friedman 1957). For this reason, you should consider the comments in this box as more of a cautionary tale rather than a deep econometric assessment of differences in estimating techniques. The issue is apparent in Table B4.2, which shows time-series estimates of income elasticities for edible oils for the three largest nations in Asia.

These *simple* time-series estimates for the three countries are greater than their cross-sectional estimates, which are in the 0.4 to 0.5 range. In our view, the higher time-series values result mainly from the variables omitted in the time-series analysis. In other words, the income variable, when regressed alone with per-capita consumption, is "picking up" the effects of other variables. For example, if real prices of oil are falling, a simple relationship between per-capita income and consumption also includes the effect of prices. Numerous other variables (urbanization, tastes, and so on) are correlated with income yet are not income effects per se.

Given a choice, we prefer operationally to use the cross-sectional estimates as our starting point, and then to take up other variables separately. Note that these estimates may err on the low side, however, to the extent that long-run behavioral processes are underway that may not be captured in one instant in time. Some analysts have interpreted the cross-sectional estimates as short-run elasticities and refer to "correctly measured" time series as long-run estimates. Our main concern is with the "correctly measured" proviso. All too often, casual empiricists look at simple correlations between income and consumption only, thereby overestimating the impact of income growth. For large countries growing rapidly, the difference, say, between an income elasticity of 0.3 to 0.5 versus 0.7 to 0.8 is very large when played out over several decades.

Table B4.2 Time-series estimates of income elasticities for edible oils in China, India, and Indonesia in terms of annual growth from 1990 to 2013

Country	*Per-capita oil consumption (%)*	*Per-capita gross domestic product (%)*	*Implied elasticity*
China	6.5	8.9	0.73
India	4.2	4.9	0.86
Indonesia	2.0	2.6	0.77

Source: USDA-FAS PSD and World Bank Data (http://data.worldbank.org, accessed May 5, 2015).

Health and Nutrition

Considerations related to health and nutrition form a third major determinant of vegetable oil demand. They influence both the total consumption of oils and also the specific oil being consumed. As mentioned, the issues are both complicated and confusing. Too little vegetable oil creates nutrition problems, but so does too much.

On one side, increased consumption of vegetable oils has made major contributions to the calorie supply of poor people who suffer from chronic undernourishment. According to FAOSTAT, the global supply of calories from 1991 to 2011 increased by 278 cal per person; 26% of this increase, likely an underestimate, came from vegetable oils alone. In South Asia, a region with one of the lowest average per-capita calorie consumption and the highest number of undernourished people, consumption of vegetable oils accounted for 32% of increased calorie consumption (FAOSTAT). For the poorest people in the region, whose intake of oils and fats was low to begin with, increased vegetable oil consumption is likely to have been important in improving their nutritional status. This role of vegetable oils in improving diets is largely obscured by the current debate about high fat consumption. Obesity is sometimes a problem even in poor societies, where undernutrition persists. In middle- and upper-income countries, cardiovascular diseases are the primary cause of death, and since the 1950s, dietary fat has been examined closely as a major cause of heart disease (Daud et al. 2012).

Early research focused on dietary fat and ways to reduce total cholesterol (Keys et al. 1965). These findings became the basis for numerous health bulletins recommending the limitation of total oil intake, and especially the substitution of unsaturated fats for saturated fats. Because palm oil, in particular, is composed of about 50% saturated fats, it became a dietary target. Campaigns against the use of tropical vegetable oils, especially coconut and palm oil, began during the 1980s. The initial impact was the replacement of these oils with partially hydrogenated vegetable oils, but the latter, in turn, contained trans fatty acids (Box 4.3).

A decade later, another series of studies showed these trans fatty acids created special cardiovascular risks, particularly among younger women (Oh et al. 2005). The new results prompted the mandatory labeling of trans fats in the United States and several other countries. The Food and Drug Administration in the United States also moved recently to eliminate trans fats in processed foods, indicating that trans fats are "no longer generally recognized as being safe" (US Food and Drug Administration 2015, p. 1).

Becasuse the tropical oils do not require hydrogenation (the process that creates the trans fats) for many food preparations, the ban on trans fats gives an advantage to palm and coconut oil that now sometimes find themselves promoted as health foods. More generally, the most forceful dietary promotions have been toward monounsaturated and polyunsaturated fats. The government of Singapore has gone so far as to subsidize the difference in price between lower priced palm oil and an allegedly more

Box 4.3 A Primer on Edible Fats and Oils

Saturated fats: Saturated fats are fat molecules that have no double bonds between carbon molecules because they are saturated with hydrogen molecules. They are usually solid at room temperature. It is typically recommended that less than 12% of calories come from saturated fats. Products that contain saturated fats include animal fats (high-fat cheeses, high-fat meat, butter) and plant oils (palm oils, coconut oil, cocoa butter). These fats may also be present in commercially prepared foods such as vegetable shortening, cakes, and pies.

Unsaturated fats: Unsaturated fats are fats that are not saturated with hydrogen molecules. This group includes monounsaturated fats and polyunsaturated fats. Products that contain unsaturated fats include nuts, vegetable oils, and fish.

Monounsaturated fats: Monounsaturated fats are fat molecules that have one unsaturated carbon bond in the molecule, or a double bond. These fats are usually liquid at room temperature. Monounsaturated fats can be beneficial for heart health if eaten in moderation or used to replace saturated and trans fats in the diet. Products that contain monounsaturated fats include nuts, olive oil, canola oil, groundnut oil, safflower oil, sesame oil, and avocado.

Polyunsaturated fats (also omegas): Polyunsaturated fats can be broken down into two different types of fats: omega-6 polyunsaturated fats, which provide an essential fatty acid; and omega-3 polyunsaturated fats, which provide another essential fatty acid, particularly from fish sources. Products that contain polyunsaturated fats include soybean oil, corn oil, and safflower oil (omega-6); and soybean oil, canola oil, walnuts, flaxseed, and fish (omega-3).

Trans fats: Consumption of trans fat increases "bad" cholesterol and may even contribute to decreasing "good" cholesterol. The general dietary recommendation is to keep trans fat consumption as a low percentage of oil intake. Products that contain trans fats include foods that contain partially hydrogenated oil, fried items, savory snacks such as microwaved popcorn, frozen pizzas, cake, cookies, margarine, ready-to-use frosting, and coffee creamers, although amounts of trans fat can vary within each food type.

Partially hydrogenated oil: Partially hydrogenated oil is formed when hydrogen is added to liquid oil, turning it into a solid fat. This inexpensive measure improves the shelf life, stability, and texture of a food.

Cholesterol: The fatty, waxy substance known as *cholesterol* is found in animal-based foods. Total cholesterol is the total cholesterol measured in a person's blood. "Good" cholesterol—high-density lipoprotein cholesterol —helps to carry cholesterol away from the body's organs to the liver, where it can be removed. "Bad" cholesterol—low-density lipoprotein cholesterol—is linked to a greater chance of incurring heart disease.

Source: Center for Disease Control and Prevention. 2012a. Dietary cholesterol. http://www.cdc.gov/nutrition/everyone/basics/fat/cholesterol.html (accessed August 5, 2014); Center for Disease Control and Prevention. 2012b. Polyunsaturated fats and monounsaturated fats. http://www.cdc.gov/nutrition/everyone/basics/fat/unsaturatedfat.html (accessed August 5, 2014); Center for Disease Control and Prevention. 2012c. Saturated fat. http://www.cdc.gov/nutrition/everyone/basics/fat/saturatedfat.html (accessed August 5, 2014); Center for Disease Control and Prevention. 2014. Trans fat. http://www.cdc.gov/nutrition/everyone/basics/fat/transfat.html (accessed August 5, 2014); American Heart Association. 2015. Saturated fats. http://www.heart.org/HEARTORG/GettingHealthy/FatsAndOils/Fats101/Saturated-Fats_UCM_301110_Article.jsp (accessed August 5, 2014).

healthy, but more expensive, mix of canola (rapeseed) oil and palm oil (Cheow 2014). Iran recently announced a decision to ban the use of palm oil completely (Islamic Invitation Turkey 2014).

The health effects of tropical oils do not depend solely on their cholesterol or fat content, but rather on what they add to or replace in diets of various populations and subpopulations (Micha and Mozaffarian 2010). Several recent metastudies compare the cholesterol effects of palm oil, soybean oil, rapeseed oil, sunflower oil, and olive oil. Few generalizations are evident across these studies. One group suggests vegetable oils are not substantially different in their impacts if consumed in moderation; some evidence suggests, however, that sunflower oil has a lesser effect on cholesterol (Fattore and Fanelli 2013; Fattore et al. 2014). Several authors also argue that health effect comparisons in low-income countries, where palm oil consumption tends to be highest, are made more complicated by the generally poorer quality of overall medical care in these regions (Chen et al. 2011; Basu et al. 2013).

A more recent metareview came to a different conclusion with respect to palm oil (Sun et al. 2015). It assessed 30 articles in which vegetable oil intake ranged from 12% to 43% of total calories. The authors concluded palm oil increased both total cholesterol and low-density lipoprotein cholesterol significantly, but it did not alter triglyceride coefficients significantly. The authors argued, when comparing their findings with previous studies, their review is more thorough and controls for the effects of other variables more adequately. In another recent review, Mancini et al. (2015) provided a comprehensive literature review of palm oil effects on various diseases, including cardiovascular diseases and cancer. They concluded the effects on cardiovascular disease varied by circumstances and, although suggestive, the results were not definitive. In the case of breast and prostate cancers, the literature found little association with palm oil consumption. In addition, May and Neseretnam (2015) argue forcefully that, despite its saturated fats, palm oil does not have negative effects on human health relative to other vegetable oils. And in yet another more recent review, the authors conclude "the intake of saturated fat (butter, palm oil, coconut oil, and lard) poses no risk to our health, particularly to the heart" (Natella et al. 2015, p. 4).

The fluid state of the nutritional literature, particularly with respect to palm oil, coconut oil, and butter, has given rise to many claims, counterclaims, and fads (Walsh 2014). Activists have also used various studies to promote other objectives, especially in arguments against deforestation and the loss of biodiversity (Brinkley 2013; Aubrey 2014; Pankratz 2014; Jeffries 2015). Actual consumer behavior may thus have less to do with scientific findings on nutrition, particularly if the message is complex, and more to do with what various advocacy groups put forward and what consumers perceive from conflicting advice.

The medical profession now seems in broad agreement on the need for more fats and oils for populations in a number of poor nations; the profession is also clear on the need for moderation in the total consumption of fats and oils (<10% from saturated fats). Although some evidence suggests high intake of palm oil may be related to "bad" forms of cholesterol, given the current state of knowledge, it seems doubtful that a tax on palm oil—sometimes suggested for cardiovascular reasons—is warranted in countries such as India. This conclusion is reinforced when the impacts of such a tax on income distribution are factored into the analysis (Basu et al. 2013).

How the health evidence will influence future consumption behavior is very unclear. We believe research on the health effects of various fats and oils will continue to be a work in progress, consumers will still be perplexed about what and whom they should believe with respect to the health effects of tropical oil consumption, and corporate groups will sometimes find themselves caught in the difficult position of choosing between what may be the best nutritional science and what consumers seem to believe with respect to their products. One promising area of research appears to be the enhancement of oil crops for specific traits. For example, soybeans bred to have oil high in stearic acid could potentially replace less healthy fats in food (Hunter et al. 2010; Huth et al. 2015). In short, much research remains to be done on the oils themselves and in educating consumers on the health front. No clear-cut trends currently reveal how health variables will affect the composition of oil consumption, although increasingly strong campaigns against obesity could well dampen total edible oil consumption, especially in rich countries.

Prices

That consumption of vegetable oil depends partially on its price is hardly a revelation. Yet, which price, over which time period, and over which range of products are all difficult questions. Our major concern in looking forward to 2050 is the trends in real prices, rather than month to month or year to year variability, which has been the focus of Brümmer et al. (2015) and others. If vegetable oil markets are working well and without significant trade barriers, there is little price variation at one point in time other than transport costs. The cross-sectional estimates helpful in describing income/consumption parameters are therefore not (generally) possible with respect to price elasticities.[9] For this reason, it seems appropriate to focus here on three propositions about the movement of vegetable oil prices through time that we believe are valid and important.

[9] A few countries, such as India, conduct nationally representative consumption surveys periodically. In these cases, it is possible to combine times-series and cross-sectional features into one analysis that produces both price and income elasticities (Pan et al. 2008; Kumar et al. 2011). The data requirements of this combined approach are formidable, however, and it requires sophisticated econometric techniques. The results may be very useful for policy work in a specific country, but the combined approach is largely infeasible for more global studies.

The first proposition is that real prices of vegetable oils are falling but that significant variation surrounds this downward trend. The trend itself may depend on the initial year chosen. For example, the real price of 1 t of palm oil—generally the cheapest oil on the market—has fallen by about 2% annually since 1950. Based on this long-term trend, the real price of palm oil in 2000 was only about one third what it was in 1950 (Fry 2011). In China, for example, the price of cooking oil fell about 50% relative to vegetables between 1980 and 2006, and about 25% relative to cereal (Overseas Development Institute 2015). On the other hand, if the world price trend line is started in 1990, the real oil price index is actually positive. A key question is whether the price changes since 1990 represent a bubble in food and edible oil prices or, alternatively, whether they represent the new links to the energy market that we discuss in Chapter 6. The second alternative is certainly hinted at in Chapter 7, where Figure 7.5 shows the relationships among price indices for energy, food, and vegetable oil/meal.

A second important feature of vegetable oil prices is the high correlation that exists among the four major oils through time. Palm oil typically trades slightly lower than its major competitor, soybean oil, but their price movements are nonetheless similar (Figure 4.8). Brümmer et al. (2015) suggested that movements in exchange rates are a key factor in keeping various oil prices in concert with one another.

This correlation is also another way of asserting there are widespread substitution possibilities in consumption. Olive oil is the distinct exception to this assertion; it tends to trade at higher prices and with different temporal patterns than the other oils. Yet, although olive oil is obviously important in the Mediterranean and in high-income cuisines around the world, it accounts for less than 2.5% of total edible oil consumption, so it is not a key feature of this analysis.

A third proposition concerns knowledge about the price elasticities of demand at varying levels of real per-capita GDP. As noted, the World Food Programme compilation (Figure 4.7) answers many of the relevant questions. At lower levels of per-capita GDP, own price elasticities are about –0.5, whereas at higher levels they are about –0.2.[10] The larger (absolute) value among poorer consumers reflects the necessity of changing behavior as economic conditions change, whereas among richer consumers, preferences may simply override price changes—particularly if the share of the household budget for the product in question is low. This difference in consumer price behavior at different income levels is often referred to as *Timmer's Law* (Timmer 1981).

The foregoing discussion of price movements and price elasticities of demand helps to describe human behavior, but it begs questions about the future direction of

[10] A recent meta-review by Green et al. (2013) shows fats and oils to have price elasticities of demand between –0.5 and –0.6 in poorer countries, and –0.3 and –0.4 in richer countries. Their results for meat and fish were very similar to the values reported in Table 4.3. For cereals, they show absolute values of –0.4 to –0.5 for rich countries, which is higher than we would have anticipated.

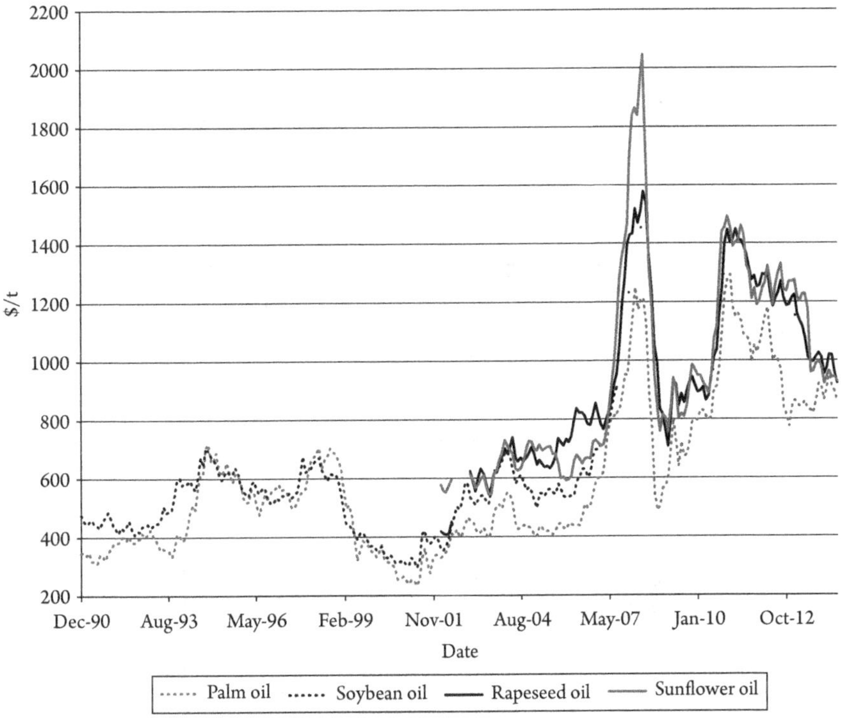

Figure 4.8 Nominal monthly vegetable oil prices, 1990 to 2014.
Source: Knoema. 2014. World Bank commodity price data (pink sheet), July 2014. http://knoema.com/WBCPD2014Jul/world-bank-commodity-price-data-pink-sheet-july-2014 (accessed May 5, 2015).

real oil prices. Our perception, developed more in the next chapters, is that real vegetable oil prices will hold about constant, albeit with significant variation, during the next 35 years.

PREDICTING GLOBAL DEMAND FOR EDIBLE OILS IN 2050

In one sense, our goal of predicting edible oils in 2050 is easy and straightforward. The task is simply to produce a table for 2050 comparable with Table 4.4 for 2013. The problem is in getting from here to there. We begin by looking at the demand for aggregate vegetable oil demand, using the determinants described previously, and then move on to examine the commodity composition of that aggregate demand. Our aims are to show relevant ranges of various variables and parameters, to offer our judgments about the most likely outcomes, and to do so transparently, so that analysts wishing to use alternative assumptions can do so easily.

Our underlying model of aggregate oil demand is straightforward. It is a simple additive model that begins with base consumption per capita in 2013 and then adds or

Table 4.4 The world of edible oils for food use in 2013

Region	*Total oils, consumption (000 t)*	*Palm oil, consumption (000 t)*	*Soybean oil, consumption (000 t)*	*Rapeseed oil, consumption (000 t)*	*Sunflower oil, consumption (000 t)*	*Other oils, consumption (000 t)*	*Population total (in millions)*	*Total consumption (kg/capita)*
World	124,621	40,094	36,534	17,512	13,914	16,567	7150	17.43
Africa	11,347	5699	2329	6	1472	1841	1100	10.32
Europe	17,452	3561	1259	2948	6536	3148	740	23.58
North America	13,265	1286	7417	3184	233	1145	350	37.90
Latin America and Caribbean	8600	1418	5791	137	869	390	610	14.10
South and West Asia	28,573	13,606	4488	3054	3544	3879	2100	13.61
East and Southeast Asia	43,284	12,955	15,296	7958	1215	5950	2210	19.59
Oceania	675	142	44	225	50	214	40	16.88

Source: USDA-FAS PSD.

subtracts percentage changes caused by the various determinants discussed earlier: population, urbanization, income, health, and price. That process gives per-capita oil consumption in 2050, which is then multiplied by the 2050 population (Eq. 4.1).

$$\begin{aligned}
&\text{Global edible oil consumption}_{2050} \\
&\quad = \left\{\text{Per} - \text{capita consumption}_{2013}\right\} \times \\
&\quad \{[\left(\%\Delta\,\text{GDP} / \text{capita} \times \text{Income elasticity}\right) \\
&\quad \pm \left(\%\Delta\,\text{Real price} \times \text{Price elasticity}\right) \\
&\quad - \left(\%\Delta\,\text{From health campaigns}\right) \\
&\quad + \left(\%\Delta\,\text{Urbanization} \times \text{Urban} / \text{rural consumption ratio}\right)] + 1\} \\
&\quad \times \left\{\text{Population}_{2050}\right\}
\end{aligned} \qquad [4.1]$$

To calculate this model, we require data for seven variables or parameters. Our judgments on each of the seven items are shown in Table 4.5, along with higher and lower estimates to cover differing views and to provide sensitivity tests for our findings.

Our "best" estimate, which assumes constant real prices for vegetable oil, is shown in column 1 of Table 4.5. It results in a projected total oil consumption of 21.7 kg/capita in 2050 compared with 17.4 kg/capita in 2013. Total global consumption of edible oils for food in 2050 is projected at 217 Mt compared with 125 Mt in 2013.

The combinations of alternative parameter choices in Table 4.5 produce 729 separate estimates for Eq. 4.1. We summarize the range of results with two additional estimates in Table 4.6. The first result comes from having chosen the parameter from

Table 4.5 Alternative projection parameters

	"Best"	*Lowest*	*Highest*
Annual real GDP growth[a] per capita %)	2.5	2.0	3.0
Income elasticity	0.25	0.20	0.30
Real price change (2050 vs. 2013) (%)	0	+15	−15
Price elasticity	−0.4	−0.4	−0.4
Health impact (2050 vs. 2013) (%)	−5%	−15%	0
Urban multiplier (urban/rural consumption) (%)	20	5	30
Population (2050)[b]	10.0 billion	9.5 billion	11 billion

GDP, gross domestic product.

[a] For varying views on long-run growth, see Carnegie (2010), Ward (2012), O'Neil (2013), Price Waterhouse and Coopers (2015), and Organisation for Economic Co-operation and Development (2012a).

[b] Population Reference Bureau (2014).

Table 4.6 Preferred, lowest, and highest estimates of total vegetable oil consumption (for food use) in 2050

Category	*Estimate (Mt)*
Preferred	217 Mt
Lowest	145 Mt
Highest	325 Mt

each set in Table 4.5 that leads to the lowest estimate for 2050 total edible oil consumption; the second reflects the choice of all parameters from Table 4.5 that results in the highest estimate.

The absolute numbers are sobering, especially if supply increases are likely to be concentrated in only one or two oil sources. Our best estimate calls for an *increase* of about 92 Mt for 2050 (about 74% higher relative to 2013), or a growth rate for edible oils of about 1.5% compounded annually. The "highest" estimate would require about 2.6% compounded annually; the "lowest" estimate, only 0.4% compounded annually. These estimates are for food uses only; biodiesel and other industrial uses have yet to be added (Chapter 7).

REGIONAL PROJECTIONS FOR 2050 EDIBLE OIL CONSUMPTION

The global model in Eq. 4.1 is useful for calibrating global growth of real GDP per capita and for examining the interactions among various variables. Because the world is far from uniform across continents when it comes to vegetable oil consumption, the logical next step is to disaggregate the global model by region. Multiple possibilities exist for selecting different areas and, in principle, models could be made for each country. We chose the regional breakdown displayed in previous tables for this projection, believing it to be both revealing and manageable.

The regional model uses the same structure as Eq. 4.1 but adds a regional dimension to each of the seven variables. This added complexity does not make for tidy summaries or easy sensitivity analyses. (Our preferred regional assumptions are shown in the Appendix A, Table A4.1 for those who wish to substitute their own assumptions for ours.) Table 4.7, which shows both per-capita and aggregate vegetable oil numbers for 2013 and 2050, presents our preferred results for the regional model. The 2050 regional results, when aggregated, are consistent with the projections from the global model discussed earlier.

Three results emerge from the regional projections in Table 4.7. First, the large percentage changes arise in Africa, where a combination of moderately high GDP growth per capita, a large income elasticity, limited health impacts, rapid rural–urban migration, and significant population growth interact in ways to triple consumption. Whether real GDP per capita in Africa grows at an annual compound rate of 2.5% is, in our view, a critical assumption. If the rate is substantially less, or if the growth is concentrated only

Table 4.7 Edible oil food consumption, regional and global, in 2013 and projected to 2050

Region	*Consumption, 2013 (kg/capita)*	*Consumption, 2050 (kg/capita)*	*Total consumption, 2013 (Mt)*	*Total consumption, 2050 (Mt)*	*Ratio of total 2050/2013*
Africa	10.3	13.9	11.3	36	2.98
Europe	23.6	24.9	17.5	19	1.04
North America	37.9	37.0	13.3	19	1.36
South America	14.1	15.3	8.6	16	1.40
South and West Asia	13.6	18.3	28.6	56	1.87
East and Southeast Asia	19.6	25.9	43.3	68	1.40
Oceania	16.9	17.7	0.7	1	1.57
World[a]	17.4	21.7	125.0	217	1.71

[a] The world and each region were estimated independently. The sum of regional consumption equals 215 Mt for 2050. The discrepancy of 2 Mt from the "world" estimated separately is the consequence of rounding several regional parameters.

within a small minority of the population, oil consumption could be considerably less. The aggregate oil consumption estimate for Africa clearly is also very sensitive to the population estimate for the region.

Second, in terms of the total oil tonnage involved, the increases for Europe, North America, South America, and Oceania seem relatively small. These regions are defined by modest growth in real GDP per capita, small income elasticities, significant health impacts, and low—sometimes even negative—population growth. The *combined* increase in edible oil consumption for these four continents is substantially less than for Africa, or for South and West Asia, or for East and Southeast Asia.

Third, in 2050, both of Asia's subregions continue to play very important global roles in edible oil consumption. Rapid growth in GDP per capita, modest income elasticities, and very large base populations foster major increases in total oil consumption for Asia.

In short, the food consumption dynamics we foresee for the decades ahead are dominated by Africa and Asia. In 2013, they constituted 9% and 57%, respectively, of the global total; by 2050, the comparable numbers are 17% and 58%, but of a much larger global total. Despite the remarkable projected growth in Africa, therefore, Asia still dominates the vegetable oil markets.

COMMODITY COMPOSITION OF EDIBLE OIL CONSUMPTION IN 2050

The final step for the 2050 projections is to translate the regional estimates into commodity-specific demands. This step is also the most difficult, primarily because of the roles that policy and consumption substitutions play in the outcome. Supply and demand interact vigorously at this stage as well, and the oil supplied domestically may also be the oil demanded (Chapter 7). In the case of imported oils, the deciding factor may be price, which would typically give palm oil the edge, or the deciding factor may be demand for the coproduct (meal). In any event, we believe the final market-share numbers must rely more on informed judgment and less on formal models.

The logical place to begin the commodity assessment is with current market shares. For the world as a whole, the share coming from palm oil has increased since 1990, soybean oil has remained constant, and all other oils have lost shares relative to total oil consumption. Market-share projections to 2050 depend critically on what is happening to oil consumption in several large countries, and on whether those forces will continue to exert their influence for the next several decades. We highlight key dimensions of oil consumption in India, Indonesia, and Nigeria—the second, fourth, and seventh most populous countries in the world. Together with the China livestock story, described in Chapter 5, they are at the heart of Asian–African consumption dynamics. They present interesting yet contrasting stories for edible oils across time and space, and they give important indications of the broader points that must be considered in establishing market shares for 2050. These country vignettes are intended to illustrate the impacts of livestock, substitutions, trade policy, and macro policy on vegetable oil consumption.

Indonesia: The Key Role of Substitution

Indonesia, the world's fourth most populous country and leading global producer of palm oil, also has the highest per-capita consumption of palm oil (21.7 kg/capita/yr). This fact is remarkable in light of Indonesia's traditional oil consumption patterns. Vegetable oil is one of several strategic consumption commodities in Indonesia. Fried (*goreng*) foods are omnipresent, both in Indonesian homes and in hundreds of thousands of tiny street food stalls. During the 1970s, coconut oil was by far the dominant oil used in food preparations, yet by 2013, a switch to palm oil was virtually complete (Figure 4.9).

During the past 30 years, production of oil palm received priority in terms of attention and investment, both from the government and from large private plantation companies; coconut did not. CPO processing units increased 10-fold between 1974 and 2013. Efficient new refining techniques made it possible to clarify palm oil and eliminate its traditional red color. The government also enacted a variable export tax system, whereby export taxes increased as world prices for palm oil increased. The result of these policies was to ensure local palm oil was both available and moderately priced. Interestingly, the substitution between oils was not induced by prices; palm oil prices remained approximately on a par with coconut prices. As Gaskell (2012) noted, the granting of near-monopolistic regional rights to favored firms to distribute vegetable oil played a powerful role in promoting domestic palm oil consumption. Now that this one-time substitution has taken place, the incredibly rapid growth of per-capita palm oil consumption for food will taper. Oil consumption will depend on income levels, relative prices, urbanization, and health concerns—not on substitution.

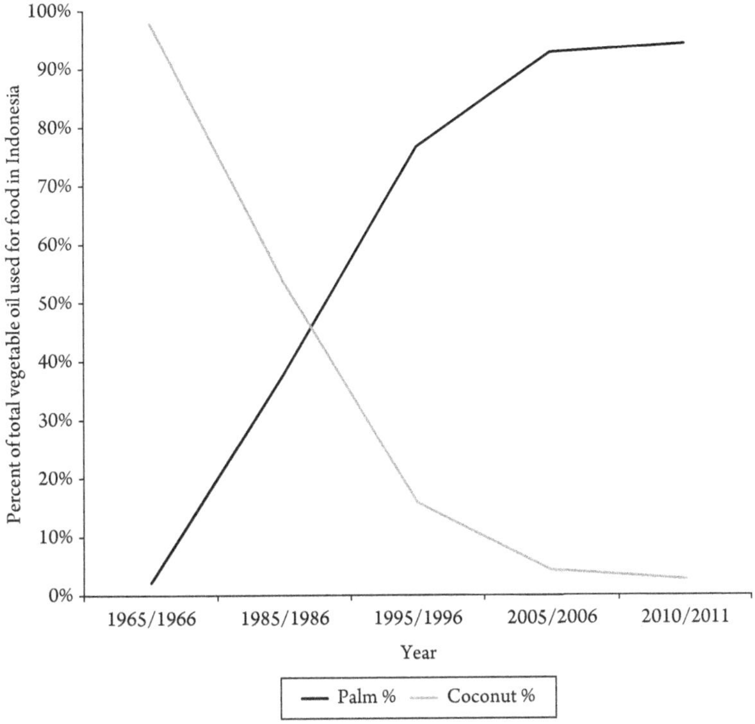

Figure 4.9 The switch from coconut oil to palm oil in food preparations, Indonesia, 1966 to 2011. *Source:* Gaskell, J. C. 2012. The palm oil revolution in Asia. PhD diss., Stanford University. http://purl.stanford.edu/zc839jm3057 (accessed May 2, 2016).

India: The Key Role of Trade Policy

India, with its population of 1.27 billion, produces an array of oils—from soybeans, cottonseed, groundnuts, sunflower, and rapeseed—and its yields per hectare for most of these crops are only about half the world average. *Ghee* (clarified butter) also contributes 8.7 Mt of fats to diets. Average oil consumption per capita in 2013 was 14.3 kg, but diets exhibit important regional differences. Across India, tastes align with production capabilities—for example, rapeseed in Uttar Pradesh and West Bengal, groundnuts in Gujarat and Andhra Pradesh, and soybeans in Madhya Pradesh and Maharashtra. Upward of 15% of oil is consumed as *vanaspati*, a semisolid product made by hydrogenating vegetable oil.

The Indian story is one of lagging domestic oil production and surging imports, which now supply about 60% of India's oil consumption. As shown in Figure 4.10, about 80% of oil imports have recently been in the form of palm oil, with soybean oil ranking a distant second.

It is clear from this figure that dramatic changes took place in India during the mid 1990s. These changes were associated with the WTO and the new rules surrounding Indian oil trade; the state monopoly that controlled oil trade was dismantled, nontariff

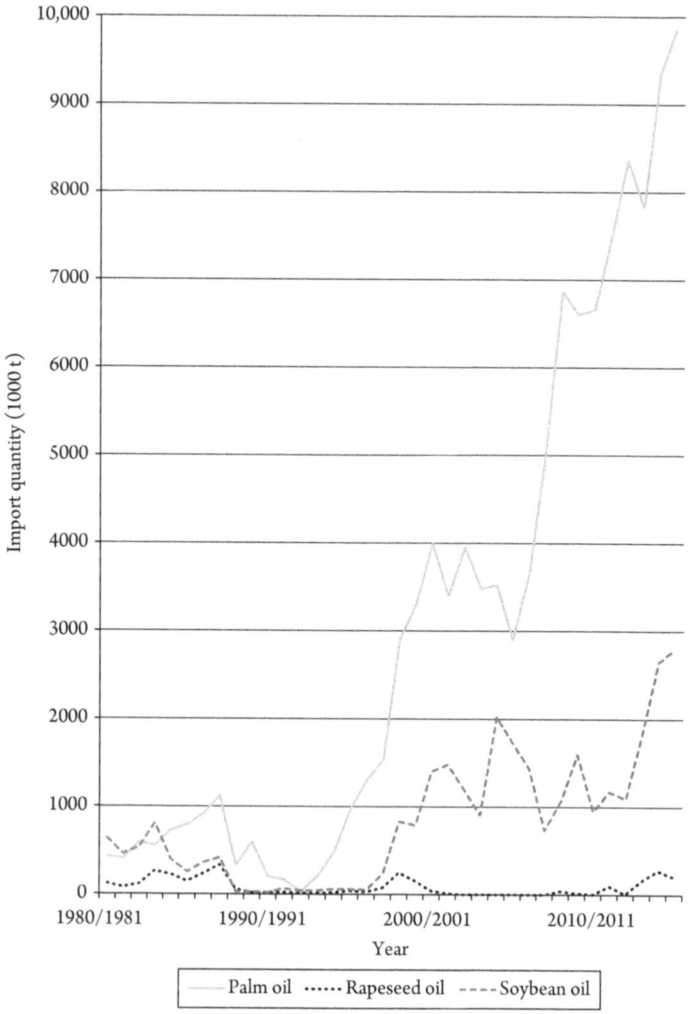

Figure 4.10 Indian oil imports, 1980 to 2015.
Source: USDA-FAS PSD.

trade barriers were abolished, and oil imports were largely turned over to private traders. The government reduced import tariffs steadily and, in 2010, it taxed imports of refined vegetable oils at a rate of 7.5%, whereas crude oils entered the country with only a 2.5% duty. To raise tax revenue and provide additional import protection for farmers, however, the Indian Minister of Finance proposed raising these rates in 2013 to 20% and 10%, respectively.

In the years ahead, consumers in India may face higher oil prices, at least relative to world prices, but unless a phenomenal breakthrough occurs in domestic oilseed production, India is likely to grow as the world's largest palm oil importer. As a consequence, the market share of palm oil in India's total oil consumption is likely to increase.

Nigeria: The Key Role of Agricultural and Macro Policy

Oil palm is native to West Africa, including Nigeria, where it still grows wild in many regions. Artisanal production and processing of red palm oil continue, and red palm stew is, in many respects, Nigeria's national dish. Given this background, it is curious the palm oil sector has fared so poorly. As discussed in greater detail in Chapter 2, Nigeria exported about 30% of the world's palm oil in 1965; in 2013, it imported one third of its consumption needs (Index Mundi 2015).

At the base of Nigeria's oil palm difficulties is a problem with the *naira*, the Nigerian currency, and its relationship to the dollar. Nigeria is a major exporter of petroleum products, which constitute 94% of Nigeria's total export earnings. The advantage of having oil, however, turns out to be a "resource curse" (also known as "Dutch Disease" after similar problems arose in the Netherlands as a result of large exports of natural gas) for labor-intensive production sectors such as oil palm. The naira is valued such that many domestic systems cannot compete efficiently with imports without substantial government assistance in the form of research, infrastructure, and tariff protection. Nigeria permits imports of CPO, but it bans imports of packaged refined oils. Domestic production of the latter has lagged, as have infrastructure and research. The oil palm system seems simply to sputter along.

Nigeria's economy has been growing robustly in recent years, and its agricultural policy is being "reformed." Perhaps a new window of opportunity will open to expand domestic production of palm oil. If not, Nigeria will become an increasingly large importer. Whether imported or produced domestically, however, palm oil seems destined to be the primary edible oil consumed.

Shares of Major Edible Oils

Overall, the experiences of these three countries point to three likely developments. First, palm oil's share in global edible oil consumption will grow. Second, soybean oil's share, because of China, will about hold its own. Third, trade policy is likely to be increasingly important in causing domestic prices in key countries to rise above world prices, thereby curbing oil demand and providing additional support to vegetable oil producers. Nigeria, for example, has recently threatened to ban palm oil imports to support prices for domestic growers of oil palm.

With these three country studies as a prelude, we construct an edible oil composition matrix for 2050. Table 4.8 provides the actual commodity shares by region for 2013, along with our assumed 2050 shares. Two key assumptions underlie our 2050 percentage: (1) growth in soybean oil consumption will tend to track livestock growth (Chapter 5) and (2) there will be a continued global shift toward greater use of palm oil.

Multiplying the estimated 2050 shares in Table 4.8 by the regional totals for edible oil consumption in 2050 (in Table 4.7) provides the global commodity totals for 2050. These totals, along with their implied compound annual growth rates from 2013 to

Table 4.8 Commodity consumption shares (measured as a percentage) by region, 2013 and 2050

Region	*Year*	*Palm oil*	*Soybean oil*	*Rapeseed oil*	*Sunflower oil*	*Other oils*
Africa	2013	50.4	20.5	0	12.9	16.2
	2050	60	20	0	10	10
Europe	2013	20.4	7.2	16.9	37.5	18
	2050	20	7	17	40	16
North America	2013	9.7	55.9	24	1.8	8.6
	2050	9	58	25	2	6
South America and Caribbean	2013	16.5	67.3	1.6	10.1	4.5
	2050	15	70	1	10	4
South and West Asia	2013	47.6	15.7	10.7	12.4	13.6
	2050	50	18	9	11	12
East and Southeast Asia	2013	29.9	35.3	18.4	2.8	13.6
	2050	32	37	17	2	12
Oceania	2013	21	6.5	33.4	7.4	31.7
	2050	20	7	35	8	30

Source: USDA-FAS PSD and authors.

Table 4.9 Projected food consumption for major edible oils for 2050, and implied compound annual growth rates from 2013 to 2050

Oil	*Consumption in 2050 (Mt)*	*Implied annual growth (%)*
Palm	80	1.9
Soybean	68	1.7
Rapeseed	25	0.9
Sunflower	22	1.4
Other edible oil	22	0.8
Total	217	1.5

2050, are shown in Table 4.9, which is for food use only. (Chapter 7 addresses nonfood uses and comparisons with other estimates.)

SUMMARY AND CONCLUSIONS

This chapter has described consumption contexts and presented data for edible oils on global, regional, and national scales. It has also analyzed key determinants of edible oil demand and provided a consistent set of consumption estimates by commodity for 2050. Many of the numbers for 2050 required important assumptions on our part, and we attempted to offer our best judgments in a transparent fashion.

Our strongest conclusion is that future vegetable oil consumption will grow, but at a slower rate than that seen during the past 30 years. The growth will be most pronounced in Africa and Asia. We see consumption of edible oils proceeding to grow at a compound annual rate of about 1.5% until 2050, with palm oil leading the way at about 2% annually. Much will happen between now and 2050, which is a very long period for forecasting. For tree crops, however, 2050 is a relevant time horizon.

The second key conclusion has to do with the land-use implications of these consumption forecasts. Given that recent increases in oil palm and soybean production have come primarily from expanding the area planted to these crops rather than from raising their yields, the issues of where producers will obtain more land for these crops and how they will manage that land will be even more crucial than they are now.

A third conclusion must take the form of a question: What forces might cause our numeric estimates to be seriously wrong? Unfortunately, that list is longer than we would like. First, much of our analysis foresees a rather booming Africa. Should that not happen, the consumption estimates for that region could be substantially overestimated. Second, vegetable oil prices have played a rather neutral role in the assessment in this chapter, in part because demand is only one blade of Alfred Marshall's price–determination scissors, and the tacit assumption has been that production would keep pace.[11] But, should oil production not grow at least by 1.5% annually, oil prices would rise, and demand would be curbed. In that scenario, price plays a crucial equilibrating role, and price variables become much more active determinants of consumption than we have assumed here. Third, health considerations, regardless of whether they are well founded in science, could have a significant bearing on oil demand, and especially on the commodity composition of that demand. Health rules and regulations also tend to spread rapidly among countries, which complicates matters further.[12]

Finally, government policy, especially on biofuels and trade, could alter outcomes significantly (Chapters 6 and 7). In mid 2015, for example, in response to foreign exchange difficulties caused by declining crude oil prices, Nigeria banned the import of 680 items including palm oil (Payne 2015). The triple pressures of increased tax revenue, more protection for domestic farmers, and curbed oil demand, separately and collectively, may cause countries to raise tariffs on edible oils if they have scope to do so within WTO rules.

[11] Mashall argued that a demand curve alone is similar to having only one blade in a pair of scissors. The single blade cannot cut a piece of paper. Similarly, a demand curve cannot determine value because there is no supply to codetermine price. (Marshall 1890)

[12] Genetically modified organisms might also become an increasingly difficult demand issue. To date, few if any genetic modifications have been used in oil palm systems, but much of the world's soybean crop is grown in systems using Roundup Ready technology, in which GM soybean varieties are grown in conjunction with Roundup herbicide.

APPENDIX A. PARAMETER VALUES BY REGION, 2050 REGIONAL VALUES, EDIBLE OILS, 2050 MODEL

Region	*2013 Base consumption (kg/capita)*	*GDP per capita growth, 2013–2050 (%)*	*GDP elasticity*	*Real price change*	*Price elasticity*	*Health impact, 2050 vs. 2013 (%)*	*Urban ratio change 2050 to 2013 (%)*	*Urban consumption as ratio of rural*	*Population, 2050*	*Consumption per capita, 2050*	*Total consumption, 2050*	*Total consumption, 2013*	*2050/ 2013*
Africa	10.3	2.5	0.33	0	−0.4	−2	16	1.2	2.5	14	36	11.3	3.19
Europe	23.6	1.5	0.2	0	−0.2	−5	9	1.05	0.75	25	19	17.5	1.09
North America	37.9	1.5	0.1	0	−0.2	−5	6	1.05	0.5	38	19	13.3	1.42
South America and the Caribbean	14.1	1.5	0.25	0	−0.3	−2	6	1.1	0.8	20	16	8.6	1.86
South and West Asia	13.6	2.5	0.33	0	−0.4	−2	18	1.15	3.0	19	56	28.6	1.96
East and Southeast Asia	19.6	3.0	0.33	0	−0.3	−2	18	1.15	2.4	29	68	43.3	1.57
Oceania	16.9	1.5	0.2	0	−0.2	−5	3	1.05	0.05	18	1	0.7	1.43

GDP, gross domestic product.

5

DEMAND FOR OIL MEAL FOR ANIMAL FEED AND THE JOINT PRODUCTION OF OIL

OIL MEAL AND THE RAPIDLY EVOLVING LIVESTOCK INDUSTRY

Because oil meal and vegetable oil are joint products, their markets are quite complex and often produce counterintuitive results. When China consumes more livestock, the resulting demand for soybean meal produces an additional supply of oil that reduces prices for vegetable oil for food and biofuel in global markets. In contrast, when the European Union imports biodiesel based on soybean oil from Argentina (Chapter 6), the supply of soybean meal on world markets increases, benefiting livestock producers and consumers in China. We return to a more comprehensive discussion of these linkages in Chapter 7.

In this short chapter, we focus on demand for soybean meal and the joint production of soybean oil.[1] Soybean meal is by far the most important oil meal used in the feed industry, accounting for more than two thirds of global oil meal use in 2014 (USDA FAS-PSD). Rapeseed and sunflower make up another 20% of the oil meal used. Together, these three crops account for nearly 90% of global oil meal production and more than half of vegetable oil production (palm oil and palm kernel oil make up a large share of all the other oil consumed).

As discussed in Chapter 3, soybean meal is favored for animal feed because of its high protein level (around 42% of soybean meal), good balance of amino acids, and ready availability at competitive prices in world markets. Soybean meal is widely used in feed concentrates, comprising up to 35% of the ration for dairy animals in the United States and Europe, 15% to 20% for poultry in Brazil and China, 20% to 25% for pigs in the United States, and about 15% for pigs in China (Wirsenius et al. 2010; Gale et al. 2012; Herrero et al. 2013). The actual share of soybean meal in feed concentrates depends, of course, on the prices of oil meal substitutes and the protein–energy balance in the diet.

Given the world's rapidly expanding livestock feed industry, the processing of soybeans for meal has produced a joint supply of soybean oil accounting for more

[1] Oil meal is 79% by weight and oil 19% by weight for soybeans. During the past 10 years, the price of oil has averaged 2.4 times the price of meal on the Chicago Board of Trade. In the short term, the share of value of the meal can vary from about 50% to 70% of the value soybean grain, but over the long term of 50 years, there has been no significant trend.

than 25% of the increased supply of vegetable oil from 1991 to 2013. In other words, without the growth of the intensive livestock industry based on animal feeds, vegetable oil prices would have been significantly higher, and food and biofuel consumption of vegetable oil correspondingly lower.

Our task in this chapter is to understand what is driving the demand for oil meal and its associated supply of vegetable oil in the context of a rapidly evolving livestock industry. Because the value of meal overshadows the value of oil only in soybeans, we focus the discussion on soybean meal and the outlook for soybeans. We first highlight briefly the main drivers of the recent upsurge in demand for livestock products and describe how that demand translates into demand for livestock feed, especially soybean meal. We then use projections of livestock demand to estimate demand for soybean meal to 2050, and the corresponding joint supply of vegetable oil.

Our assessment of the demand for soybean meal considers a complex set of drivers:

1. Demand for meat, milk, eggs, and, increasingly, for farm-raised fish; and changes in the balance among these products in human diets.
2. The relative importance of three major production systems for each animal product: extensive grazing, mixed crop–livestock, and intensive (also referred to as *industrial*) systems (Herrero et al. 2013). Grazing systems use feed concentrates to a limited extent only, whereas intensive systems depend almost entirely on concentrates.
3. The feed conversion efficiency (FCE) in intensive systems and for each livestock product, and differences in FCE by country and over time.
4. Shifts in the composition of feed concentrates—both the share of oil meals in concentrates and the share of soybean meal in the oil meal component.

The lack of data is a clear constraint, and it is no surprise to find that other assessments consist of only partial analyses of past trends and there are only a handful of projections to the future. To simplify the task, we focus on pigs and poultry (the so-called *monogastrics*), which consume an estimated 74% of all feed concentrates supplied to livestock globally (this figure increases to 78% if farm-fed fish are included) (Alltech 2014; Msangi and Batka 2015).[2] Dairy animals consume much of the remainder. About 5% of soybeans are fed unprocessed to livestock, mainly to ruminants.

DEMAND FOR LIVESTOCK PRODUCTS

Similar to the logic outlined in assessing the food demand for vegetable oils in Chapter 4, demand for livestock products is driven by population growth, incomes,

[2] These calculations exclude the small share (3%) of feed concentrates consumed by household pets and horses.

cultural and regional differences in preferences, and health concerns. Income growth is by far the most important driver, because the income elasticity of demand for most livestock products is relatively high—at least double that for food staples such as rice and wheat (Table 5.1 and Chapter 4). The strong association of livestock consumption with rising incomes is well documented (Delgado et al. 1999). Figure 5.1 shows the rapid increase in livestock consumption with rising income levels.

As income growth took off in emerging economies during the 1990s, so did livestock consumption, in what Delgado et al. (1999) label a "livestock revolution." From

Table 5.1 Approximate income elasticities for livestock products

Product	*Mean*	*Median*	*Minimum*	*Maximum*
Meat	0.69	0.70	0.27	0.86
Fish	0.80	0.80	0.31	0.96
Dairy	0.76	0.76	0.30	0.96

Source: Data are from World Food Programme. 2007. *PDPE market analysis tools: Price and income elasticities.* Rome: World Food Programme.

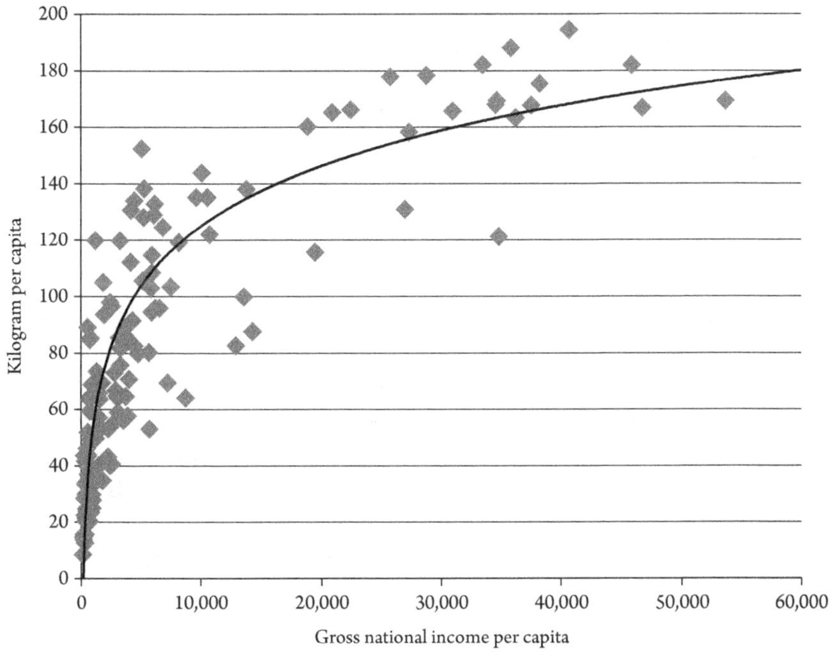

Figure 5.1 Per-capita annual consumption of livestock, milk, and fish products plotted against gross national income per capita, 2009. Milk products are multiplied by 0.25 to convert to the protein equivalent per unit weight of meats.

Source: Data are from Keats, S., and S. Wiggins. 2014. *Future diets: Implications for agriculture and food prices.* London: Overseas Development Institute.

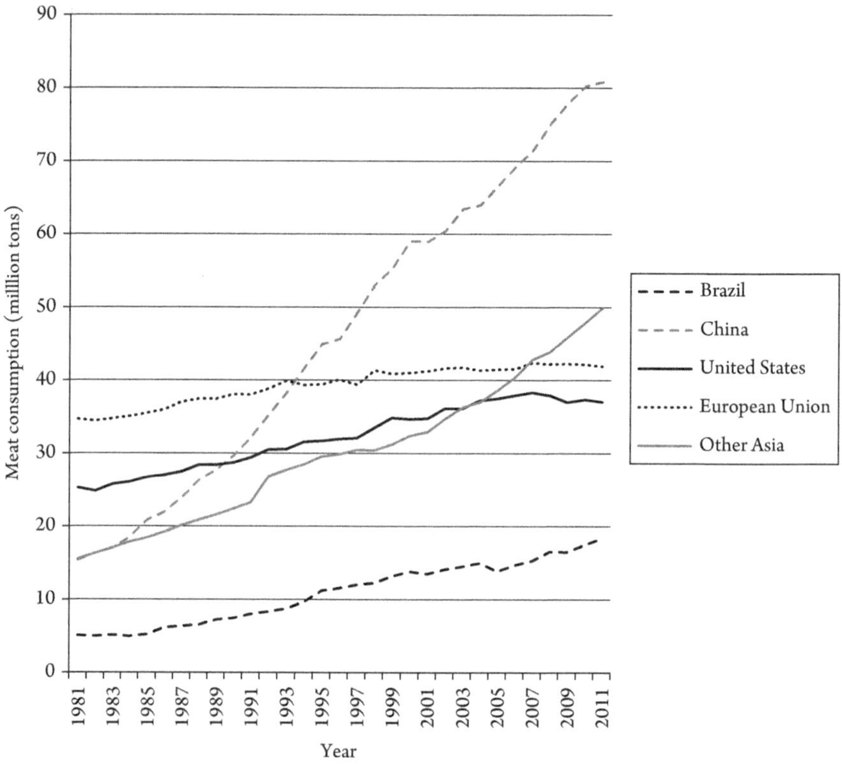

Figure 5.2 Trends in the annual total consumption of all meat by major producing country/region, 1971 to 2011.
Source: FAOSTAT.

1981 to 2001, per-capita meat consumption grew by 1.0% per year in the world as a whole, but in East Asia it grew by 4.9% per year. China now accounts for 28% of global meat consumption (Figure 5.2). The importance of culture as well as income is illustrated in Table 5.2, which shows the diversity of consumption patterns by region. In South Asia, annual meat consumption is only 4 kg/capita, much less than even sub-Saharan Africa, in part because a large share of the Indian population is vegetarian. Higher milk consumption in India means animal protein consumption is greater in India, however.

The composition of livestock products matters, as well, for our calculations. Each product depends on a different mix of production systems, and FCEs vary widely among product type, even in the intensive system. In part because real prices have declined (the result of gains in productivity) and in part for cultural and health reasons, poultry production has expanded much more rapidly than production of any other type of livestock product (by 4.4% from 1981 to 2011), followed by pig meat (2.3%) (mostly in China) and beef (1.1%). Poultry and pig meat now account for more than 70% of global meat consumption. The developing world consumes two thirds of poultry and pig meat.

Table 5.2 Diversity of livestock consumption by region and growth rates

Region	*Per-capita consumption (kg/yr), 2011*						*Growth in meat consumption per capita (%/yr)*	
	Beef	*Pig*	*Poultry*	*All meat*[a]	*Fish*	*Milk*	*1981–2001*	*2001–2011*
Oceania	39	22	42	116	27	203	0.0	0.7
North America	36	28	50	115	22	253	0.5	-0.2
Western Europe	18	44	20	87	23	261	0.0	0.0
South America	31	11	35	78	10	140	1.8	0.9
Eastern Europe	11	28	22	65	16	174	0.0	0.0
East Asia	5	34	13	57	35	34	4.9	1.2
Southeast Asia	4	12	12	29	33	18	2.8	0.0
Sub-Saharan Africa	6	2	5	17	11	32	−0.2	0.0
South Asia	2	0	3	7	6	86	0.8	0.8
World	9	16	15	42	19	91	1.0	0.6

[a] Includes sheep and goat meat.

Source: Calculated from FAOSTAT.

SHIFT TO INTENSIVE SYSTEMS AND CHANGING FEED CONVERSION EFFICIENCIES

The increased consumption of meat from monogastrics—pigs and poultry—favors disproportionately the growth of intensive systems based on feed concentrates. In some regions, smallholder systems that use a range of feeds to produce monogastrics are important; globally, they account for about 20% of pig and poultry production. Yet, intensive production systems are responsible for most of the growth in the market for pig and poultry meat (Herrero et al. 2013). In many countries, the rise of intensive systems coincides with the rapid rise of the feed industry (Box 5.1).

How the shift to monogastrics affects demand for feed concentrates depends on a number of sometimes countervailing factors:

1. Production systems for monogastrics depend much more on feed concentrates than production systems for ruminants (which consist largely of grazing and mixed crop–livestock systems).
2. In intensive systems, monogastrics are much more efficient than ruminants in converting feed concentrates into meat. In Figure 5.3, for example, the FCE (measured by the ratio of plant protein input to animal protein output) is about 0.4 for poultry, relative to 0.3 for pigs, and 0.05 for ruminants.
3. FCEs are improving for each species and even more in the aggregate as more efficient species gain a larger share of production (Figure 5.3). Some improvement

Box 5.1 The Global Feed Industry

The global feed industry produced about 1 billion tons of feed concentrates in 2014, valued at about US$400 billion. The feed is supplied from an estimated 28,000 mills with an average annual throughput of 34,000 t. The mills mix the bulk ingredients, depending on local supplies, but frequently they combine maize and soybean meal with other nutrients and vitamins, depending on the species and age of the livestock. Mills may be owned individually, but large agribusiness operations such as Thailand's Charoen Pokphand increasingly dominate the feed business. These vertically integrated multinational operations supply other inputs, especially genetic stock, as well as veterinary and other technical services, generally through contractual operations with producers. They also process and market the output (Delgado et al. 2008).

China is now the largest producer of feed concentrates, providing 20% of world supply, followed by the United States with 18%, the European Union with 17%, and Brazil with 7%. The use of feed concentrates varies across countries, but the dominant share usually goes to poultry, ranging from 73% in India to 51% in the United States (only 19% in France). Pigs are the second most important consumer of concentrates globally, but their production is centered in northern Europe, East Asia, and North America. Dairy animals account for more than 10% of feed use in Europe and the United States. In China, aquaculture has emerged as an important user of feed concentrates, although fish meal is still needed to provide a balanced diet to carnivorous fish species.

Source: Alltech. 2014. *2014 Alltech global feed survey summary.* http://www.alltech.com/sites/default/files/alltechglobalfeedsummary2014.pdf (accessed 31st Jan 2016).

is also evident as livestock operations move to a larger scale, as in China (Wang et al. 2015).

4. FCEs vary by region. Because FCEs are higher in developed countries, the overall FCE is reduced by the increasing share of consumption in developing countries.

Extreme care is needed in interpreting FCEs. Based on the second factor listed here, it is commonly asserted that eating beef consumes five times more grain than eating poultry, because the FCE of beef is so much lower (Smil 2002). Beef is commonly quoted as having a feed-to-meat ratio of 10 to 12 expressed by weight (e.g., see Smil [2002] and Msangi and Batka [2015]). In practice, much beef is not fed on grain, and when it is, feed concentrates are used only in one or two stages—usually for rearing young stock and for fattening prior to slaughter. For example, in the United Kingdom, where mixed grazing and feeding systems are used, the feed-to-beef ratio was estimated at 2.7–4.6 for the most common systems (Wilkinson 2011).[3]

[3] The ratio was 8.8 for a cereal-fed beef production system, but this system accounted for only 6% of production.

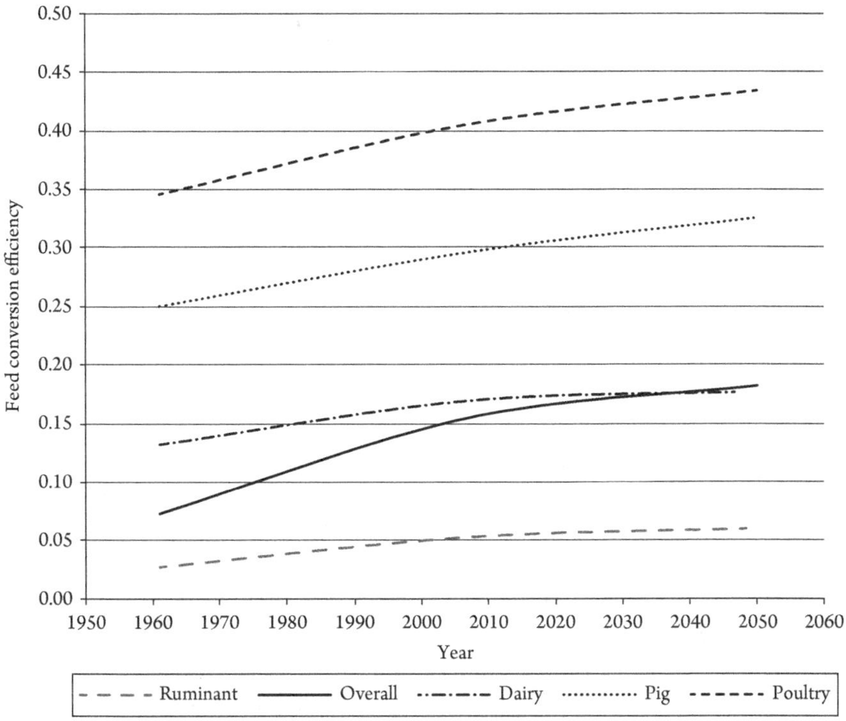

Figure 5.3 Feed conversion efficiencies (in kilograms protein in meat to kilograms protein in feed) estimated over time and in aggregate. Feed conversion efficiency increases for all products but slows from 2007. Overall efficiency increases most rapidly because the share of more efficient products (poultry, for example) increases in the product mix.
Source: Animal Change. 2012. "Storylines for the Livestock Sector Scenarios in EU, Studied SICA Regions and Global Level." *Animal Change*: 1–24.

The composition of concentrates also changes as the livestock industry develops. When the industry is in the early stage of development, intensive systems use many crop by-products such as brans as well as food wastes; but, over time, the growth of the industry exhausts these sources and the industry depends increasingly on using cereals and oil meals in feed. For example, the use of feed concentrates tripled from 2001 to 2012, when the Chinese livestock industry entered this "modernization phase" (Gale 2015). The relative price of cereals and oil meal also matters. During the past 20 years, the ratio of the prices of soybean meal to maize in world markets has ranged between 2 and 3, with an average of about 2.5.

Together, these trends have caused the use of oil meal in feed to grow faster than meat production, whereas the use of cereals in feed has grown more slowly than meat production (Table 5.3). Ignoring the relatively small share of feed that goes to ruminants, oil meal has expanded faster than total poultry and pig meat.[4] The major factor explaining

[4] Note that in China, more than 5 Mt of soybean meal used in aquaculture is not included in Table 5.1 (De Haan et al. 2010).

Table 5.3 Global feed use and meat production in million tons, 1991 to 2013

Year	*Oil meal feed*[a]	*Soybean meal feed*[a]	*Cereal feed*[a]	*All meat*[b]	*Beef*	*Pork–poultry*
1991	142	72	624	148	49	99
2013	325	185	850	254	58	196
Ratio 2013 to 1991	2.29	2.60	1.36	1.72	1.12	1.98

[a] Includes oilseeds and cereals fed directly without processing.

[b] Excludes sheep and goats.

Source: USDA-FAS PSD.

this trend appears to be a sharp shift toward a higher share of oil meal in feed concentrates. In North America, the ratio of oil meal (mostly soybean meal) to cereal feed grains doubled from 0.16 to 0.33 between 1991 and 2013. During the same period in China, this ratio (again, mostly soybean meal) increased even more dramatically from 0.06 to 0.40 (note that China uses a lot of noncereals such as sweet potatoes in its concentrates) (USDA-FAS PSD). These changes represent an effort to improve FCEs through more balanced diets, the substitution of oil meal for animal by-products and fishmeal in feed concentrates, and a push for leaner meat, which requires a diet higher in protein.

The final element in soybean demand is the substitution of soybean meal for other oil meals. Between 1991 and 2014, soybean meal increased steadily as a share of oil meal consumption, rising from 62% to 68%. This substitution accounts for about 14% of the growth in soybean meal consumption during the period (estimated by assuming the share of soybean meal in total meal consumption in 2014 remained constant at the 1991 share). Soybean meal became the dominant protein source in animal feeds for several reasons: the European Union ban on animal by-products in animal feeds (imposed because of the mad cow disease scare), the readily available supplies of soybeans and soybean meal from Latin America, and the decision in 1999 by China to liberalize soybean imports and reduce to a minimal tariff to stimulate its livestock sector.

OUTLOOK TO 2050

Looking to the future, most observers believe the spurt in intensive livestock growth has peaked and they expect future demand for oil meal to expand much more slowly. The latest estimates available from the FAO and the International Food Policy Research Institute have global meat consumption expanding at 1.3% annually to 2050 (from a 2005/2007 base for the FAO and a 2010 base for the International Food Policy Research Institute), which is only about half the rate of expansion since 1990 (Alexandratos and Bruinsma 2012; Rosegrant et al. 2012). The global shift to poultry is expected to continue, with demand for poultry meat expanding by 1.8% annually, faster than for any other meat (Alexandratos and Bruinsma 2012).

Growth of meat consumption has already slowed sharply in China since 2001, with consumption now surpassing 50 kg/capita (plus 33 kg of fish and seafood), and this slowing trend is expected to continue. Indeed, the fastest growth in demand in China will be for beef, dairy, and mutton, and, as noted, production systems for these products use feed concentrates much less (Dong et al. 2015).

Although China has set the pace for increasing meat consumption during the past two decades, future growth will shift to sub-Saharan Africa, South Asia, and some other Asian countries, notably Indonesia. Some 2.7 billion people in these regions live in countries where the annual per-capita meat consumption is less than 20 kg. Based on the FAO projections in Table 5.4, it is this bottom 35% of the population that will account for about 40% of the increase in world meat consumption to 2050.

Taking into account some of the complexities discussed earlier in relating growth in the livestock industry to growth in feed demand, particularly oil meal demand, Alexandratos and Bruinsma (2012) project that the use of oil meal will increase from 220 Mt in 2005/2007 to 390 Mt in 2050. If soybean meal continues to increase as a share of oil meals—for instance, if its share reaches 75% in 2050—and the share of soybeans processed increases to 90%, then 470 Mt of soybeans will be needed in

Table 5.4 Projection of total meat consumption to 2050

Region	*Per capita meat consumption, 2005/2007 (kg/yr)*	*Growth rate, meat consumption, 2005/2007–2050 (%/yr)*	*Total meat consumption, 2005/2007 (Mt)*	*Total meat consumption, 2050 (Mt)*	*Total increase, 2005/2007–2050 (Mt)*
Sub-Saharan Africa	10.0	3.0	7.3	27.3	20.0
South Asia	4.5	4.2	6.7	42.5	35.8
Southeast and East Asia, excluding China	31.6	2.0	19.0	45.8	26.8
China	50.0	1.2	68.0	115.3	47.3
West Asia and North Africa	23.8	2.3	10.3	28.3	18.0
Latin America	62.2	1.3	34.6	61.3	26.7
Developing	28.1	1.8	145.9	320.6	174.7
Developed	81.0	0.4	109.4	130.5	21.6
World	39.0	1.3	255.3	451.0	196.3

Source: Calculated from Alexandratos, N., and J. Bruinsma. 2012. World agriculture towards 2030/2050: The 2012 revision. ESA Working Paper no. 12-03. Rome: Food and Agriculture Organization of the United Nations.

2050, representing an annual growth rate of 1.5% from 2013.[5] This rate of growth is far slower than the 4.6% growth rate seen between 1991 and 2013, but it is still faster than the expected increase in meat consumption. Note that recent projections to 2030 for China also estimate the use of protein meal will grow much more slowly, at 1.8% annually, in part because of slower growth in meat consumption and in part because consumption will shift toward less feed-intensive species (Dong et al. 2015; Gale 2015).

Obviously, any projections to 2050 are fraught with uncertainty, especially projections of feed use, given the complexities of the livestock industry and the many substitutes among production systems and types of feed. Indeed, projections of supply and demand for soybeans made by international agencies hugely underestimated their extremely rapid growth during the last 15 to 20 years. Actual consumption was more than 25% higher than the projected level (Chapter 7).

Our own estimates may well be too conservative for a number of reasons. Demand projections for livestock products may be on the low side, according to Keyzer et al. (2005), who suggest that many consumers in middle-income countries and most consumers in low-income countries have not reached a minimum threshold of income at which livestock consumption takes off. The FAO projections may underestimate the pace at which meat consumption will grow among the bottom 40% of the population—countries referred to earlier, the populations of which consume less than 20 kg of meat per capita—especially if income growth in Africa and South Asia maintains its recent momentum. These countries have huge potential to increase their use of feed concentrates, because current per-capita consumption of oil meals is less than one quarter of the world average (Figure 5.4).

The future trajectory of meat consumption in India is particularly uncertain. The FAO projections (Table 5.4) estimate the most rapid growth for livestock products will be in South Asia (4.2% annually), but it is starting from a very low base. This rate of growth assumes many Indian consumers will make a significant switch away from a vegetarian diet—a cultural trend that is much debated (Keyzer et al. 2005).

Another source of uncertainty is the rapidly growing aquaculture industry, especially in Asia (Naylor et al. 2009). Although only some fish are herbivores, Msangi and Batka (2015) suggest aquaculture will use about 12 Mt of soybean meal by 2030, which is still small in relation to global use. Given the industry's very rapid expansion, aquaculture could be a significant player in global oil meal markets by 2050.

In rich countries, demand for livestock products may decrease as consumers move to diets they perceive are healthier and more sustainable. Beef consumption has already fallen in the United States and Western Europe. Much scope also remains to improve FCEs in developing countries through a better balance of feed nutrients and by operating at scale (Dong et al. 2015). New and more efficient sources of synthetic protein derived from single-cell proteins and modified algae may also become available before 2050 to substitute for soybean meal and even some meat products (Mattick and Allenby 2013).

[5] Assumes soybean meal is 80% of processed soybeans by weight and 90% of soybeans are processed in 2050 (the rest are mostly consumed directly for food and feed).

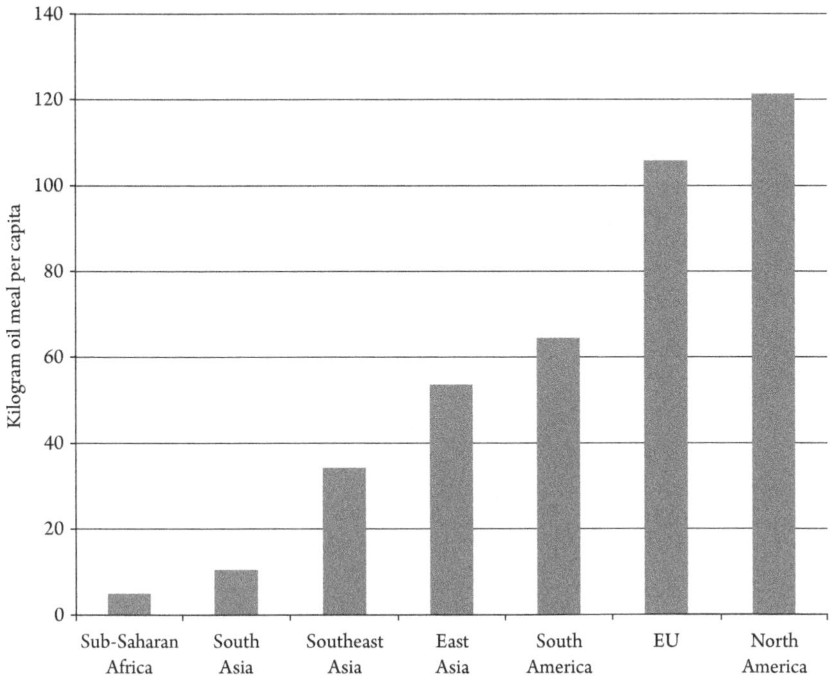

Figure 5.4 Per capita annual oil meal consumption in 2014. The world average per-capita meal consumption is 40 kg/yr. EU, European Union.
Source: Calculated from USDA-FAS PSD and FAOSTAT.

CONCLUSION

Oil crops produce joint products, oil meal and oil, and, in the case of soybeans, the meal (used to feed livestock) is the most valuable product. Demand for livestock products from intensive production systems that rely heavily on feed concentrates—especially poultry and pig meat, but also fish—has accelerated in recent decades, led by China. The use of soybean meal expanded much faster than meat production because of the shift to feed-intensive poultry, pigs, and fish; the sharp increase in the share of protein in feed rations relative to energy; and the growing share of soybeans in the market for oil meals.

Projections of demand for soybean meal are especially uncertain. We are confident, however, that the recent surge in soybean meal consumption has peaked and that future growth will be aligned much more closely with growth in the livestock industry. Growth of that industry will also be slower than in the past, and China's extraordinary spurt in soybean meal consumption and imports may be ending. At the same time, 2.7 billion people with rapidly increasing incomes have yet to consume significant quantities of meat and other livestock products, which leaves plenty of room for the livestock sector to grow.

Another outcome of the livestock industry's demand for soybean meal is the increase in joint production of soybean oil, which has made an important contribution to the world vegetable oil supply. We estimate the soybeans processed for animal feed between 1991 and 2013 generated an additional 30 Mt of oil that supplied about 28% of the increase in vegetable oil consumption from 1991 to 2013. Although this contribution to oil supplies will surely slow, reasonable projections to 2050 suggest that the supply of soybean oil could add 0.5% annually to edible oil supply, or one third the estimated annual growth in food demand for oils (Chapter 4). As in the past, processing of soybeans for animal feed will continue to be a significant factor in vegetable oil markets.

6

BIODIESEL
A SOURCE OF GROWTH AND UNCERTAINTY IN VEGETABLE OIL MARKETS

The use of biodiesel in the global energy sector presents considerable uncertainties with regard to future patterns of production, consumption, and trade in vegetable oils. Oil processed from these crops contributes directly to about 85% of biodiesel; the remainder is manufactured from animal fats and recycled vegetable oils (Table 6.1).

Widespread use of biodiesel in the ever-expanding transportation industry of developing nations could raise biodiesel demand, leading to sharp increases in vegetable oil prices and significant growth in the production of tropical oil crops in the coming decades. If, on the other hand, biodiesel consumption stagnates and is outpaced by other sources of energy for transportation as a result of market forces or policy actions, vegetable oil prices will likely level off, remaining relatively constant in real terms.

Looking out to 2025, the dynamics of the global biodiesel market will depend largely on government policies in a wide range of industrialized and developing countries aimed at (1) bolstering demand and prices of vegetable oils and efficient use of coproducts, (2) increasing the share of renewable (nonfossil) fuels in overall energy use, (3) reducing the net climate impact of energy use, (4) securing domestic energy supplies, and (5) supporting rural development. Predicting long-term trends in the biodiesel market beyond 2025 requires a closer look at the fundamentals of the energy market, because policy dynamics are almost impossible to anticipate a decade or more in advance. Regardless of which time period one examines, the biodiesel story is both fascinating and complex, given the substitution in demand for different vegetable oils and the interplay between agriculture and energy markets throughout the world.[1]

A large body of literature has emerged since the mid 2000s on policies surrounding the development of biofuels (ethanol and biodiesel) and their effects on agricultural markets, land use, GHG emissions, rural development, and food security.[2] The interesting questions for our purposes in this chapter concern how and where markets

[1] This chapter draws on work by Joanne Gaskell (2012) during her graduate training at Stanford University and also benefited greatly from the research assistance of Matthew Higgins. A complementary short-run outlook for the global biodiesel sector out to 2020 can be found in Naylor and Higgins (2016).

[2] For references, see material reviewed in Mitchell (2011), German et al. (2011), Moschini et al. (2012), Naylor (2014); Langeveld et al. (2014), and de Gorter et al. (2015).

Table 6.1 Estimated shares of biodiesel production by feedstock, 2014

Country	*2014 Biodiesel production (billion L)*	*Estimated feedstock share (%)*[a]							
		Rapeseed oil	*Soybean oil*	*Palm oil*	*Sunflower oil*	*Other virgin vegetable oils*	*Recycled vegetable oils*	*Animal fats*	*Unknown*
European Union	12.7	55	8	15	3	1	14	4	—
United States	4.8	14	64	1	—	13	—	7	—
Brazil	3.5	—	77	—	—	1	—	21	1
Argentina	2.9	—	100	—	—	—	—	—	—
Indonesia	3.3	—	—	100	—	—	—	—	—
Thailand	1.2	—	—	100	—	—	—	—	—
China	1.1	—	—	—	—	—	100	—	—
Colombia	0.6	—	—	100	—	—	—	—	—
Malaysia	0.5	—	—	100	—	—	—	—	—
Canada	0.3	60	—	—	—	—	26	14	—
India	0.1	-	-	-	-	61	36	3	—
Total	31.0	25	32	24	1	3	10	5	<1

[a] Not all feedstocks are produced domestically (e.g., palm oil is not produced in the European Union or the United States).

The data reflect estimated feedstocks used in biodiesel as reported by the US Energy Information Administration, not biomass-based biodiesel as defined by US biodiesel standards.

Source: Data from US Department of Agriculture. 2015. Published GAIN reports. http://gain.fas.usda.gov/Recent%20GAIN%20Publications/Forms/AllItems.aspx (accessed June 23, 2015);and US Energy Information Administration. Table adapted from Gaskell, J. C. 2012. The palm oil revolution in Asia. PhD diss., Stanford University. http://purl.stanford.edu/zc839jm3057 (accessed January 8, 2014).

for biodiesel in particular have developed during the past decade, what direction the biodiesel industry will take during the coming decades, and which oil crops will be affected most by changes in energy demand.

We begin with a description of the emerging biodiesel market and its links to various oil crops, then turn our attention to the policy context of biodiesel development. Biofuel policies play an important role in determining vegetable oil demand and prices, but they are not always an exogenous factor in the dynamics of vegetable oil markets. Policymakers in many large agricultural economies have strong interests in boosting farm incomes and supporting various parts of the agricultural supply chain. As a result, policies promoting biodiesel growth typically focus on domestically produced oil crops and tend to be introduced during periods of relatively low crop prices. These patterns reflect an endogenous policy response to agricultural surpluses; when policies supporting biofuels are in place, agricultural surpluses (and hence the slack in agricultural markets) are reduced.

Policies promoting biodiesel are also responsive to fossil fuel prices, creating a link between real energy and agricultural prices, at least in the short run. The nature and strength of this coupling depends on which types of policies are used to support biodiesel use (quantity vs. price instruments) and on the market dynamics of coproducts for different oil crops, ranging from soybeans (with their high protein meal content) to oil palm (with its high oil content).

The feedbacks among crop prices, energy prices, and policy responses create substantial analytical challenges for assessing connections between vegetable oil and biodiesel markets, particularly over the long term. As the biodiesel industry has expanded, the creation of excess plant capacity has allowed production to increase when profits have been favorable and to decline when profit margins have dipped. Subsidies often have been essential for keeping biodiesel plants in operation. In the third section of this chapter, we examine how price and policy interactions determine the profitability of biodiesel production with different feedstocks (crop inputs). Given the importance of policy—but also the difficulties associated with predicting specific policy measures far into the future—we end the chapter by presenting scenarios of energy demand and biofuel mandates to explore how the biodiesel industry might affect global vegetable oil markets.

THE EXPANSION OF BIODIESEL MARKETS

Global biodiesel production increased sixfold between 2005 and 2013, and by some estimates could grow by another 50% by 2025 (Organisation for Economic Co-operation and Development 2014; Food and Agriculture Organization of the United Nations 2014). Box 6.1 describes the basic properties of biodiesel and other renewable diesel fuels. The expansion of biodiesel markets has been responsible for roughly half the increase in vegetable oil consumption since 2003. The share of vegetable oils consumed in biodiesel worldwide is expected to remain relatively constant at 14% to 15% until 2025, but several countries—including Colombia, Indonesia, Brazil, the European Union, Thailand, and Argentina—are expected to

Box 6.1 The Basics of Bio-Based Diesel Fuels

Biodiesel is a liquid transportation fuel that substitutes for its petroleum-based alternative, diesel, and is used primarily to power cars, trucks, and, increasingly, ships. Biodiesel also shows some promise for wider use in stationary applications (heating and power) and in aeronautics (Rosillo-Calle et al. 2012). By volume, biodiesel contains ~93% of the energy content of petroleum diesel (US Energy Information Administration 2007). The energy content of ethanol, by comparison, is only 66% that of gasoline. In terms of vehicle kilometers/L (KPL), biodiesel also outperforms ethanol. The KPL of biodiesel is about 91% that of fossil diesel (whereas ethanol gets only 70% the KPL of gasoline), and diesel engines achieve approximately 37% higher KPL than gasoline engines (de Gorter et al. 2015). From a mileage efficiency standpoint, therefore, biodiesel is a relatively good bet for the future, particularly for countries that anticipate rapid growth in transportation fuel demand.

Converting vegetable oil into biodiesel is a relatively simple, low-cost process typically involving a chemical reaction between vegetable oils and methanol in the presence of sodium hydroxide (a transesterification process) to produce fatty acid methyl esters (biodiesel) and glycerol (Meher et al. 2006). The biodiesel industry is versatile and innovative in terms of processing technologies, inputs, and vehicle compatibility. Several alternative technologies exist to convert fats and oils into commercial-grade transportation fuel (Marchetti et al. 2007). One example of a promising fuel technology is hydro-treated vegetable oil (HVO, also called *renewable diesel*), which is made from the same feedstocks as biodiesel. Hydro treatment uses catalytic hydrogenation to convert vegetable oils and animal fats into long-chain hydrocarbons (paraffins), and the resulting HVO-based fuel is identical chemically to traditional diesel fuel.[a]

The specific vegetable oil feedstock used to produce biodiesel depends largely on relative prices of vegetable oils, markets for oilseed coproducts, and policy incentives. Global biodiesel feedstocks in 2014 were composed of soybean oil (32%), rapeseed oil (25%), palm oil (24%), recycled vegetable oils (10%), animal fats (5%), and other vegetable (and unknown) oils (4%).[b] Biodiesel from these feedstocks can be blended with petroleum diesel in any ratio. Only during the past decade or so have engine and vehicle manufacturers tested and approved the use of biodiesel blends specifically. Although recommendations vary, B20 (20% biodiesel blended with 80% fossil-based diesel) is the most commonly approved blend level. In practice, biodiesel blends are often less than B20.

The performance of biodiesel derived from different sources, particularly in cold environments, also plays an important role in the selection of feedstocks. Similar to petroleum diesel, biodiesels can form wax crystals at a certain low temperature point (typically referred to as the *cloud point*), causing the fuel to become too viscous to travel through a vehicle's fuel system (Radich 2004). Cloud points differ for biodiesel made with different feedstocks and processing technologies (Aatola et al. 2008). Palm-based biodiesel has a higher cloud point than rapeseed- or soy-based diesel, for instance. A higher cloud point makes biodiesel from palm oil less suitable for use during the winter in temperate regions.

Solutions to this problem include the use of special engine heaters, chemical additives, hydro treatment, and biodiesel blends.

[a] Neste Oil, located in Singapore, is currently the leading producer of biodiesel and HVO fuels (Neste n.d.).

[b] See Table 6.1. Much of the information in this section comes from the Renewable Energy Policy Network for the 21st Century (2014), the US Department of Agriculture (2015c), and the US Energy Information Administration (n.d. d).

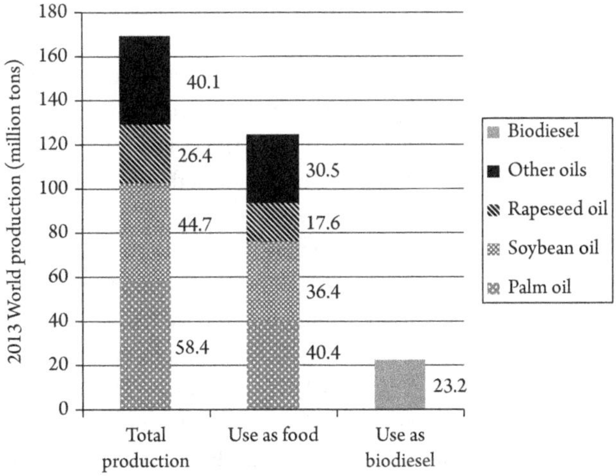

Figure 6.1 Global production and use of vegetable oils, 2013. Industrial uses of vegetable oils, waste, and changes in stocks account for most of the difference between total production and uses in food and fuel.

Sources: US Department of Agriculture. 2015. Oil crops yearbook. http://www.ers.usda.gov/data-products/oil-crops-yearbook.aspx (accessed June 23, 2015); and Renewable Energy Policy Network for the 21st Century. 2014. *Renewables 2014 global status report*. Paris: Renewable Energy Policy Network for the 21st Century. http://www.ren21.net/Portals/0/documents/Resources/GSR/2014/GSR2014_full%20report_low%20res.pdf (accessed January 8, 2016).

use from one third to one half of their vegetable oil production to produce biodiesel (Organisation for Economic Co-operation and Development 2014; Food and Agriculture Organization of the United Nations 2014). If these predictions hold true, the biodiesel industry could have significant impacts on regional and global vegetable oil markets.

As noted in the previous chapter, about 75% of vegetable oil demand comes from the consumption of vegetable oils for food (Figure 6.1). That figure was 90% as recently as 2000, however, revealing how the biodiesel sector has assumed a greater role in shaping the dynamics of the vegetable oil market at the margin. Whether the demand for biodiesel continues to expand and outpace the growth in demand for edible oils on a regional basis has important implications for vegetable oil prices and investments in particular oil crops.

The trajectory for biodiesel demand depends most importantly on regional per-capita income growth, energy consumption, and relative prices for vegetable oil and diesel commodities. Theory and empirical evidence show that as incomes increase, expenditures on food typically level off (a concept known as *Engel's Law*, discussed in Chapter 4).[3] In contrast, direct and indirect expenditures on energy continue to increase with income. Beyond a minimum income level, the additional dollar one earns is no longer spent on basic food commodities; instead, it might be used to purchase foods shipped or flown in from across the world, to buy a new car, or to travel abroad (Lobell et al. 2014). As a result, income growth is expected to be far more important than population growth in stimulating the demand for vegetable oils via biofuels.

What differentiates the biodiesel story from the story of edible oils told in Chapter 4 is that policy—not income, population growth, or prices per se—largely drove growth in demand during the past decade. The next section reviews the types of policies used to stimulate biodiesel production and their effects on vegetable oil markets, but it is worth noting at the outset that many countries have introduced mandates for renewable fuel use, which in many cases cause the demand for biofuels to increase regardless of price (Abbott et al. 2009; Naylor 2014). Global production of biodiesel has expanded significantly since the mid 2000s with the introduction of renewable fuel policies in the United States, Europe, and several other countries, reaching 31.0 billion L in 2014 (Figure 6.2). The regional distribution of biodiesel production is diverse (much more so than ethanol, which is dominated by the United States and Brazil), with the United States, Germany, Brazil, Argentina, Indonesia, and France all contributing sizeable volumes. Several other countries, including Thailand, Poland, and Colombia, are also gaining shares in the biodiesel market.

International trade in biodiesel is only 10% to 15% of global production (USDA, 2015c). A number of countries that produce biodiesel, including Brazil, Colombia, India, Thailand, and China, report virtually no trade in biodiesel. Other countries, including Indonesia, Malaysia, and Argentina, export but do not import biodiesel. The United States and the European Union are two of the largest importers, and they also export biodiesel. Overall, the trade data indicate a relatively high degree of self-sufficiency in biodiesel production, underscoring the important role of biofuels in supporting domestic agriculture (and hence feedstock selection), energy, and rural interests.

[3] Engel's Law holds up empirically with both time-series and cross-sectional data and states the income elasticity of demand for food in the aggregate is less than one and declines toward zero with income growth (Timmer et al. 1983). By comparison, the income elasticity of demand for vehicle fuels ranges from 0.5 to 1.6 (mean 1.2) (Victoria Transportation Policy Institute 2014). For a review of studies on elasticities for transportation fuel, see Johansson and Schipper (1997), and Lipow (2008).

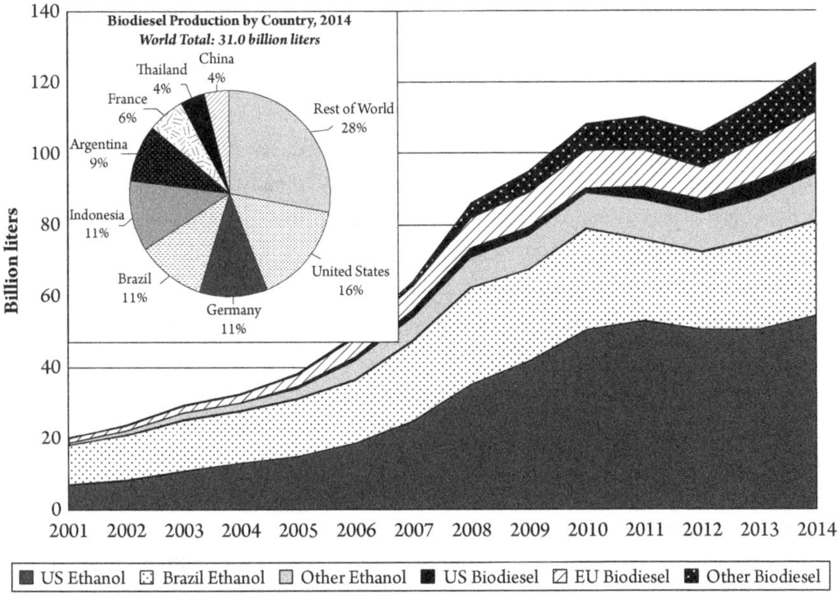

Figure 6.2 World biofuel production 2001 to 2014. The figure shows ethanol and biodiesel production by region from 2001 to 2014. Biodiesel production by country in 2014.
Source: Data for the large figure are from US Energy Information Administration. n.d. International energy statistics database. http://www.eia.gov/cfapps/ipdbproject/IEDIndex3.cfm (accessed June 23, 2015); and data for the inset figure are from the Renewable Policy Energy Network for the 21st Century (2014). http://www.ren21.net/Portals/0/documents/Resources/GSR/2014/GSR2014_full%20report_low%20res.pdf (accessed May 5, 2016).

THE POLICY CONTEXT

With the exception of Brazil, the global biofuel boom has been largely a 21st-century phenomenon and, by 2013, more than 50 countries had implemented policies promoting the development of biodiesel and ethanol (Renewable Energy Policy Network for the 21st Century 2014; Smeets et al. 2014). The main policies have included price incentives (subsidies, tax exemptions) for biofuel use, binding or nonbinding quantitative targets on the blending of biofuels with conventional fossil fuels, and trade policies. Other policies used to facilitate biofuel development have included the provision of subsidized credit to the biofuel industry, the provision of public land for feedstock production, and direct budgetary spending on biofuel support. These types of policies contributed to Brazil's establishment as a leader in the market for biofuels during the 1980s and 1990s (Valdes 2011a; Valdes 2011b) and are now allowing other countries to gain a market share. The global cost of biofuel policies—calculated on the basis of direct budgetary spending, tax relief, import duties, and indirect market price transfer—was estimated at US$22 billion in 2010 (Gerasimchuk et al. 2012; International Energy Agency 2011).

In the United States and European Union in particular, several circumstances impelled the widespread introduction of biofuel policies during the mid 2000s. High energy prices and international conflicts involving oil-producing rogue states were an incentive to move away from oil imports. Renewable fuel agendas aimed at reducing GHG emissions stemmed from a growing awareness of the potential negative effects of climate change on society and ecosystems. Finally, a long history of crop supports, agricultural surpluses, and declining (real) prices for agricultural commodities led policymakers to identify new sources of crop demand. Just as in earlier periods when a decline in real cereal prices induced the grain-fed livestock sector and the corn (maize) fructose industry to develop, low crop prices during the late 1990s and early 2000s helped usher in a new era of growth for biofuels (Naylor and Falcon 2011).[4] Production gains in oil crops (rapeseed, soybeans, oil palm), combined with developments in coproducts (oil, meal) and supply chains, also fostered growth in the biodiesel industry.

Developing countries marshaled their own incentives to support biodiesel as conditions in agriculture and energy markets changed. Indonesia, the world's largest exporter of palm oil, introduced policies in 2015 in response to falling international prices and declining export tax revenues. These policies, which we discuss further in Chapter 7, included an immediate increase in the blending mandate for biodiesel to B15, a targeted mandate of B20 beyond 2016, and a biodiesel subsidy supported by an export levy for CPO (Pakiam and Rusmana 2014; Ministry of Energy and Natural Resources, Republic of Indonesia 2015). Colombia has also adopted ambitious policies supporting biodiesel based on palm oil. The case studies that follow illustrate a range of strategies for supporting the biodiesel industry and show the variable effects of these strategies on vegetable oil markets from 2000 to 2015.

The US Renewable Fuel Standard

The US Renewable Fuel Standard (RFS) was introduced through the Energy Policy Act of 2005 and strengthened through the Energy Independence and Security Act of 2007 (Naylor and Falcon 2011). The RFS mandates the minimum quantities of renewable fuels that must be used in gasoline blends in the United States. According to the 2007 law, the minimum target for 2015 was set at 56.7 billion L (15 billion gal) of conventional (corn or other first-generation) ethanol in gasoline blends. The RFS also stipulates that by 2022, the use of advanced biofuels (made from waste, agricultural, cellulosic, or algal materials) must reach a minimum of 79.4 billion L (21 billion gal). Within the advanced biofuel mandate, at least 3.8 billion L (1 billion gal) must consist of biomass-based diesel (including biodiesel and renewable diesel). The

[4] The US ban on the use of methyl tertiary butyl ether (MTBE) as an oxygenate in fuels stimulated the demand for ethanol blends in gasoline (Naylor and Falcon 2011). de Gorter et al. (2015) show that the boom in the US ethanol production also stemmed from increasing crude oil prices during the mid 2000s that activated existing ethanol tax credits and thus made the biofuel industry viable economically.

US biodiesel industry has benefited from a US$0.26/L ($1/gal) blenders tax credit (subsidy) throughout most of the period from 2005 to 2015, and biodiesel production has exceeded the RFS mandate. The biodiesel tax credit has expired but has been reinstated on more than one occasion in recent years, and was reinstated once again in 2015 (Kotrba 2015).

The United States has had a complicated history of trade policy in biodiesel, especially in relation to the European Union. Incentives by US companies to profit from the biodiesel blenders subsidy has resulted in a pattern of "splash and dash"—in other words, adding biodiesel to a fuel blend and exporting it—that resembled dumping.[5] The European Union has responded by imposing antidumping and countervailing duties on biodiesel imports from the United States (de Gorter et al. 2015). The trade loopholes in US blending credits for biodiesel have been closed, but incentives to use biodiesel from domestically produced feedstocks have remained intact, particularly as first-generation biodiesel produced from soybean oil now qualifies for the broader scope of the overall renewable fuels mandate. The incentives for biodiesel production and use in the United States have induced the area planted to canola (which has a higher oil content than soybeans) to rise in the United States and have reinforced biodiesel imports to the United States from exporting countries such as Argentina.

The US Environmental Protection Agency (EPA), which oversees the RFS, announced a revised set of mandates in December 2015 that further supports the biodiesel sector.[6] The mandate for biodiesel was raised by more than 20%, from 6.2 billion L in 2014 to 7.6 billion L in 2017. Largely as a result of uncertainty in policy targets and regulations, new investments in cellulosic fuels have declined, and success in meeting existing cellulosic fuel targets have lagged expectations. As a result, biodiesel has played an increasing role in meeting the advanced biofuel mandates over time. The EPA's new targets set the stage for the 2016 presidential elections, that began with the Iowa caucus on February 1, 2016.

The RFS mandates in the United States have an important GHG stipulation: conventional biofuels must be 20% lower in GHG emissions than petroleum-based transportation fuels (based on 2005 petroleum carbon intensity), and advanced biofuels must be at least 50% lower in GHG emissions for noncellulosic material and 60% lower for cellulosic material than gasoline and diesel (calculated through a life cycle analysis).[7] Based on these GHG criteria, palm oil-based biodiesel does not qualify within the RFS because, according to the EPA, it reduces GHG emissions by only

[5] The biodiesel blenders credit of US$1/gal applied to all fuels, whether imported or produced domestically. The blend was often 0.1% diesel and 99.9% biodiesel, and blending credits could be earned even by enhancing imported biodiesel from Europe with additional US-produced biodiesel and then reexporting it to Europe (de Gorter et al. 2011).

[6] For more information on the 2015 renewable fuels mandates, see US Environmental Protection Agency (2015). The RFS standards have been in flux since 2014 as a result of industry litigation, which has led to great uncertainty surrounding the future of the US biofuel industry. The Obama administration remains highly supportive of renewable fuels.

[7] For further information on the RFS, see US Environmental Protection Agency (2014).

17% relative to conventional (fossil-derived) diesel. Soy-based biodiesel, on the other hand, clearly passes the bar, with estimated GHG reductions of 76% (Pradhan et al. 2012). Some countries, most notably Colombia, are attempting to prove to the EPA that GHG reductions from palm-based biodiesel can be far greater than 20% relative to conventional fuels, provided the oil palm is grown on degraded land and not in pristine forests or in high-carbon soils. Even if they succeed, a daunting, if not impossible, challenge for exporting countries will be to certify compliance with US renewable biomass requirements, which (among other tough provisions) prohibit net agricultural land expansion for biodiesel production.

The EU Renewable Energy Directive

The European Union passed legislation in 2009 through its Renewable Energy Directive (RED) that required 10% of all transportation energy to come from renewable resources by 2020, thus expanding the region's use of biofuels (European Commission 2015). The policy tools used to achieve RED reflect a transition within the European Union from earlier tax and trade incentives and indicative consumption targets to mandatory targets, with trading permitted among states. In 2010, EU biofuel consumption was made up of 80% biodiesel and 20% ethanol, and biodiesel blends had reached B5 (Flach et al. 2013). Although feedstocks differ by country within the European Union, rapeseed is the main feedstock in biodiesel production for the region as a whole particularly for Germany and France (Table 6.1). The heterogeneity among countries in feedstock use and energy technologies makes the EU biodiesel industry a particularly interesting case study.

The increasing demand for biodiesel in the European Union since the introduction of RED provides an excellent example of the connections among energy use, crop substitutions, and trade, as shown in Table 6.2. Biodiesel consumption in the European Union increased from around 0.81 billion L in 2000 to 13.8 billion L in 2011. Given the European Union's use of rapeseed as a dominant feedstock, the region's rapeseed production rose by 70% during this period, from 11.3 to 19.2 Mt, and the area sown to rapeseed also expanded significantly from 4.2 to 6.8 Mha. Rapeseed has traditionally been a rotation crop with wheat in the European Union, and with the introduction of RED, it has increasingly displaced other major rotation crops such as potatoes and sugar beets (International Council on Clean Transportation 2013).

Despite impressive growth in rapeseed production, the European Union's rapidly expanding biodiesel industry has increased its imports of vegetable oils for both food and fuel. Between 2000 and 2011, vegetable oil imports by the European Union more than doubled from 10.6 billion L to 21.6 billion L, with ~40% of the increase comprised of palm oil imports. Biodiesel imports, primarily from Canada and the Ukraine (based on oilseed feedstocks such as rapeseed and sunflower) have also increased (International Council on Clean Transportation 2013). The European Union has a range of trade policies designed to protect domestic rapeseed and biodiesel producers, including a 3.5% import duty on biodiesel blends of B30 and under, and a 6.5% import duty on B30 to B100 fuels (Flach et al. 2013). The European Union has

Table 6.2 Biodiesel, rapeseed production, and vegetable oil imports in the European Union, 2000 and 2011

	Year	
	2000	*2011*
Biodiesel production (billion L)	0.87	10.6
Biodiesel consumption (billion L)	0.81	13.8
Biodiesel imports (billion L)	0.0	3.2
Rapeseed production (Mt)	11.3	19.2
Rapeseed area harvested (Mha)	4.2	6.8
Palm oil imports (billion L)	3.8	8.1
Other vegetable oil imports (billion L)	6.8	12.5
Distillate fuel imports[a] (billion L)	83.6	133.8
Approximate diesel fuel imports (billion L)	75.2	120.4

[a] Distillate fuel is one of many petroleum products that can be produced from a barrel of crude oil, and diesel currently accounts for about 90% of all distillate fuel production. Distillate fuels are made from longer carbon chains than gasoline, but the two have a substantial overlapping range, and refineries can optimize between the two fuels in real time. One barrel of crude oil (42 gal) produces roughly 12 gal distillate fuel (equating to about 10 gal of diesel) and 19 gal of gasoline. Global distillate fuel production in 2010 was 9.3 billion barrels, equivalent to 8.4 billion barrels of diesel. For further details, see US Energy Information Administration (n.d. c).

Assumes 1136 L/mt of vegetable oil. As mentioned, diesel currently makes up about 90% of all distillate fuels.

Source: Data from the US Energy Information Administration (EIA). n.d. International energy statistics database. http://www.eia.gov/cfapps/ipdbproject/IEDIndex3.cfm (accessed June 23, 2015); Independent Petroleum Association of America. Diesel fuel: The unsung hero. http://oilindependents.org/diesel-fuel-the-unsung-hero/ (accessed June 23, 2015); FAOSTAT (n.d.); National Renewable Energy Laboratory. 2013. International trade of biofuels. US Department of Energy National Renewable Energy Laboratory. http://www.nrel.gov/docs/fy13osti/56792.pdf (accessed January 7, 2016); and US Department of Agriculture. 2015. Published GAIN reports. http://gain.fas.usda.gov/Recent%20GAIN%20Publications/Forms/AllItems.aspx (accessed June 23, 2015).

also implemented antidumping tariffs on biodiesel imports from the United States, Canada, Argentina, and Indonesia (de Gorter et al. 2015).

A key component of the EU directive is the sustainability criteria for its feedstock sourcing. RED requires that biofuels used under the mandate lead to a 35% reduction in GHG emissions relative to conventional fossil fuels (gasoline and diesel) on implementation, and that the reduction in GHGs be scaled to 50% for existing plants by 2017 and 60% for new installations (European Union Renewable Energy Directive, 2015). The directive also provides a double mandate credit for the use of second-generation biofuels (derived from cellulosic material such as switchgrass, animal fats, distillers' corn oil, and recycled vegetable oils). A 2015 amendment to RED stipulates that calculations of indirect land-use change associated with biodiesel production must be transparent in reports of GHG emissions by fuel suppliers.[8] The 2015 amendment also limits the share of biofuels made from edible crops to 7% of all transportation energy use, and thus ensures the use of food crops (such as rapeseed) in biofuel production increases only at a rate proportional to total transportation fuel consumption (Saikkonen et al. 2014).

Brazil's Biodiesel Industry

Brazil's experience with biodiesel dates to the 1920s, but it was not until the 1970s that the country's biofuel industry began to develop at scale. (For a thorough review of Brazil's biodiesel industry, see Bailis [2014]). Brazil's biofuel industry, like that of the United States, is dominated by ethanol (in Brazil's case, sugar-based ethanol) (Figure 6.2). Serious public and commercial interest in biodiesel development surfaced during the early 2000s, and as a result of a series of policy initiatives, Brazil now ranks as the world's third-largest biodiesel producer. Rapid growth in the country's biodiesel industry can be traced to the 2002 election of President Lula da Silva, who ran on a platform of social justice and wealth redistribution (Flexor et al. 2011). At that time, the Ministry of Science and Technology introduced a new program, PROBIODIESEL, which set a blending target of B5 by 2005 and B20 by 2020. An important goal of this program was to source more feedstock produced by small-scale farmers in poor agricultural regions. The president advocated a flexible approach in terms of technology options, engagement of farmers throughout the country, and voluntary targets. In 2004, however, that initiative led to the establishment of the National Biodiesel Production and Use Program (Programa Nacional de Produção e Uso de Biodiesel), and not long thereafter Brazil implemented nonvoluntary mandates for biodiesel blends.

Although PROBIODIESEL was intended to engage farmers in Brazil's poorest agricultural regions (the north, northeast, and semiarid regions), achieving the mandated blends of biodiesel required the scale and efficiency of feedstock production present in the wealthier soybean-producing regions. By 2012, soy-based biodiesel, produced mainly in the south and central–western regions, accounted for more than 85% of the cumulative biodiesel output under the program. More than 100,000 family farms were participating in the program at the time, and roughly 60 cooperatives had been formed. Yet, the large-scale soybean farmers—those with a long history of commercial soybean production—reaped most of the benefits (Bailis 2014).

[8] This amendment reinforces the use of indirect land-use change in reporting but not specifically in compliance with the EU sustainability criteria. The indirect land-use change reporting was previously voluntary. For more information on RED amendments, see European Parliament (2014).

Gains in rural employment were not as substantial as the program initially intended. Soybeans cover ~35% of Brazil's area under annual crops, but soybeans are one of the country's least labor-intensive crops, accounting for only 8% of employment in annual crop farming. Employment in refineries also failed to meet the original development goals. More than 50 biodiesel refineries were in operation throughout the country by 2011, but only 10 of them were in the north or northeast. Those 10 refineries produced only 15% of national biodiesel output.

Despite these regional disparities, Brazil's biodiesel program provides a template for how a country might engage small-scale farmers. The government provides tax breaks for biodiesel producers who procure feedstocks from small-scale farmers, with larger tax breaks accruing to producers in the country's least developed north, northeast, and semiarid regions. In addition, the program establishes formal contracts with family farms or cooperatives of small-scale farmers, and it provides them with technical assistance to produce biodiesel feedstocks.

Brazil's northern regions produce a wide range of potential inputs for the biodiesel industry, including beef fat, castor oil, cottonseed oil, native oilseeds, soybean oil, and palm oil. Between 2005 and 2011, the area under oil palm expanded by 23%, and production of palm oil rose by 44%, mostly in the states of Pará and Bahia (Instituto Brasileiro de Geografía e Estatística 2012).[9] Despite the high oil yield of oil palm, however, oil palm does not contribute to the region's total biodiesel output. The high perishability of palm fruit requires farmers to be located close to palm oil mills—not an easy requirement for smallholders to meet in Brazil's vast northern region. In addition, high palm oil prices make it more advantageous to sell processed palm oil into the market for food (vs. fuel), and farmers face a range of constraints on oil palm production, such as bud rot and high soil potassium requirements.

Ultimately, Brazil's expansion of biodiesel will depend on the country's energy policies and energy markets. Fuel subsidies and fluctuating crude oil prices influence the profitability of biodiesel expansion, as discussed later in this chapter. The Brazilian government has a long history of subsidizing fuel to curb inflation and garner voter support. It also maintains a trade policy that protects domestic biofuel producers (14% tariff on biodiesel imports and 20% tariff on ethanol imports) (Barros 2013). Brazil's macroeconomic policies and stance on fuel subsidies will continue to have spillover effects on the biodiesel industry and on farmers who grow biodiesel feedstocks.

Colombia's Biodiesel Expansion

Similar to Brazil's biodiesel strategy, Colombia's palm oil and biodiesel expansion programs have the publicly stated goals of promoting rural income growth and stability, expanding rural employment, and diversifying sources of energy (for further information, see Gilbert and Pinzon [2013], and Pinzon [2012]). Colombia began promoting

[9] The state of Pará accounts for more than 80% of Brazil's palm oil production (Instituto Brasileiro de Geografía e Estatística 2012).

a "new energy paradigm" in 2001, and by 2004 the government had established policy guidelines for biodiesel blends in transportation fuel, with an emphasis on palm oil-based biodiesels (Castiblanco et al. 2015). The government's agenda to develop oil palm is focused highly on peacemaking in an attempt to rid the countryside of insurgency, coca production, and violence. Colombia's strategy, particularly since 2008, has emphasized sustainable feedstock sourcing, with the goal of producing most of its oil palm on existing pasture and agricultural land as opposed to clearing pristine forests and savannahs (as discussed in Chapter 2). To meet these objectives, the government introduced a variety of subsidies and other incentives at various points along the biodiesel supply chain, such as tax-free zones for biofuel production, cultivation incentives for oil palm (for example, agricultural income, crop insurance, and price stabilization schemes), and public–private financial partnership programs (Castiblanco et al. 2015). The government also established a blending mandate for biodiesel of B10. Colombia plans to expand the biofuel industry at a steady pace in the coming decades, with most (if not all) of the growth coming from biodiesel production through oil palm expansion.

The country's biodiesel industry has developed rapidly since 2008, and today nearly half its palm oil production is used as fuel (Figure 6.3). Some large plantation companies have integrated vertically to first-stage processing as well as biorefineries and other second-stage processing sectors. Colombia's biodiesel refineries are operating at capacity, and several new refineries are under development. The government has pledged to increase its biodiesel mandate to B20 (and later to B30) as additional capacity goes online, but this pledge has yet to be formalized or signed into law. FEDEPALMA (the National Association of Oil Palm Producers) estimates that palm oil production from the country's existing plantations and farms is already sufficient

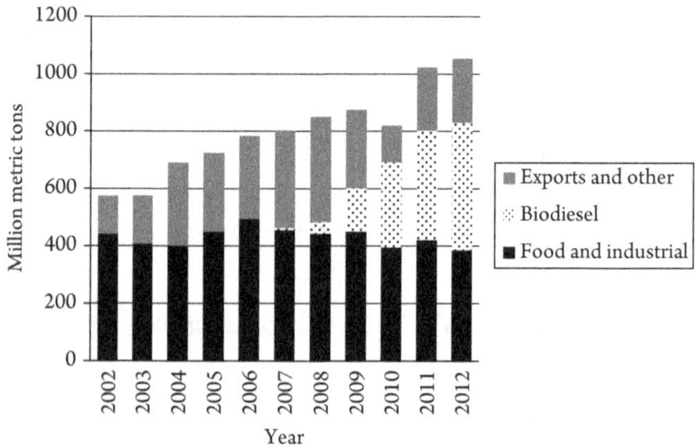

Figure 6.3 Use of palm oil and palm kernel oil, Colombia (measured in million metric tons).
Source: Martinez Pelaez, G. 2013. Panorama de la Agroindustria Palmera—Retos y Oportunidades. Paper presented at the seminar La Agroindustria de la Palma de Aceite: Un Negocio Sostenible e Inclusive. Bogota, Colombia. October 2013. Also available online at web.fedepalma.org/sites/default/files/files/Fedepalma/Panoramaagroindustriapalmeraretosyoportunidades_opt.pdf (accessed February 16, 2016).

to satisfy a B20 mandate, as soon as the recently planted oil palm plantations mature (Pinzon 2012). Increasing the biodiesel mandate to B30 could require an estimated expansion in oil palm area of 480,000 ha, however (Johnson and Franco 2009). As discussed in Chapter 2, FEDEPALMA has a palm oil production target of 3.5 Mt by 2020 and a goal of establishing a biodiesel export industry.

The social benefits and costs of Colombia's palm oil and biodiesel policies remain open to serious debate. Operating under the stated goals of rural economic growth, poverty alleviation, rural stability, and energy diversification, these policies have also served to advance the interests of large landholders and vertically integrated biodiesel producers. Land distribution in Colombia is among the most unequal in the world. Biodiesel entrepreneurs and large land owners have been the main beneficiaries of policy support as the industry has become capitalized and integrated more vertically (Chapter 2). Palm oil production and biodiesel development do not appear to have reduced inequality and poverty in Colombia since the policies were initiated during the early 2000s.

Colombia's biodiesel policies also create large social costs in the form of fiscal expenditures, foregone exports, and reduced foreign exchange balances (Johnson and Franco 2009). For much of the recent history of biofuels, palm oil has sold at a higher price than biodiesel on international markets, making it advantageous for Colombia to sell palm oil on the market rather than to convert it to biodiesel. To offset the disadvantage to its domestic biofuel industry, the government has established a floor price for biodiesel—set at the export price for palm oil—which ensures that returns from domestic biodiesel production remain at least as high as, if not higher than, returns from palm oil sales.

Colombia does not yet trade significantly in biofuels. The government's goal, however, is to become a biodiesel exporter as soon as its production capacity expands and outpaces its mandated demand (Gilbert and Pinzon 2013). The country's potential to export biodiesel, at least in terms of the US market, depends importantly on GHG emissions from its palm oil production, as we discussed in our case study of the United States.[10] Most of Colombia's new palm plantations are expected to be situated in current agricultural areas or on current and degraded pasture land. Less than 13% of new palm plantings will replace natural vegetation (forest, scrubland, and savannah) (Castiblanco et al. 2015). Based on this strategy, the Colombian Ministry of Energy has determined its palm oil-based biodiesel can reduce GHG emissions by 83% relative to fossil diesel, but analysts in the United States challenge this assertion. As noted, the Ministry of Energy is petitioning the US EPA to recognize the sustainability of Colombian biodiesel so it can qualify under the US biofuel mandate (Gilbert and Pinzon 2013), but the United States retains a firm policy against palm-based biodiesel or palm oil imports for biodiesel production.

[10] It is worth remembering that the 2015 amendment to RED limits the share of biofuels derived from edible oils in transportation fuels and thus curtails Colombia's potential exports to the European Union.

POLICY INSTRUMENTS, PRICE EFFECTS, AND PROFITABILITY

The main instruments of biodiesel support, as seen in the case studies, are mandates, blending credits, and tax exemptions. The effect of biodiesel policies on vegetable oil demand and prices depends on which policies are implemented in any given country (de Gorter et al. 2013). Subsidies and tax exemptions allow biodiesel to become competitive with conventional fossil fuels and thus provide a substitute in the market. When petroleum prices rise, fossil fuel consumption falls and biodiesel consumption rises, leading eventually to an increase in the price of biofuels and associated feedstocks. When mandates are in place, on the other hand, biodiesel acts as a complement to fossil fuels because it is used in a targeted fuel blend. In this case, when petroleum prices rise, the consumption of the blended transportation fuel declines and hence the demand for biodiesel and its feedstocks also declines.

By ensuring a targeted share of use within the transportation sector, biofuel mandates create a relatively reliable market for the development of the renewable fuels sector. The use of mandates worldwide transfers the burden of costs from governments to consumers through fuel and food markets, but mandates still come at a high cost to many governments, especially in countries where public investments are required to develop agricultural supply chains or refining and transportation infrastructure. Table 6.3 lists biodiesel mandates and the extent to which various countries have achieved them.

If mandates are sufficiently high (at B10 or higher, for example), they have the potential to distort agricultural prices more than conventional subsidies or tariffs and can thus have significant impacts on food security. This outcome is less likely, however, when the biodiesel feedstock used to satisfy the mandate is high in protein coproducts (meal), such as soybeans in the United States and Brazil. In this case, the increasing demand for soybean oil to meet the biodiesel mandate will lead to excess supplies of soybean meal, unless the demand for animal protein rises at a rate equal to or above that of biodiesel demand. Declining prices caused by an excess supply of soybean meal offset the increase in soybean price that would otherwise arise from a mandated shift in demand for biodiesel.

Mandates can also affect the variability of crop prices. Strictly enforced mandates essentially create an additional and inelastic level of demand for crops used as feedstocks, up to the point where the mandate is binding (Abbott et al. 2009; Abbott 2013). With this new demand, any supply shock (such as a drought) is amplified in the market, causing a larger price hike than would be the case without the mandate. Price spikes are likely to be even greater if agricultural stocks decline because of expanded biofuel mandates. When the mandated blend of biofuel is reached, the marginal demand for biofuels and their associated feedstocks falls toward zero. With a more elastic demand for the fuel, the reverberations in crop prices caused by exogenous supply shocks are reduced.

In the context of biodiesel, an asymmetric relationship exists between petroleum prices and agricultural prices. Petroleum prices have a potentially significant influence

Table 6.3 Biodiesel mandates and actual blending rates in various countries, 2014

Country	*Biodiesel mandate*	*2014 Biodiesel blending*
Brazil	B7	5.8%
Canada	B2; many provinces have B4 mandates	2.3%
Colombia	B10	7.4%
European Union	20% renewables in energy use overall, 10% of energy use in transport sector by 2020	5.3%
Indonesia	B15; to B20 by 2016	6.3%
India	B20 target by 2017; expected to become a binding mandate as production capacity increases	0.1%
Malaysia	B10	5.0%
Thailand	B7; mandate temporarily cut to B3.5 in 2015 because of palm oil shortage	5.6%
United States	RFS mandates 136 billion L total renewables by 2022; 3.8 billion L biodiesel mandated now and in 2022 (36 billion gal and 1 billion gal, respectively); in mid 2015 the US Environmental Protection Agency proposed a revised set of mandates, increasing biodiesel requirements to 7.2 billion L by 2017	2.5%; consumed 5.4 billion L biodiesel compared with a mandate of 3.8 billion L

All mandates other than the United States and European Union are in volumetric terms and are based on a percentage of biodiesel blending. The US mandate is also volumetric but is based on the biodiesel quantity, not rate of blending. The European Union's mandate is unique in that it is based on energy equivalence rather than volume. EU policy states that 10% of energy used in transportation must be renewable; each country within the European Union can achieve this target however it wishes and blending rates may thus vary between countries depending on the energy technology in place.

Source: Information on mandates from US Department of Agriculture. 2015. Published GAIN reports. http://gain.fas.usda.gov/Recent%20GAIN%20Publications/Forms/AllItems.aspx (accessed June 23, 2015); Global Renewable Fuels Alliance. n.d. Global biofuel mandates. http://globalrfa.org/biofuels-map/ (accessed June 23, 2015); Wisner, R. 2013, February. Biofuels mandates outside the US. *AgMRC Renewable Energy and Climate Change Newsletter*. http://www.agmrc.org/renewable_energy/biofuelsbio-refining_general/biofuels-mandates-outside-the-us/ (accessed June 23, 2015); Sapp, M. 2015. Thailand slashes B7 mandate by half due to low palm oil supplies. *Biofuels Digest*, January 21. http://www.biofuelsdigest.com/bdigest/2015/01/21/thailand-slashes-b7-mandate-by-half-due-to-low-palm-oil-supplies/ (accessed June, 23 2015); and US Environmental Protection Agency. 2015. EPA proposes renewable fuel standards for 2014, 2015, and 2016, and the biomass-based diesel volume for 2017. Report no. EPA-420-F-15-028. United States EPA Office of Transportation and Air Quality. http://www.epa.gov/otaq/fuels/renewablefuels/documents/420f15028.pdf (accessed Accessed May 5, 2016). Data updated as of January 7, 2016. Blending statistics from US Department of Agriculture. 2015. Published GAIN reports. http://gain.fas.usda.gov/Recent%20GAIN%20Publications/Forms/AllItems.aspx (accessed June 23, 2015) (for international data) and US Energy Information Administration (n.d. b) (for US data).

over oil crop prices, but the opposite is not true, because the scale of the renewable fuels sector is small relative to the nonrenewable transportation energy sector. In 2013, only 14% of global vegetable oil supplies were used to produce biodiesel (Figure 6.1), and global biodiesel production was equivalent to only 1.5% of total diesel demand (Renewable Energy Policy Network for the 21st Century 2014). Even if all vegetable oils in the world market were converted to biodiesel—an impractical assumption at best—biodiesel would represent only 12% to 13% of global diesel consumption.[11]

In the short to medium term, the extent to which major biodiesel-consuming countries adjust and enforce their mandates—especially for crops grown to produce oil as the primary product, such as oil palm—will determine the strength of the energy–agriculture linkages. Indonesia's implementation of an export levy on palm oil in 2015, along with its support of the palm oil-based biodiesel sector through subsidies and mandates, is one example. Without the enforcement of aggressive mandates, the connection between energy and agricultural markets could dissipate over the longer run—particularly if crude oil prices are low—leaving market dynamics within each sector to determine the respective price (Myers et al. 2014). Nonetheless, irreversible investments in biofuel processing technology by fuel blenders are likely to perpetuate some connection between the two sectors regardless of crude oil prices.

The policy environment for biofuels is important not only because of the linkages it creates between agriculture and energy markets, but also because biodiesel has rarely been profitable to date without direct policy support. The profitability of biodiesel depends primarily on the relative prices of diesel and vegetable oils that serve as biodiesel feedstocks. Production cost estimates for biodiesel vary by location and feedstock, but in virtually all cases, feedstock costs comprise 85% to 90% of total (fixed and variable) costs. Box 6.2 presents an illustrative budget for soy-based biodiesel production in the United States.

Figure 6.4 shows break-even prices for US biodiesel production derived from soybeans, rapeseed, and palm oil in relation to conventional diesel prices.[12] The figure plots monthly vegetable oil prices on the *y*-axis and monthly diesel prices on the *x*-axis in real terms (2005 prices) from 2000 to mid 2015. The break-even lines between biodiesel derived from the three feedstocks and diesel *without subsidies* are shown along the gray dashed line (without fixed costs) and along the black dashed

[11] This estimate is based on the following calculation: Edible oil production in 2013 was 165 Mt (US Department of Agriculture 2015b), which generates 161.7 Mt of biodiesel after the 98% conversion rate (Fukuda, Kondo, and Noda 2001). Using a density conversion for biodiesel of 1136 L/t (Renewable Energy Policy Network for the 21st Century 2014), total biodiesel production was 183.7 billion L, or 163 to 171 billion L of diesel equivalent based on either the lower (89%) energy equivalence value (Radich 2004) or the higher (93%) energy equivalence value (US Energy Information Administration 2007). In 2013, petroleum-based diesel production (derived from distillate fuels, of which ~90% is diesel) was ~8.4 billion barrels, or 1344 billion L (US Energy Information Administration n.d. b; IPAA 2015).

[12] Glycerol, a by-product of the chemical manufacturing process, might add revenue (Radich 2004), but the glycerol market is sensitive to volumes, so the value of this coproduct may disappear at large production scales (Gaskell 2012).

Box 6.2 Model Budget for a Typical Soy Biodiesel Plant in Iowa

This illustrative monthly budget for soy biodiesel production in Iowa (displayed in Table B6.1) is based on a standard soy biodiesel plant built in 2006/2007. The model used to develop the budget includes the following parameters and assumptions:

Table B6.1 Representative production costs for a biodiesel plant

Production costs (¢/L)	
Feedstock costs	56–123
Chemical ingredients	1.51
Repairs and maintenance	0.79
Transportation	2.64
Water	0.18
Electricity	0.79
Other	0.79
Total variable costs	6.70
Depreciation	2.41
Interest	1.71
Labor and management	1.42
Marketing and procurement	1.06
Property taxes, insurance, etc.	0.31
Total fixed costs	6.91
Feedstock costs	56–123
Total variable costs	6.70
Total fixed costs	6.91
Total costs	69.61–136.61

- 113,550,000-L (30,000,000-gal) nameplate capacity; plant operates at 100% capacity
- Construction cost: US$5.94/L of nameplate capacity
- One liter of biodiesel requires:
  - 0.91 kg soybean oil
  - 0.052 m^3 natural gas
- One liter of biodiesel produces 0.11 kg glycerin coproduct
- Fifteen-year estimated life of factory
- Financing: 50% debt,10-year loan, 8.25% interest

Production efficiency, capacity, and feedstock vary substantially across the US biodiesel industry, but soybean oil is the most common feedstock (Table 6.2). With this representative budget, biodiesel is rarely profitable, even on a pretax basis, without the tax credit. Fossil diesel fluctuated between UD.32/L and US.95/L (in 2005 US$) throughout the same time span.

Source: Hofstrand, D. 2014. Tracking biodiesel profitability. Iowa State University Extension and Outreach. https://www.extension.iastate.edu/agdm/energy/html/d1-15.html (accessed May 5, 2016); and Irwin, S. 2015. The profitability of biodiesel production in 2014. *FarmDoc Daily* 5: 16. http://farmdocdaily.illinois.edu/2015/01/profitability-of-biodiesel-production-in-2014.html (accessed May 5, 2016).

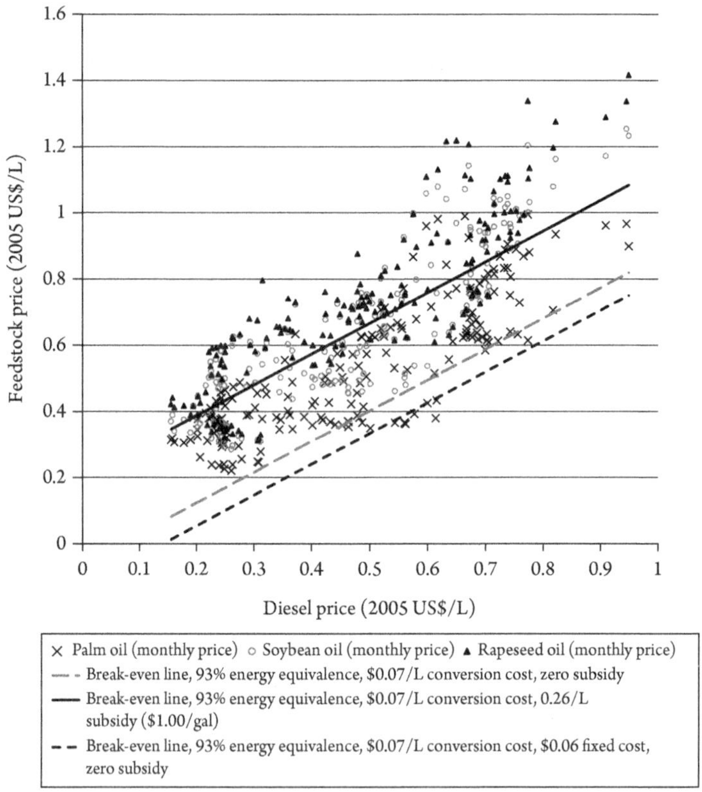

Figure 6.4 Profitability of biodiesel from three vegetable oil feedstocks, with and without subsidies, 2000 to May 2015. This figure shows monthly prices for palm oil (x), rapeseed oil (triangle), and soybean oil (circle) along the *y*-axis, in comparison with diesel prices along the *x*-axis, using real (2005) prices. Three break-even lines are drawn to indicate the profitability of biodiesel based on these feedstocks relative to diesel with and without subsidies. Each line assumes a variable cost for producing biodiesel (apart from feedstock and energy costs) of US$.07/L (US.24/gal [From Haas, M. J., A. J. McAloon, W. C. Yee, and T. A. Foglia. 2006. A process model to estimate biodiesel production costs. *Bioresource Technology* 97: 671–678.]), and a 93% energy equivalence between biodiesel and diesel (U.S. Energy Information Administration, 2007). All dots below the lines indicate that biodiesel from that particular feedstock is profitable. When fixed costs are factored into the analysis, biodiesel is almost never profitable.

Source: Figure adapted from Gaskell (2012). Other sources include World Bank. 2015a. Data. http://data.worldbank.org (accessed May 5, 2015); World Bank. 2015b. GDP deflator (base year varies by country). http://data.worldbank.org/indicator/NY.GDP.DEFL.ZS (accessed June 23, 2015); US Energy Information Administration. n.d. Petroleum and other liquids. http://www.eia.gov/petroleum/data.cfm (accessed June 23, 2015); International Monetary Fund. n.d. IMF primary commodity prices. http://www.imf.org/external/np/res/commod/index.aspx (accessed June 23, 2015); and US Department of Labor. n.d. Bureau of Labor Statistics. http:// http://www.bls.gov (accessed September 13, 2015).

line (with fixed costs). The break-even line *with subsidies* is shown along the black solid line. We see clearly in this figure that vegetable oil prices have rarely been low enough for biodiesel to break even with conventional diesel in the absence of subsidies. Palm oil is the only feedstock that experienced sufficiently low prices in some months during 2000 to 2015 to allow biodiesel to compete with diesel without subsidies. These profitability points are apparent only when we consider variable costs alone; when we add fixed costs to the equation, profits disappear—even with palm oil as a feedstock. On the other hand, when we add a US$0.26/L (US$1.00/gal) subsidy on biodiesel (as in the United States in the past, through the RFS), biodiesel becomes more competitive with diesel, particularly when palm oil is used as a feedstock. With the subsidy, soy-based biodiesel has also been competitive with diesel on numerous occasions during this time period, whereas rapeseed has proved to be competitive less frequently.

It is important to note that biodiesel has 93% of the energy content of fossil diesel, and that the variable costs of converting vegetable oils to biodiesel are around US$0.07/L. Variable costs are more important than fixed costs in short-run profitability assessments because excess capacity still exists in biofuel operations in many countries, such as Malaysia (which uses less than 15% of its existing capacity), Indonesia (~43%), the United States (~60%), and Europe (~42%) (US Department of Agriculture 2015c; US Energy Information Administration n.d. b). Using the full capacity of existing biodiesel plants and adding more production facilities would require either a major expansion in government policies to support biodiesel, or a substantial hike in diesel prices with an increasing demand for transportation fuel, or both. A longer view of energy prices and energy demand, presented in the next section, provides a more realistic assessment of long-run demand in the biodiesel sector.

BIODIESEL OUTLOOK TO 2050

The role of policy in shaping biodiesel markets is clear, but it raises considerable challenges for attempts to predict demand for biodiesel and, in turn, for vegetable oil feedstocks. The task of estimating future vegetable oil demand for fuel becomes easier if we break it down into two parts: What is the projected growth in demand for transportation fuel, particularly for diesel? And how would different scenarios of policy mandates affect future demand for biodiesel given the projected growth in overall demand for diesel for transportation? It would be far more difficult to assess scenarios of biodiesel subsidies and trade policies, which vary widely among countries. As a result, we confine the following discussion to mandates, because they remain a key policy instrument in most countries that produce and consume biodiesel.

Table 6.4 shows projections of biodiesel demand to 2050 with a B10 mandate in place (and enforced) for the countries that accounted for more than 90% of global biodiesel production in 2013. China and India, both large and growing consumers of transportation fuel, are not included in the projections because they are large vegetable oil importers that rely mainly on recycled vegetable oil and other (unknown) feedstocks for biodiesel (Table 6.1). Africa is also excluded because the continent has not yet played a significant role in the biodiesel sector and it faces a looming import deficit in vegetable oils. The Philippines, the Ukraine, and Canada, all net exporters of vegetable oils, are also absent because of their relatively small market share, although they could be potentially significant players in the future. With these omissions, our projections are conservative—but in our view still realistic—in terms of future demand for vegetable oils to produce biodiesel. Another consideration is that few, if any, countries have achieved biodiesel blending at or above B10 to date, and pulling all the major fuel-consuming countries up to that level is likely to be a major challenge.

Underpinning the data in Table 6.4 are estimated annual average growth rates for transportation fuel demand of 2.7% in Southeast Asia, 0.9% in Latin America, –1.2% in the United States, and –1.9% in the European Union (Organization of the Petroleum Exporting Countries 2014). The share of diesel in transportation demand is expected to increase at various rates in all countries; for the world as a whole, diesel's share in transportation fuel is projected to rise from ~36% to 50% (Exxon Mobil 2014; Green Car Congress 2014). Growth in demand for nontransportation diesel is not included in our analysis, which again leads to a conservative estimate of biodiesel and vegetable oil feedstock demand in 2050.

Table 6.4 Projections of growth in demand for biodiesel and vegetable oil feedstock for biodiesel, 2014–2050

Country	*Transport biodiesel demand in 2014*[a] *(billion L)*	*2014 blend rate*[a] *(%)*	*Transport biodiesel demand in 2050 (B10)*[b] *(billion L)*	*Growth in global biodiesel production required to meet B10 in 2050*[c] *(%)*	*Vegetable oil feedstock needed to produce biodiesel in 2050 (B10)*[d] *(Mt)*
Argentina	1.10	8.2	2.34	4.0	2.1
Brazil	3.46	5.8	8.43	16.0	7.6
Colombia	0.61	7.4	0.95	1.1	0.9
Indonesia	1.64	6.3	7.54	19.0	6.8
European Union	10.40	5.3	15.12	15.2	13.6
Malaysia	0.26	5.0	1.61	4.4	1.5
Thailand	1.18	5.6	4.10	9.4	3.7
United States	5.36	2.5	12.97	24.5	11.7

[a] US Department of Agriculture (2015c).

[b] Transport fuel demand growth projected based on values in *World Oil Outlook 2014* (OPEC 2014); diesel market share growth based on *The Outlook for Energy 2014 Report* (Exxon Mobil 2014).

[c] New biodiesel production as a share of 2014 production (31.0 billion L).

[d] Assumes a conversion efficiency of 98% by weight (Fukuda et al. 2001) and 1136 L biodiesel/t (Renewable Energy Policy Network for the 21st Century 2014). World production of vegetable oils was 169 Mt in 2013 (Figure 6.1). Global production of CPO, the most competitive feedstock (Figure 6.4), was 58.4 Mt in 2013 (Figure 6.1).

Two main observations stand out from these projections. First, the amount of biodiesel production required to meet the expected demand in 2050 under a B10 mandate for countries that currently comprise the bulk of the market is 53 billion L, roughly double the volume of biodiesel produced in 2013. Second, the amount of vegetable oil required to supply the biodiesel output in 2050 is estimated at 48 Mt, which accounts for 28% of current global vegetable oil supplies (biodiesel currently uses 14% of those supplies; see Figure 6.1).

A doubling of biodiesel production is significant in terms of future vegetable oil demand, yet it is not overwhelmingly large when spread over the 35-year period. As noted in Chapter 1, palm oil production has doubled *each decade* since 1970. Even if some of the countries omitted from our analysis were included in our projections, the share of global vegetable oils used in biodiesel production would likely top off at about one third or—at the upper limit—one half. These projections are substantially less than Koh's (2007) frequently cited prediction that biodiesel production capacity will need to grow about 100-fold by 2050 to meet demand based on a B20 mandate for the entire world. In our view, it is unreasonable to assume that a B20 mandate for biodiesel would be adopted successfully worldwide by 2050.

CONCLUSION

During the past 10 to 15 years, the biodiesel industry has been a wild card in vegetable oil markets. Before the turn of the 21st century, few analysts predicted that biodiesel would play a major role in boosting global vegetable oil demand and prices. The case studies in this chapter demonstrate wide variation in motivations and strategies for biodiesel development from one country to the next, with differing impacts on specific crops. Looking ahead, we believe the influence of biodiesel growth on vegetable oil demand and price formation remains equally uncertain at the national or global scale. One aspect of this energy–agriculture nexus is fairly certain, however: policies have played, and will continue to play, a dominant role in the profitability of the biodiesel industry. Without national policies mandating the use of biodiesel in fuel mixes, or incentivizing its use via subsidies, the industry might just fade away.

Is it possible that a renewable fuel such as biodiesel could stagnate without government support in an era of rising per-capita incomes, escalating demand for transportation fuel, and growing concern over climate change? We believe the answer is yes, mainly because a number of other competitive transportation fuels are coming online, ranging from compressed and liquid natural gas to electric vehicles. Vehicle efficiencies are also improving at impressive rates and are likely to continue improving in the coming decades. Despite these alternative fuel options, the reality in terms of biodiesel and vegetable oil feedstocks is that governments in most countries are prone to support domestic farm interests, particularly interests that have become entrenched in the policy process. The endogenous response of biofuel policies to low agricultural commodity prices is an important factor bound to keep biodiesel in the transportation fuel mix—at least in countries that have strong interests in particular oil crops, such as Indonesia, Malaysia, and Colombia in the case of oil palm, and the United States, Brazil, and Argentina in the case of soybeans.

How the crop-based biodiesel story plays out in the future depends to a large extent on innovation within the biodiesel sector, markets for protein meal coproducts, and societies' willingness to trade off food for fuel. We have already mentioned the limit announced by the European Union in 2015 on using edible crops in biodiesel, and other countries, including China, have declared that domestic staple crops are off limits for fuel. The future of crop-based biodiesel also depends on its life cycle GHG emissions relative to fossil fuels, which will likely reduce the use of vegetable oils linked to tropical forest clearing, at least by the European Union and the United States. As the biodiesel industry expands, the social concerns surrounding food security, climate change, and biodiversity conservation will undoubtedly grow.

For all of these and other reasons, the biodiesel story is far from over. Specific countries, such as Indonesia and Colombia, still have big plans for biodiesel. The biodiesel industry continues to make technological advances, including the development of fuels based on HVOs. The airline and shipping industries continue to look toward bio-based fuels and fuel processes. These demand factors have high margins of uncertainty, but no higher than the uncertainty surrounding future demand for palm oil as a food in sub-Saharan Africa. Whether the impacts of biodiesel demand will be confined to national markets, or whether biodiesel will increasingly shape international markets and global vegetable oil prices, depends importantly on trade—the focus of our discussion in the next chapter.

7

VEGETABLE OIL TRADE AND MARKETS

MARKET AND TRADE LINKS IN THE GLOBAL VEGETABLE OIL SYSTEM

The preceding six chapters focused on production systems and supply chains for various vegetable oils as well as the food and biodiesel demands for them. In this chapter, we weave these strands into a synthesis discussion of vegetable oil trade, marketing, and prices. Our goal is not to analyze trade and markets with a formal econometric model, but rather to show the broad tapestry of vegetable oil dynamics unfolding across time and space.

On any given day, month, or year, multiple forces combine to determine vegetable oil prices and trade. Land clearing and area expansion in particular countries have been key variables on the supply side. New technologies have altered end uses for commodities. Palm oil is a prime example; once largely a home-produced "red" oil, it is now a ubiquitous, industrially refined oil used in hundreds of food and nonfood products. New technologies have also permitted vegetable oils to become an important transport fuel during the past 50 years, thereby linking vegetable oil and petroleum markets in complicated ways. Similarly, the spectacular growth of soybean as a world crop has added substantially to the amount of protein meal produced, thus tying soybean to the livestock sector. To complicate matters further, government policies—on biofuel mandates, taxes, interest rates, foreign exchange rates, subsidies, and trade—come and go as prices, politics, and public opinion change.

Figure 7.1 presents our simplified version of how the global vegetable oil system works, especially with respect to price formation. We covered some aspects of this system in previous chapters. Here, in attempting to gauge real vegetable oil prices for the future, we are faced with making judgments on a series of additional determinants. Given all the complexities, it seemed wisest to begin with a survey of how several key components of the vegetable oil sector have operated during the past few decades. Later in this chapter, we offer both analyses and judgments on how vegetable oil uses, sources, and prices are likely to play out in 2050.

THE DYNAMICS OF VEGETABLE OIL TRADE

A key feature of the vegetable oil complex is its magnitude in international trade. Table 7.1 shows that 28 countries had trade interests in oilseeds, oils, and meals surpassing US$1 billion in 2011, and seven nations had imports or exports surpassing

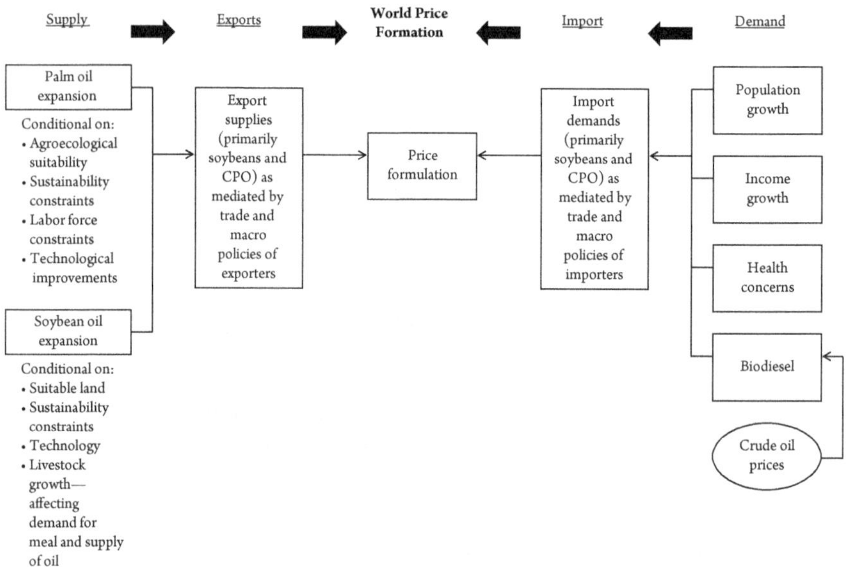

Figure 7.1 Future world prices for vegetable oils, as driven by palm oil and soybean oil outcomes. For simplicity, the arrows showing price formation indicate causality going only in one direction. Clearly, there are feedbacks from prices running in the opposite direction as well. CPO, crude palm oil.

US$15 billion. In value terms, the vegetable oil complex—palm oil plus oilseeds—forms the largest component of all traded agricultural commodities. Although we focus on vegetable oils in this book, the discussion on livestock in Chapter 5 underscores the complexities added by oilseed meals, especially soybean meal. In China, for example, the composition of vegetable oil consumed by humans is determined in part by soybean imports, which are intended largely to feed livestock, but add substantial amounts of soybean oil for food consumption. And the livestock products exported from some countries, notably Brazil, come from animals that have consumed feed produced from domestically grown oil crops. In this sense, livestock products substitute in part for direct exports of soybeans or soybean meal.

In physical terms, the system is growing rapidly. In 1970, total exports of all vegetable oils (shipped as oil) were only about 4 Mt. By 2014, exports had reached about 71 Mt, implying a 9% annual growth rate for the 43-year period (USD-FAS PSD). In fact, the growth in vegetable oil trade turns out to be even larger when soybean data are modified appropriately. A high proportion of "soybean oil" shipped internationally is sent as beans that are crushed in the destination country. In 2014, for example, about 118 Mt of whole beans were exported globally, which embodied about 22.3 Mt of soybean oil and 94.4 Mt of soybean meal (Table 7.2). With this addition, the "augmented" total of soybean oil shipments is about three fourths of all palm oil shipments. With a total of 92 Mt, the ratio of global vegetable oil trade to total oil consumption was about 55%. In the case of grains, in contrast, the ratio of global trade to global grain consumption in 2014 was less than 15% (USDA-FAS PSD).

Table 7.1 Major importers and exporters of oil crops and their products (meal and oil) categorized by value, 2011

Net exports >US$1 billion	*Net exports >US$1 billion but import meal*	*Net exports >US$1 billion but import oil*
Brazil (US$23 billion)	Indonesia (US$18 billion)	United States (US$16 billion)
Argentina (US$23 billion)	Malaysia (US$15 billion)	
Canada		
The Ukraine		
Paraguay		

Net imports >US$1 billion	*Net imports >US$1 billion but export meal*	*Net imports >US$1 billion but export oil*
China (–US$41 billion)	India	Thailand
European Union (–US$27 billion)		
Japan		
Mexico		
Pakistan		
Iran		
Egypt		
Korea		
Turkey		
Vietnam		
Bangladesh		
Taiwan		
Saudi Arabia		
Algeria		
United Arab Emirates		
South Africa		
Venezuela		
Russia		

Includes olive oil. Countries with trade flows exceeding US$10 billion shown in parentheses.

Source: Calculated from FAOSTAT.

Palm oil, palm kernel oil, and "augmented" soybean oil constituted 85% of all vegetable oil exports in 2014 (Table 7.2). Quite remarkably, just five countries—Indonesia, Malaysia, Brazil, the United States, and Argentina—provided 75% of all the vegetable oil exported in 2014 (Figure 7.2). Other countries were involved in smaller ways, but it is clear that vegetable oil exports are dominated currently by palm oil from Southeast Asia and soybean oil from the Americas. The Ukraine is a major supplier of sunflower oil, and the European Union, Australia, and Canada supply much of the rapeseed oil.

Table 7.2 Exports of vegetable oils and of oilseed meal, 2014

Vegetable oil	*Exports (Mt)*
Palm oil	44.3
Soybean oil	9.7
Soybean oil from beans	22.3
Sunflower oil	6.9
Rapeseed oil	4.2
Palm kernel oil	3.1
Coconut oil	1.7
Olive oil	0.8
Groundnut oil	0.2
Cottonseed oil	0.1
Total	92.3
Oilseed meal	*Exports (Mt)*
Soybean meal	62.8
Soybean meal from beans	94.4
Sunflower meal	6.1
Rapeseed meal	5.8
Palm kernel meal	6.5
Copra meal	0.8
Groundnut meal	0.1
Cottonseed meal	0.4
Total	176.9

In addition to soybeans, a combined total of about 5 Mt of copra, cottonseed, groundnut, and sunflower was traded internationally in 2014 as "whole" commodities. These products are used in a variety of ways and their oil content is not included. The table does not include the 14 Mt of uncrushed rapeseed shipped internationally, which would embody about 6 Mt of rapeseed oil and 8 Mt of rapeseed meal.

Source: USDA-FAS PSD.

Vegetable oil imports were also concentrated among countries, although less so than on the export side (Figure 7.3). Six countries/regions—China, India, European Union, the United States, Pakistan, and Egypt—accounted for 64% of oil imports in 2014, with China alone taking about 26%. The number of key importers is relatively small, yet they exhibit significant differences in the role of imports in their respective vegetable oil sectors and in whether those imports are sourced sustainably. China's massive soybean imports (representing 65% of the global soybean

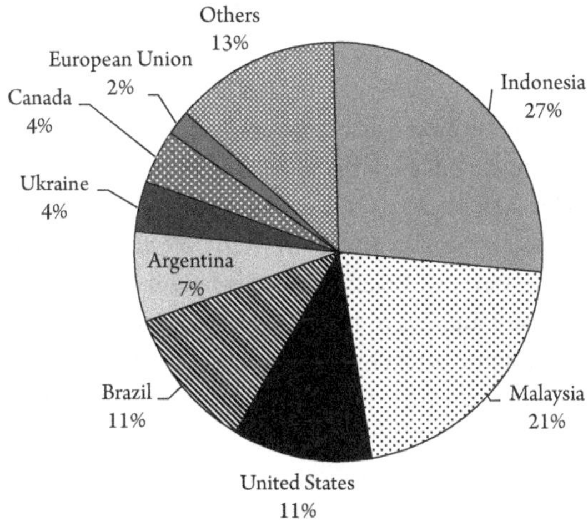

Figure 7.2 Major vegetable oil exporters, 2014. Includes oil traded as whole beans (19% by weight). *Source:* USDA-FAS PSD.

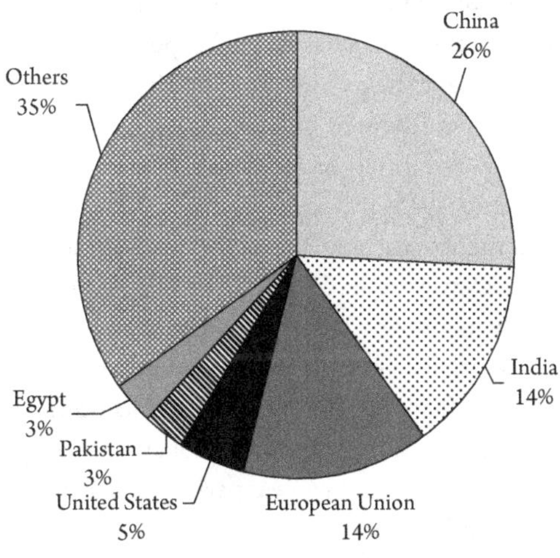

Figure 7.3 Major vegetable oil importers, 2014. Includes oil traded as whole beans (19% by weight). *Source:* USDA-FAS PSD.

trade in 2015) are associated with the need for soybean meal to support the country's rapidly growing livestock sector, as discussed earlier in Chapter 5.

TRADE POLICIES

Exports

International trade in vegetable oils, relative to most other agricultural commodities, is not beset with widespread trade-distorting policies (US Department of Agriculture 2015d). Trade was liberalized substantially during the 1990s, particularly for soybeans, as major exporters such as Brazil removed export barriers and dropped export taxes to become competitive internationally.

Three of the five major exporting countries in Figure 7.2—the United States, Malaysia, and Brazil—have placed relatively few restrictions on either soybean or CPO exports, but domestic policies still play a role in the determination of exports. For example, the tightening of sustainable sourcing rules altered Brazilian export supplies, corn ethanol polices in the United States had the indirect effect of shifting soybean production and exports to Brazil, standards for renewable fuels in the United States also reduced export supplies of soybean oil, and Malaysia's and Indonesia's promotion of biodiesel has come partially at the expense of their palm oil exports.

As of 2015, however, one major soybean exporter, Argentina, was taxing exports heavily, with trade-distorting consequences for the global soybean oil market (Deese and Reeder 2007; Gonzalez 2015). Export taxes were imposed for several reasons: to raise revenue, stabilize internal prices, ensure higher profits for crushers and millers, and compensate for inadequacies in macro policy.

Argentina renewed its use of export taxes shortly after the peso was delinked from the US dollar in 2000. In 2002, the government essentially imposed a flat tax of 10% on all exports. Initially, the export duties were intended to stabilize internal prices (Nogués 2014), but before the 2007 election, when federal expenditures rose sharply, the government turned to export taxes to help cover fiscal deficits—first raising them to 27.5%, and later to 35% for soybeans. The new taxes, coupled with currency controls and a depreciating peso, led soybean farmers to rebel, and they stopped selling soybeans for a period in 2008. As of mid 2015, the soybean export tax continued at 35%, but to provide incentives for domestic crushing, the export taxes on soybean oil and soybean meal were set at 32% (US Department of Agriculture 2012a). There seems little doubt, however, that, historically, tax policy has curtailed soybean production and trade in Argentina. In looking forward, the key uncertainty is whether Argentina will become an even larger soybean supplier than it is now by reforming its trade and macroeconomic policies. In December 2015, Argentina's new president announced plans to reduce soybean export taxes from 35% to 0% over a 6-year period (Dube 2015).

Uncertainty also surrounds Indonesia's long-term intentions with respect to taxing CPO exports. For some years, Indonesia has operated with a variable export levy; in other words, when the world price of CPO exceeds a reference value, export taxes rise. This tax policy has provided important revenue for the government and has helped to stabilize the domestic price of a key food staple. The policy's recent origins lie in the instability of the world price of palm oil during the early 1990s. In 1998, the Indonesian currency (rupiah) suffered massive depreciation against the US dollar as a consequence of the Asian financial crisis. The domestic price signals created by the depreciation suggested that virtually *all* of Indonesia's tradable goods should be exported. The scarcity of vegetable oil became a domestic food crisis, and the export tax became an even more entrenched feature of Indonesian oil palm policy. This brief history provides two important reminders. First, Indonesia became the world's leading exporter of crude palm oil despite the export tax. In large part, this achievement occurred because producer and trader profits were relatively high, and the government did not "overtax" the sector into poverty. Second, financial crises and sharp changes in exchange rates were (and are) powerful determinants of economic development, and of commodity exports.

The likely impacts of a new Indonesian export tax on oil palm are not clear. In May 2015, the government of Indonesia announced the imposition of a new export tax of US$50/t to be applied whenever the global price of CPO fell to less than US$750/t (US Department of Agriculture 2015a; World Bank 2015c). (Under the prior export levy system, export taxes went to zero when world prices were US$750/t or less.) The decrease in world vegetable oil prices to US$600/t in mid 2015 meant that no export tax revenue was coming into the government under the old scheme, and this fiscal factor was significant in the change in policy.[1]

Proceeds from the new export tax could generate up to US$1 billion annually. These funds are to be earmarked in part to improve the productivity of the oil palm sector, but mostly to subsidize the production of biodiesel. In conjunction with the B15 and B20 biodiesel mandates now being discussed for transportation and energy generation, respectively, the impacts of this tax policy could be very significant (Chapter 6). First, the effects of the new tax, given the structure of the world vegetable oil market, will be felt primarily by domestic oil palm producers.[2] Consumers will benefit, but how seriously the tax will threaten overall incentives to producers remains to be seen. Second, it is not clear whether a significant portion of the revenue will reach smallholder producers, directly or indirectly.

Third, the tax/mandate combination could result in a much greater portion of Indonesia's CPO being used domestically for fuel. One of Indonesia's primary fiscal worries currently is the amount of the budget going toward petroleum subsidies, and the country is looking at biofuels as a serious energy alternative. It is also unclear whether the users of CPO for biodiesel will share the same concern for sustainable sourcing that the internationally traded food sector has recently prompted (Chapter 9). The world price implications of lower quantities of palm oil exports from Indonesia also pose an interesting issue that we have not assessed. The optimal export tax literature suggests the reduced export volume might be compensated for in part by higher prices in the short run, but almost surely not in the longer run as other countries are induced to bring additional vegetable oil supplies to the world market (Yilmaz 1999; Rifin 2010). How these tax and trade policies are implemented will be key. They have the potential to be helpful to consumers and to stimulate the domestic biodiesel industry. If done poorly, or if the tax is levied at increasingly high rates in the future, this same set of policies has the potential to thwart incentives for oil palm production, to promote an inefficient and environmentally damaging form of renewable energy, to decrease CPO exports, and to encourage smuggling.

[1] If global palm oil prices return to a level greater than US$750/t of CPO, the US$50 tax would apply, and the prior export levy would also go into effect (*Indonesia-Investments* 2015).

[2] If global palm oil prices are significantly less than soybean oil prices, parts of the Indonesian tax might be passed on to importers of Indonesian palm oil, thereby decreasing the global soy–palm price margin.

New institutional arrangements constitute a final source of uncertainty for palm oil exports. In November 2015, Malaysia and Indonesia announced formation of an Organization of Petroleum Exporting Countries–like Oil Palm Council (*Times of India* 2015). How forcefully this council will be promoted is unclear. Although the two countries supply about 85% of global palm oil exports, they supply less than half of total vegetable oil from all sources. Given the close substitution among vegetable oils, the limited storage life of CPO, and the nature of oil palm production (a 30-year tree life with more or less continuous production of fruit bunches after the fourth year), we are very doubtful the council will be an effective mechanism either for stabilizing or supporting world CPO prices.

Imports

Taxation of vegetable oil (and whole soybean) imports throughout the world has been quite modest. In part, these low duties are the result of WTO negotiations and in part they represent the growing recognition by many countries that low import duties are advantageous for key consumer products.

Table 7.3 illustrates these points for three large importers. For both soybean oil and CPO, most-favored nation duty rates are quite moderate. China's rates, for example, are 3% for soybeans, 9% for soybean oil, and 5% soybean meal—a rate structure designed to encourage domestic crushing. In India, the 12.5% rate for palm oil and soybean oil is a recent phenomenon (2015) stemming from government efforts to collect additional revenue and give domestic oilseed producers more protection at a time of relatively low world prices for vegetable oils (Bhardwaj 2015). Overall, however, import duties, unlike export taxation, have not been a major destabilizer of future world trade in vegetable oils; however, this situation may be changing. Nigeria, in response to a foreign exchange crisis in 2015, banned imports of 680 items, including palm oil (Payne 2015). And in early 2016, France threatened to impose a substantial tax on palm oil (and its derivatives), ostensibly to protect biodiversity, but we suspect also to protect domestic oilseed producers (*Business News* 2016).

Table 7.3 Import duties, selected countries, 2015 (%)

Country	*Soybean tariff*	*Soybean oil tariff*	*CPO tariff*
China[a]	3.0	9.0	6.0
India	15.0	12.5	12.5
European Union	9.6	6.5	1.7

[a] Five percent tariff on soybean meal. CPO, crude palm oil.

Source: Duty Calculator. 2015. Duty calculator. Pitney Bowes. http://www.dutycalculator.com (accessed May 2, 2016 ,) and Rodenburg, J. 2012. Iowa Soybean Association Asian trade mission—recap. http://www.iasoybeans.com/chinablog2012/ (accessed Mar 2, 2016).

THE DYNAMICS OF VEGETABLE OIL PRICES

Fundamentally, this chapter is forward-looking, focusing on vegetable oil prices, their determinants, and their likely future movements. But before looking ahead, it seems prudent to look backward, and to learn the price lessons of vegetable oils during the past four decades.

Historical Lessons

Multiple narratives can be written about the movement of real vegetable oil prices during the past 50 years. Conclusions about both the direction and magnitude of prices turn out to depend greatly on the time period of analysis. Figure 7.4, using soybean oil and palm oil as representative price indicators, suggests a general U-shaped real-price pattern for the entire period between 1970 and 2015. The graph also indicates substantial amounts of price variation. Since 1990, however, real vegetable oil prices have increased, also with substantial variation. Put in slightly different terms, there have been periods when increased supplies of vegetable oils, particularly the surge in soybean production in Latin America and the palm oil output in Southeast Asia, exceeded demand, and other periods when growth in (primarily) Asian food demand and expanded biodiesel use exceeded the growth in vegetable oil supplies. Finally, Figure 7.4 suggests that a simple linear projection of past vegetable oil prices is unlikely to provide insight into or be relevant to the future.

A second lesson from history is that vegetable oil prices among commodities are linked very tightly via substitution, and actions and reactions in any of them are soon reflected in the markets of others (Chapter 4, Figure 4.8). At the same time, the price movements are not identical, indicating that substitutions are neither perfect nor instantaneous. These points are reflected in Table 7.4, which shows high correlations among prices of the four major vegetable oils during the 30 years from 1985 to 2015 (after linear time trends have been removed from all the series). At a monthly timescale, there is no statistical evidence suggesting either palm oil or soybean oil was *the* leader in setting prices.[3]

The power that general economic conditions exert over all price series, not just vegetable oils, provides a third lesson on prices. Figure 7.5 shows the movement of price indices for oils and meals, grains, and energy from 1970 to 2015. Quite strong correlations appear among the indices, and the price movements in the three series are especially striking since 2000. As discussed in Chapter 6, government policy on biofuels contributed to the new linking of energy and vegetable oils.[4] Figure 7.5 is also

[3] Monthly price data for palm oil and soy oil were analyzed for January 1970 to December 2014. Tests were run on stationarity, co-integration, and Granger "causality" (Granger 1969). There was no statistical evidence that suggested palm oil prices "led" soy oil price or vice versa. It is possible that weekly, daily, or hourly price data might produce a different conclusion.

[4] Sanders et al. (2014) show, using vector autoregression techniques on data from 1980 to 2010, changes in energy prices did not "cause" changes in palm oil prices.

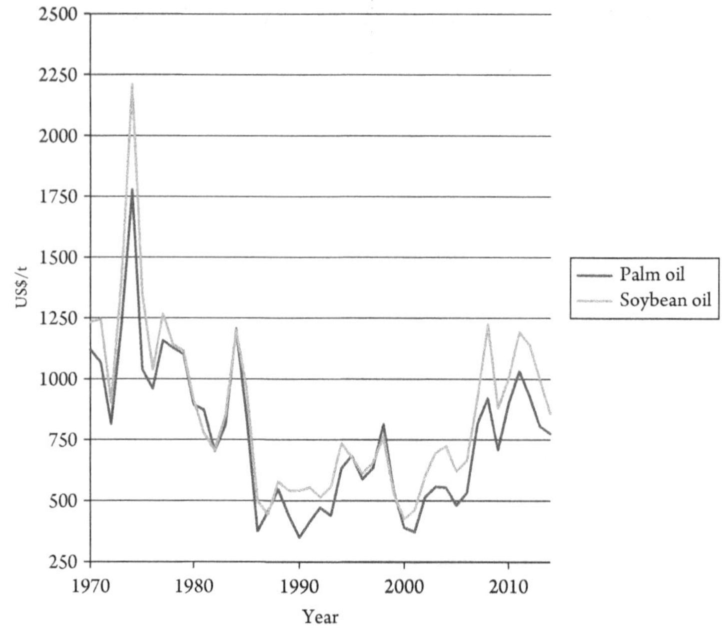

Figure 7.4 Real prices of soybean oil and palm oil 1970 to 2015 (2010 US$).
Source: World Bank commodity price data (pink sheet), July 2014. http://knoema.com/WBCPD2014Jul/world-bank-commodity-price-data-pink-sheet-july-2014 (accessed May 5, 2015).

Table 7.4 Correlations among monthly prices for soybean, palm, rapeseed, and sunflower oils, 1985 to 2015, nominal prices with linear time trends removed

Oil	*Soybean oil*	*Palm oil*	*Rapeseed oil*	*Sunflower oil*
Soybean oil	1.00			
Palm oil	0.85	1.00		
Rapeseed oil	0.89	0.79	1.00	
Sunflower oil	0.88	0.67	0.79	1.00

a sobering reminder that it is possible to know everything about vegetable oil prices but still to miss the macro and policy decisions outside the vegetable oil sector that drive many of the fundamental price relationships more broadly across the economy.

Price Discovery

The preceding figures indicate clearly that vegetable oil prices are linked closely across commodities and time. The market forces that drive these linkages are powerful and almost mysterious. Thousands of public and private agencies are involved in collecting

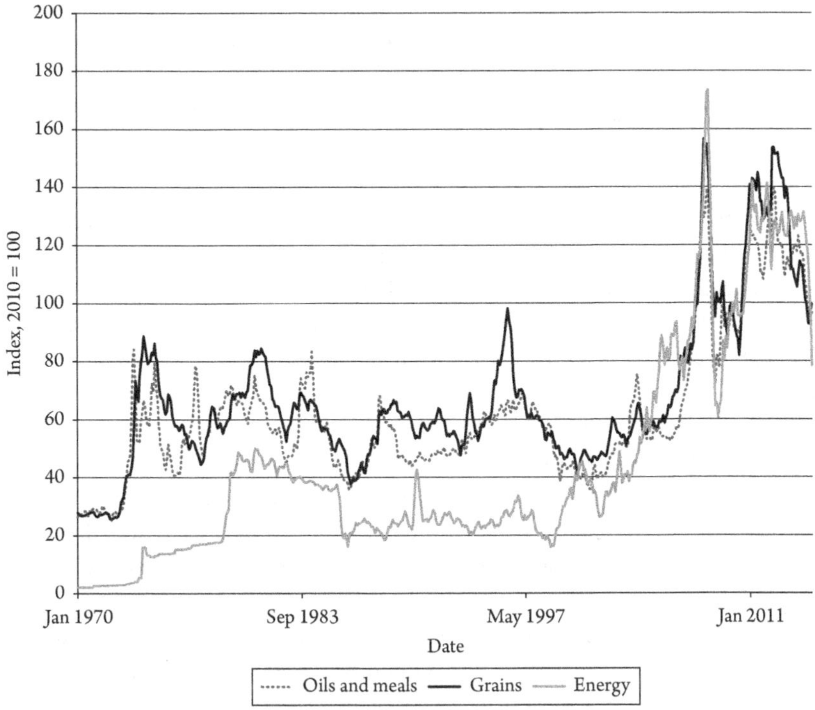

Figure 7.5 Price indices for oils and meats, grains, and energy, 1970 to 2015.
Source: World Bank commodity price data (pink sheet), July 2014. http://knoema.com/WBCPD2014Jul/world-bank-commodity-price-data-pink-sheet-july-2014 (accessed May 5, 2015).

price data, and many organizations also engage in producing forecasts. Nevertheless, it is organized futures exchanges that provide the glue holding all the markets together. It is at these exchanges where most price "discovery" takes place for key commodities, including vegetable oils.

The vast body of literature on futures markets is well beyond the scope of this volume. Futures contracts are sufficiently important for pricing and trade issues, however, that a few comments about them are essential to set the context for the discussions that follow. We again focus on soybean/soybean oil and palm oil to illustrate the importance and operations of these markets as well as the differences among commodities.[5]

Soybean futures are traded in many futures markets around the world, but there is one primary center: the Chicago Mercantile Exchange Group (CME) in Chicago.[6] The CME (in prior years, the Chicago Board of Trade) has multiple contracts for

[5] Chicago Mercantile Exchange (2006) provides a good general introduction to futures markets. For a readable version of how the global soybean export market operates, see HighQuest Partners and Soyatech, LLC (2011). For a good review of the details surrounding trade in oil palm products, see International Trade Centre (2012).

[6] The dominance of the CME exchange means a large proportion of international sales contracts (in physical terms) are priced using quotations from Chicago.

soybeans, soybean meal, and soybean oil, for which price quotations are found easily in financial newspapers or on multiple websites. Buyers and sellers come together, at least electronically, to form prices for various future time periods.

Physical trading does not take place in Chicago; rather, what is being bought and sold are contracts—for example, for 5000 bushels (~136 Mt) of No. 2 yellow soybeans—that call for (potential) delivery in one of seven specific months during the year. In principle, delivery could take place between the seller and buyer of the contract; in fact, most contracts are not completed by shipping soybeans from seller to buyer. The contracts, instead, are closed by buying or selling an offsetting futures contract.

Well-functioning exchanges ensure low-cost contract compliance. To work efficiently, however, futures markets need the liquidity provided by large numbers of buyers and sellers. This need is one reason why there is typically one predominant exchange for each commodity contract; the lack of liquidity is what causes contracts for many commodities to fail. On any given day (or minute) successful markets assimilate what is known about future market conditions. This knowledge does not mean the price quotation for a future contract will eventuate; rather, it shows what traders know about future market conditions at an instant in time. Conceptually, this is what price discovery means and how futures markets provide this service.

Most trading is done by firms and specialists, not by individual producers or consumers. Although futures markets play a dominant role in vegetable oil marketing, they do not solve all problems or cover all special situations. For instance, futures markets are less adept at handling the inclusion of more specialized characteristics (such as products certified to sustainability standards), so direct contracting of the commodity between producers and consumers frequently results. Exchanges sometimes monitor this "over-the-counter trading," but it is called more properly *forward contracting* rather than *futures trading*.

Soybean trading provides a very useful example. The CME offers futures contracts in soybeans, soybean oil, and soybean meal. The markets thus coordinate across commodities as well as time in the "soybean complex"—best thought of as a three-legged stool where each leg can influence the price of the other two. If, for example, a large weather shock occurs, future soybean supplies may be down, and prices of beans, oil, and meal are all likely to rise. However, what happens to absolute and relative prices among the three contracts if bean supplies are normal, but China's demand for soybean meal, for example, suddenly increases? Meal prices are likely to go up, which might drive bean prices up, but may also drive the price of soybean oil downward relative to the price of soybean meal. As explained in Chapter 6, these dynamics also explain why the development of biodiesel has been of such complementary importance to the soybean industry.

Soybean processors (crushers) are also affected directly by what is going on in this complex. Their crushing margins for a ton of soybeans are bounded by the price of the beans less the soybean oil yield (~19%) multiplied by the oil price, less the soybean meal yield (~79%) multiplied by the meal price. Those margins cannot be negative over the longer run, although in the short run they can certainly be less than zero. They vary considerably by time period and by country, depending largely on tax structures.

Using futures data for the same date for beans, meal, and oil, US soybean crushing margins from 2010 to 2015 ranged from US$1.43/t (April 2011) to US$232.84/t (September 2014). In Argentina, the range was from –US$35.02/t (April 2015) to US$90.97/t (April 2011).[7]

Further complexity is added to soybean oil trading by what is going on in the palm oil (and other oil) markets. Palm oil has a much less mature futures market, so it is instructive to see the difference in how it operates. One major difference is that direct physical contracts between buyers and sellers are much more common in trading CPO (International Trade Centre 2012). In the early days, virtually all contracts were in physical terms, although some hedging of palm oil occurred via futures contracts in soybean oil. It was only in 1980 that the Kuala Lumpur Exchange (now Bursa Malaysia) began trading palm oil futures in 1980 (Corley and Tinker 2003).

In an important development in 2009, Bursa Malaysia entered a new strategic alliance with the CME Group in Chicago. CME customers are now able to trade US dollar-denominated palm oil contracts, as opposed to the Malaysian ringgit-denominated contract in Kuala Lumpur. This new alliance provides convenience and adds liquidity to the palm oil market. It eliminates the need to worry about the dollar/ringgit exchange rate and creates easier opportunities for cross-trading in other vegetable oils.[8] Palm oil contracts will no doubt encounter some difficulties along the way forward—for example, in distinguishing certifiably produced palm oil from non-certified oil. However, we believe the use of palm oil futures contracts, in particular, will become an increasingly important instrument for managing palm oil price risks as margins narrow and as the industry expands regionally.

The Role of Stocks

Futures markets are central to the price discovery process. For certain storable commodities, such as grains, oilseeds, and vegetable oils, futures markets for a particular commodity are themselves linked through time via storage (Houthakker 1959). Suppose, for example, that all of a product was produced/harvested during just 1 month of the year. If consumption is to be equalized throughout the year, storage becomes vital, yet storage incurs carrying charges—costs of warehousing, possible physical damage, and the opportunity cost of the capital tied up in the product. Because of these costs, even completely competitive markets exhibit seasonal price movements. These movements are a strong determinant of the varying prices of futures contracts within a marketing year.

Storage also links prices between marketing years. Typically, a key variable in estimating next year's price for a storable product is the amount of stocks available at the

[7] Margin data are all in metric tons and obtained from personal communication with Paula Savanti, Rabobank, on August 4, 2015.

[8] For further details about futures contracts, as well as a side-by-side comparison of soybean oil and palm oil futures contracts, see Chicago Mercantile Exchange Group (2010).

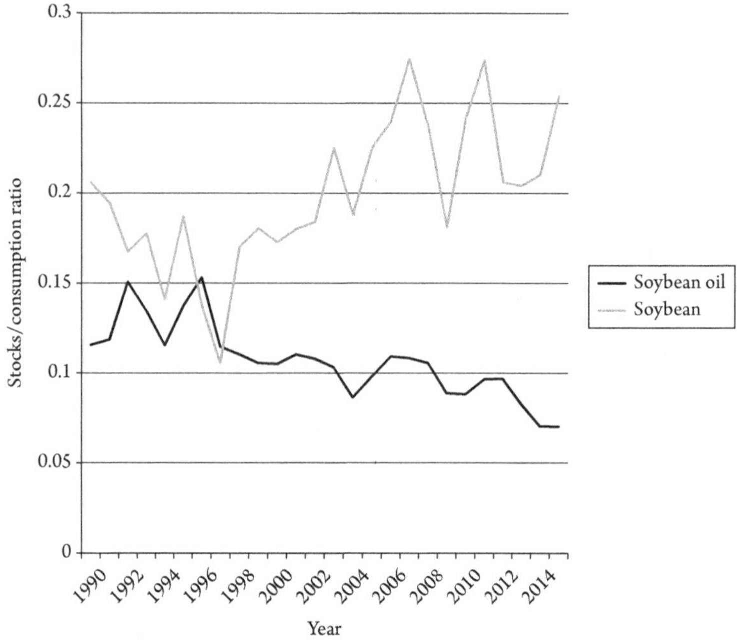

Figure 7.6 Global stock-to-consumption ratios for soybeans and soybean oil.
Source: Authors' calculations from USDA-FAS PSD.

end of the current year. In particular, ratios of stock to consumption (so-called *stock/use ratios*) are often used in price predictions.[9]

Figures 7.6 and 7.7 show two patterns of stock/use ratios from 1990 to 2014. Several things about these patterns are worth noting. For soybean products, it is clear stocks are held increasingly in the form of beans rather than soybean oil, both for ease of storage and for greater product flexibility.

In an important econometric contribution from 1960 to 2012, Baffes and Dennis (2013) show that the elasticity of the stock/use ratio to prices in the succeeding year is –0.2; in other words, a 10% increase in the end-of-year soybean stock/use ratio gives rise to a –2% fall in soybean prices the following year. This relationship is significantly smaller in magnitude (and of less good fit statistically) than the elasticities of –0.6 and –0.5 they found for maize and wheat, respectively.[10] This difference suggests the soybean market is less sensitive to stock levels, especially low stock levels, than markets

[9] Vegetable oils can be stored for months, provided temperatures are neither very hot nor very cold, and the oils are stirred in storage to prevent sedimentation.

[10] Irwin and Good (2015) indicate the Chicago soybean prices are correlated tightly with stock/use ratios for stocks of beans held in the United States. They note, however, that global stock/use ratios do not correlate well with Chicago prices. Part of the difficulty arises from adding ending stocks for the northern hemisphere (United States and China) with stocks from the southern hemisphere (Brazil and Argentina), which have different marketing years. They also note that more than 90% of

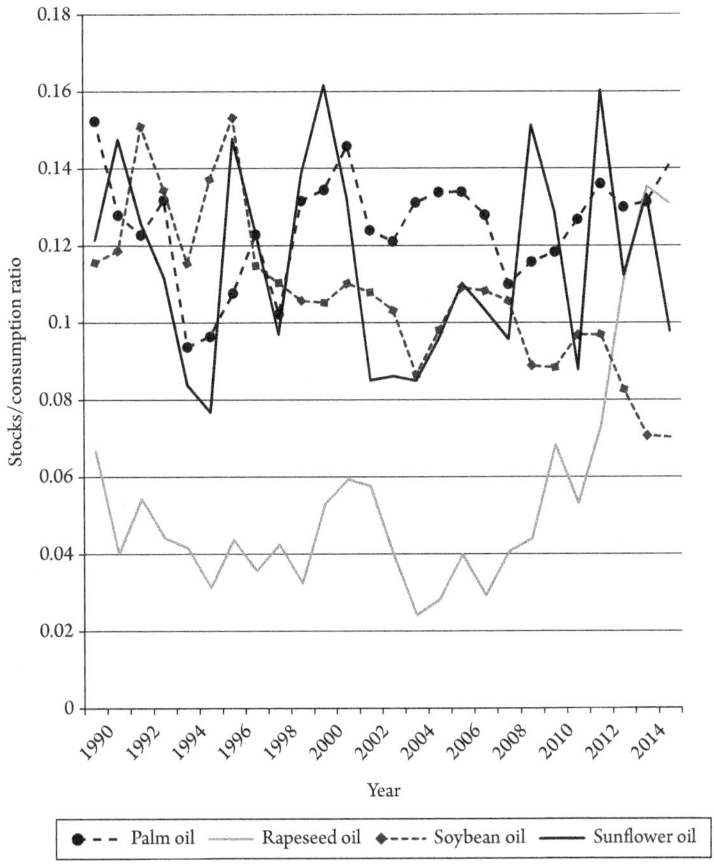

Figure 7.7 Global stock-to-use ratios for palm oil, rapeseed oil, soybean oil, and sunflower oil, 1990 to 2015.
Source: Authors' calculations from USDA-FAS PSD.

for grains. We believe lower sensitivity for soybeans has to do with the greater number of substitutes for both meal and oil.

Figure 7.7 shows stock/use variation of the four leading vegetable oils through time. Sunflower ratios exhibit the greatest variability, reflecting temperature and moisture shocks in the Ukraine—the major sunflower oil exporter. Recent political turmoil in that country has no doubt also affected sunflower production and marketing. Palm oil ratios show less sharp variation, reflecting the more continuous harvesting of oil palm bunches throughout the year. As a tree crop, oil palm is less prone to annual production fluctuations; however, Southeast Asia is still buffeted by El Niño events that tend to disrupt oil palm production about every 5 years (Ling 2014). Rapeseed

ending stocks in recent years, as defined by USDA-FAS PSD, are held outside the United States—mainly in Brazil, Argentina, and China—and that their stock positions may be related more to trade and other strategic variables than to Chicago prices.

ratios, initially very low, have shown sharp increases during the past 5 years, mostly the result of very good harvests and the leveling off of biodiesel production in the European Union.

More generally, vegetable oil production has surged since 2005, whereas vegetable oil demand has lagged as a result of slower economic growth rates in several large countries, and as a consequence of the near completion in Indonesia of the substitution of palm oil for coconut oil (Chapter 4). In 2004, for example, global year-end stocks of all vegetable oils totaled 9.4 Mt, whereas in 2014 the total was 19.4 Mt. During that decade, supplies grew faster than demand, and stocks increased.

THE LONG-RUN DRIVERS OF VEGETABLE OIL PRICES

Futures markets help to portray and explain price movements in the shorter run. Introducing stocks into the discussion is helpful in understanding price movements in the short to medium run. But, looking toward vegetable oil prices in 2050 requires us to consider longer run, more fundamental variables. The central question, of course, is quite simple: Will the growth in demand for tropical oils exceed the growth in their supply? If it does, prices will rise; if it does not, vegetable oil prices will fall. And if the discrepancy is significant or ill timed in either direction, vegetable oil prices will themselves provide an important equilibrating mechanism, curbing demand and inducing supply with rising prices, and expanding demand and curtailing supply with falling prices.

The interdependence between prices and quantities outlined here—the so-called *identification problem*—does not permit completely independent analyses of supply and demand. On the other hand, our vision of what is likely to happen does permit some separation. We believe demand for tropical vegetable oils, driven mainly by population growth, income growth, and policy mandates, will "lead" the pricing process. Supply will likely "follow," mainly via changes in the areas harvested. The expansion of area for oil palm, a tree crop, and soybean and other annual oilseeds will affect different locations, but both will be important. Price spikes and price dips will undoubtedly occur along the way, but we foresee no major changes in real prices during the decades ahead. [11]

The foregoing view can certainly be contested, but it is informed by previous chapters in this volume and by the discussion that follows. Drawing on Chapter 4, we summarize three demand variables: population growth, income growth, and the income elasticities associated with vegetable oils used for food. Based on Chapter 6, we tabulate the growing use of vegetable oils in manufacturing biodiesel, as conditioned by petroleum prices and by renewable fuel mandates and prices. On the supply

[11] IFPRI's IMPACT model, in contrast, shows a 70% increase in real vegetable oil prices by 2050 under business as usual assumptions (Rosegrant, M. W., personal communication to W. P. Falcon, Setember 26, 2014).

side, we draw from Chapters 2 and 3 to examine prospective yield growth as well as likely area expansion, focusing mainly on oil palm and soybeans. Chapter 5 provides us an estimate of likely soybean oil supply given livestock demand for meal. Given the commodity substitutions that prevail in the vegetable oil market, we write initially as if there was one vegetable oil price in 2050 before taking up the outlook for individual vegetable oils.

GROWTH IN DEMAND FOR VEGETABLE OILS

Virtually everyone agrees on the importance of future population, income levels, and government policies as primary determinants of vegetable oil demand in 2050. Little consensus exists on which particular numbers are most logical, however.

Our view toward 2050 builds on a 2013 base. As shown in Table 7.5, the 165 Mt of vegetable oil used in 2013 was split among food (125 Mt), biodiesel (23 Mt), and "other uses," such as lubricants and cosmetics (17 Mt).[12] Our view of 2050 is also shown in Table 7.5. The column for 2050 reflects a population of 10 billion, a per-capita growth of real GDP per year of 2.5%, an income elasticity for food of 0.25, and "other" increasing at the rate of per-capita GDP growth. We also estimate that 48 Mt of vegetable oils will be devoted to biodiesel, using the eight-country B10 scenario of Chapter 6. In total, these demands show a global use of about 307 Mt. This sum implies a growth in demand for vegetable oil for all uses of about 1.7% per year for 2013 to 2050.

Table 7.6 shows the effects of altering individual assumptions. The various rows indicate the *additional* effects on total use in 2050 for each of the changed parameters.

If the "high" assumptions of Table 7.6 were used in *all* four instances (and ignoring interaction terms), the combined increment of +65 Mt would raise total use in 2050 to 372 Mt. This level of annual use would imply an annual growth of about 2.2% between 2013 and 2050.

The task now at hand is to determine whether this set of projected demands is consistent with our projections on the supply side, and then to examine what supply and demand together imply for the real prices of vegetable oil in 2050.

GROWTH IN SUPPLIES OF MAJOR OILSEEDS AND PALM OIL

Considerable talk in the literature, some of it science based and some of it not, concerns the need to double agricultural output during the 21st century. Despite the talk, the number of systematic attempts to project agricultural output—"even" to 2050—is quite small, especially with respect to vegetable oils. Part of the difficulty with vegetable oil projections derives from the need to combine annual oilseed improvements

12 We have spent little time analyzing industrial, lubricant, soap, and cosmetic uses, and assume their growth would equal the growth of per-capita GDP.

Table 7.5 Projected vegetable oil uses, 2013 versus 2050 (estimated)

Use	*2000*	*2013*	*Annual growth, 2000–2013*	*2050 (estimated)*	*Annual growth, 2013–2050*
Food (Mt)	78	125	3.6%	217	1.5%
Food per capita (kg)	12.8	17.1	2.3%	21.7	0.6%
Biofuel (Mt)	N/A	23	N/A	48	2.0%
Other (Mt)	10[a]	17	10.7%[b]	42	2.5%
Total	88	165	4.8%	307	1.7%

[a] Includes a small amount of biofuels.

[b] Growth rate for biofuels and other combined.

N/A, not available.

Source: Table 5.1 and authors' calculations from USDA FAS-PSD.

Table 7.6 Sensitivity analysis, 2050 vegetable oil uses

Changed assumption	*Effect on total use (Mt)*
Raising population estimate from 10 to 11 billion	+21
Raising gross domestic product per-capita growth rate from 2.5% to 3.0% annually	+16
Increasing the income elasticity of demand for vegetable oil for food from 0.25 to 0.35	+15
Raising the eight-country biofuel blending coefficient from B10 to B12.5	+13

with a 30-year life cycle for a tree crop that produces virtually nothing until 4 years after planting.

With these complications in mind, it seems prudent to begin with history—specifically, with two questions. First, how does a future demand projection of 1.7% per year compare in magnitude with recent growth on the supply side? Second, how does recent growth disaggregate into yield-increasing and land-expanding components?

Table 7.7 provides data on growth in the production of palm oil and the three leading oilseed crops from 2000 to 2013. Except for sunflowers, area expansion has been the dominant component of production growth and overwhelmingly so for oil palm (Dallinger 2011; Malaysian Palm Oil Board 2011; US Department of Agriculture 2011, 2013).

Two points about Table 7.7 are of particular interest and reinforce findings from the first three chapters. First, unlike the green revolution for wheat and rice, which was based overwhelmingly on yield increases, the vegetable oil revolution has occurred mainly via area expansion. Second, as seen from a recent historical perspective, the growth of oil palm and major oilseeds has been impressive in its magnitude. The

Table 7.7 Growth of palm oil, soybean, sunflower, and rapeseed production, 2000 to 2013

Commodity	*Production (%/yr)*	*From area (%/yr)*	*From yield (%/yr)*
Palm oil	7.3	6.3	1.0
Soybeans	3.4	2.7	0.6
Sunflower	4.6	1.6	3.0
Rapeseed	5.6	3.8	1.8

Growth is calculated as $\log_e y = a + b$ time. All trends for production and area are statistically significant, as are all yields except for soybean yields.

Source: Authors' calculations from USDA-FAS PSD and FAOSTAT for oil palm.

historical supply numbers seem to indicate that meeting future growth in demand of 1.7% or even 2.2% per year represents a sharp slowdown compared with this recent history.

Oilseeds

Since 2000, growth in soybean yields in the United States, Brazil, Argentina, and China—the four largest producers—has been modest at under 1% annually. In looking forward to 2050, none of the frequently cited projections show much optimism or suggest major yield breakthroughs. Yield gaps in the three major soybean producers are already quite small, as we saw in Chapter 3. Alexandratos and Bruinsma (2012) imply that soybean yields will increase by 0.7% per year from 2005/2007 to 2050, Masuda and Goldsmith (2009) indicate an increase of 0.5% per year from 2020 to 2030, the Organisation for Economic Co-operation and Development (2012a) suggests yield improvements of 1.0% per year for all oilseeds until 2024, and Sands et al. (2014) suggest an increase of 0.9% per year for all oilseeds until 2050. None of these projections hints at major yield breakthroughs or the emergence of revolutionary technologies. Based on the projections in this chapter, as well as the extensive literature reviewed in Chapter 3, we believe there is potential for yield increases on the order of 0.6% to 0.8% per year for the major oilseeds.

We have little doubt that additional land will be planted to oilseeds in Latin America and other regions, but nothing on the horizon begins to compare with what has happened during the past 35 years—the tripling of Brazil's soybean area between 1990 and 2015, for instance. Although significant areas in sub-Saharan Africa resemble Brazil's Cerrado, Africa's poor soils, undeveloped land markets, and weak governance militate against their rapid development. Based on the review in Chapter 3, a potential global area expansion of 1% to 1.5% annually for soybean and other oilseeds with minimal environmental costs seems feasible, especially through conversion of pasture land in South America. Together with the estimated yield growth of 0.6% to 0.8% annually, this should meet demand growth, even our higher growth estimate.

Oil Palm

As noted, the outlook for oil palm is complicated by the 30-year yield profile for a given tree, and by the 10-year lag from planting to full production. Producers are likely to replace existing oil palms only once during the next 35 years. The large "stock" of trees at any given time relative to plantings means there is great stickiness in year-to-year changes in average yields.

Even establishing a global base yield for palm oil is not easy. Because the sector is growing so rapidly in Southeast Asia, the large number of young trees greatly influences the measurement of current "average" yields. And West and Central Africa have millions of hectares of semiwild oil palm groves; their yields are "guessed" to be less than 1 t/acre (US Department of Agriculture 2014).

Perhaps the best yield data come from Malaysia, which is also where some of the best research on oil palm technology is now taking place. Data from the Malaysian Palm Oil Board (MPOB) indicate yields from mature trees currently average 4 to 5 t/ha of CPO, but yield growth since 2000 has been slow—around 0.7% annually (Chapter 2).[13] Virtually all commercial oil palms now in the ground will be replanted before 2050—assuming oil palm remains a vibrant economic activity. This 37-year replanting with higher yielding varieties and clones would result in the equivalent of about 1% per year of yield growth from genetic gains in yields (Chapter 2). A concerted effort to raise smallholders' yields in Asia and Africa through improved management of existing planted area or through incentive systems to replant with high-yielding genetic stocks would also enable higher overall growth in yield.

We have saved the most crucial and most difficult piece of the projection puzzle until last. The central question is whether the phenomenal growth in area that has characterized oil palm during 1990 to 2013 will persist for the next 35 years. More precisely, if yields grow at ~1% per year, will oil palm area grow by 1% to 1.5% per year at the real price level that existed in 2013 to 2015? Can the land, labor, and capital be mobilized within an environmentally sensitive framework to ensure the needed investments (Chapter 9)? For if area expands less than 1% annually, real prices will probably increase, and the higher vegetable oil prices would feed back on investments. With a 5-year maturity lag, a different price equilibrium will be reached, but a price spike will probably occur along the way. In sum, area response is where the price–quantity simultaneity problem is most severe in our analysis. Land is also particularly exposed to the vagaries of public policy, which dominates the granting of concessions and rights as well as patterns of land use (Chapter 9).

With these considerations in mind, what is likely to happen to oil palm area in Southeast Asia, Latin America, and West and Central Africa? Malaysia is putting increased constraints on land being devoted to oil palm. Indonesia is rethinking its strategies on sustainable production, making it highly unlikely that Indonesia will continue

[13] Yields per hectare can be increased both by increasing the weight of fruit bunches and by increasing the efficiency with which fruit bunches are milled into CPO. The gain in milling efficiency, in turn, involves timelier processing of fruit, and better engineered and operated mills.

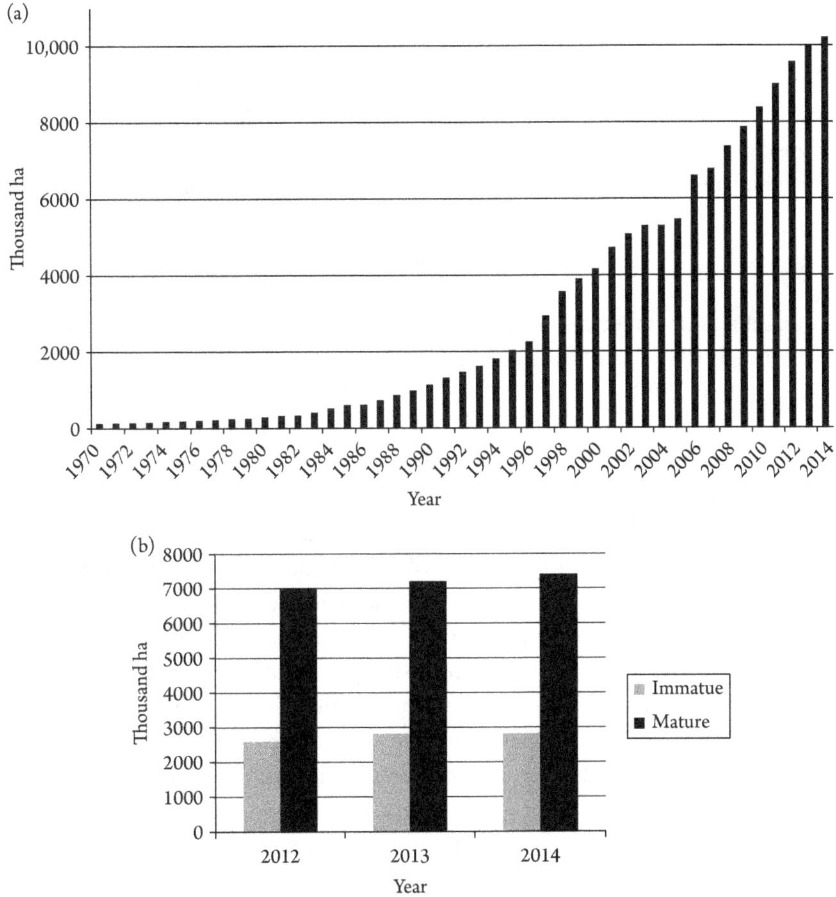

Figures 7.8 Total hectares of oil palm in Indonesia, 1970 to 2014.
Source: Tree Crop Estate Statistics of Indonesia. 2013. *2012–2014 Kelapa Sawit (palm oil)*. Jakarta: Directorate General of Estate Crops.

the trends shown in Figure 7.8. On the other hand, part of the future "area expansion" has already been planted. Together, Indonesia and Malaysia have more than 3.5 million "immature" hectares that will soon begin to produce significant yields.[14] This is a large area, comparable with the approximately 11 Mha now considered mature in the two countries.

Thailand, Colombia, and possibly Brazil will probably plant more area. During the next 35 years, Cameroon and other countries in West Africa project expansion as well. West and Central Africa may still require major pioneering investments to become significant participants in palm oil trade. Land and forest policies could be crucial,

[14] Ling (2014) shows the 30-year yield profiles for oil palm trees. He indicates that 18% of Indonesia's oil palms are 3 years of age or less, and that only 15% are 19 years of age or older.

possibly leading production units for oil palm to tend toward medium-size units rather than the dual plantation/smallholder landscape that characterizes Southeast Asia.

New plantings of oil palm will surely decelerate from their recent ~4.5% annual rate (2000–2013), but we believe that area expansion of 1.5% could occur easily. Coupled with yield growth, this expansion in area would suffice to keep real prices constant. In fact, we think real vegetable oil prices are much more likely to fall than to rise.

PROJECTIONS

In interpreting our conjectures about the future, it is useful to look at the recent dynamics of vegetable oil markets. In doing so, we note that earlier projections made by the FAO and the IFPRI for 2000 to 2015 or 2020 have underestimated growth consistently in vegetable oil and oilseed supply and demand by a wide margin (Rosegrant et al. 2001; Bruinsma 2003). A summary of those projections, plus some of the factors that accounted for the "misestimates" are shown in Box 7.1.

In looking forward, our general conclusion is that growth of the oil crop sector will slow sharply from 2013 to 2050 relative to 2000 to 2013. Increments to global population will be smaller in the years ahead, China's economic growth can be expected to slow along with its consumption of vegetable oil and meat, and further substitutions through imports—even if they are possible—are unlikely to be acceptable politically, as in India. Biodiesel mandates in the European Union and the United States are also less likely to be a driving force (Chapter 6). These slowdowns on the demand side will probably be matched with less rapid growth in supplies arising from increased concerns related to sustainable sourcing and more restrictions on area expansion for both soybeans and oil palm. These changes may not be abrupt, but we now see a definite shift to growth in the oil crops sector that is more in line with growth of the agricultural sector growth as a whole.

We thus see a substantial expansion of the vegetable oil sector, yet at only a little more than third of the 4.8% annual growth from 2000 to 2013. Our base projection indicates a growth rate of 1.7% annually between 2015 and 2050, and it would not surprise us very much if the rate was 2% per year. These growth rates would put us at the middle to high range of annual growth rates currently being projected: IFPRI, 1.2% (Rosegrant 2015); FAO, 1.4% for food (Alexandratos and Bruinsma (2012); Organisation for Economic Co-operation and Development (2012a), 1.6% for oilseeds; National Soybean Laboratory, 1.8% for soybeans (Masuda and Goldsmith (2009); Corley (2009), 2.2% for food; and US Department of Agriculture, 1.9% (Sands et al. (2014). Note that the end period for these projections varies between 2025 and 2050.

Whatever rate is suggested, all sources see a continued shift in shares toward palm oil. We see palm oil leading the way in Africa, where a combination of population and income growth make that region a more important actor in vegetable oil markets in the future than in the recent past. We see continued global substitution for, and erosion of, shares going to "other minor" oils. And as described in Chapter 5, we see a tapering off of livestock growth in China, slower growth in soybean imports, and a

Box 7.1 Problems with Prior Projections

The tropical oil crops revolution was largely unanticipated, at least by the major international agencies charged with making long-run projections of the world food economy: FAO (Bruinsma 2003) and International Food Policy Research Institute (IFPRI) (Rosegrant et al. 2001). Around 2000, both agencies published projections to 2015 or 2020. Although their projections for rice and wheat were quite close to realized production in 2013, they underestimated production of oilseeds and vegetable oils by a wide margin.

It is instructive to review these projections to understand why consumption and production were so much higher than anticipated. FAO published detailed projections for 2015 from a 1997/1999 base, using simple bottom-up approaches based on demand and supply for each commodity and consistency checks. IFPRI made detailed projections to 2020 of all major crops to 2020, with 1997 as the base, using their multicommodity partial equilibrium model, IMPACT. IFPRI does not provide detailed crop breakdowns for oil crops except for soybeans. Notably, the report of more than 200 pages does not mention palm oil. Projections for 2013 were interpolated linearly from the 2020 projection by IFPRI, and the 2015 projection by FAO. FAOSTAT data for 2013 were used as the actual (IFPRI's baseline also used FAOSTAT).

The main differences apparent in these projections and what actually happened can be summarized as follows:

1. Production of vegetable oils was hugely underestimated by both sets of projections. In FAO projections, palm oil production turned out to be nearly double that projected. All the major oilseeds (soybeans, rapeseed, sunflower, and groundnuts) exceeded their FAO projections by at least 25%. In IFPRI projections, production of fats and oils as a category was underestimated by 31%. The surge in vegetable oil use for biodiesel to about 18 Mt explains part of the underestimation for vegetable oils—a surge that almost nobody anticipated in 2000. However, it seems both sets of projections underestimated the rapid growth in food demand for vegetable oils, especially palm oil.
2. In the case of soybeans, FAO's estimate was 57 Mt and IFPRI's was 78 Mt less than actual for 2013—a staggering underestimate in the space of only 15 years. Because soybeans have a low oil content (in volume), we must look on the livestock side for the reasons for this major underestimate. Meat production was underestimated by a significant amount especially for poultry (FAO, 14%; IFPRI, 25%). The higher than expected growth in global income explains in part the underestimate of meat consumption as a result of its relatively high-income elasticity. However, neither set of projections anticipated the extent of the rapid shift to feed-intensive poultry meat in the developing world and the growing share of soy meal in feed concentrates globally (Chapter 5).

Table 7.8 Vegetable oil supplies and uses, 1990, 2013, and 2050 (estimated), and implied growth rates, 2013 to 2050

Use and supplies	*1990 (Mt)*	*2013 (Mt)*	*2050 (estimated) (Mt)*	*Implied growth, 2013–2050 (%/yr)*
Total	56.2	165.4	307	1.7
Oil use				
Food	50.7	125.1	217	1.5
Biodiesel	N/A	23.0	48	2.0
Other	5.5[a]	17.3	42	2.5
Oil supply				
Palm oil/palm kernel oil	12.5	63.5	140	2.0
Soybean oil	15.4	45.0	80	1.6
Rapeseed oil	8.7	25.0	35	1.8
Sunflower oil	8.2	15.0	25	1.3
Other oils	11.4	16.9	22	0.5

[a] Includes small amounts of biodiesel.

Food uses of vegetable oils fall from 76% in 2013 to around 70% in 2050.

N/A, not available.

Source: USDA-FAS PSD and authors' calculations.

corresponding slowing of growth in soybean oil consumption. We believe palm oil will have a price advantage relative to soybean oil, and palm oil will therefore also increase its share in China's consumption bundle.

Our assessment suggests further that a significant increase in real prices is unlikely in the decades ahead, although price spikes and dips might occur through variations in economic growth on the demand side and from El Niño events and other climate shocks on the supply side. We think long-run growth in supply is more likely to push ahead of growth in demand than to lag it, thereby putting downward pressure on real vegetable oil prices. We do not think that downward price pressures will be severe or last long because of what happens to palm oil production. Assuming that countervailing policy incentives are not put in place, lower prices would delay investment in new plantings and replacement planting, and the growth in palm oil production would slow.[15] We find this ratchet-like process for prices to be a quite likely outcome for the future.

We regard what happens to oil palm planting as the key and most uncertain variable in the price discussion. The palm oil sector rests on a rather narrow genetic base, and a disease outbreak in Asia could devastate palm oil supplies. Government policies, especially as they relate to biodiesel mandates, could have a similarly dramatic effect on

[15] Low replanting could be counterbalanced by policies to provide incentives to replant during periods of low prices, as Malaysia has done in the past and continues to do.

the demand side. What happens in China and India will also be crucial. China's overall rate of growth will affect total vegetable oil demand and supply in important ways, especially the rate at which its livestock consumption grows (Chapter 5). In India, will the newly announced US$1.5 billion program to produce oil palm (*Economic Times* 2015) be implemented and prove successful, given the marginal production environment and land scarcity? And will domestic Indian vegetable oil producers be politically powerful enough to curtail imports through higher tariffs? On the consumer side, will new efforts at sustainability certification take on sufficient force to restrict the expanding area of oil palm everywhere, and not just in forested regions, to protect local land rights? We are not sure which, if any, of these events will transpire, but we are sure the world of palm oil will have its surprises in the years ahead. Our numerical projections, and those of all others, should thus be read as having significant error bars attached to them.

Our summary description of the past and future is best shown in Table 7.8. The data on production, uses, and shares for 1990 and 2013 are facts insofar as we have been able to determine them. The estimates for 2050 reflect our best judgments about the future.

8

CONTRIBUTIONS TO GROWTH, JOBS, FOOD SECURITY, AND SMALLHOLDER DEVELOPMENT

INTRODUCTION

Oil crops are a big business in their major producing and exporting countries, such as Indonesia and Brazil, where they generate revenues in the tens of billions of dollars and affect many other sectors of the economy indirectly. Given the phenomenal growth and large size of this "oil crop pie," the distribution of industry benefits matters a great deal in the economies where they are produced. In this chapter, we review the limited evidence on how oil crops have affected economic growth, job creation, and contributions to food security. Because the participation of smallholders is a major determinant of inclusive growth in the sector, we pay special attention to business models that partner companies with smallholders, especially oil palm producers.

ECONOMIC GROWTH

Oil palm has generated huge economic benefits during the past two decades, adding around US$40 billion to world agricultural output and accounting for a significant share of increased foreign exchange earnings of major exporters. Oil palm now accounts for nearly half of Indonesia's agricultural export earnings. Likewise, soybeans provide more than US$50 billion in export earnings to South American producers and about one third of the agricultural export earnings of the largest producer: Brazil.

The indirect benefits of oil crop production are also potentially very large. They accrue through the stimulation of upstream industries (farm inputs, for example) and downstream industries (soap, margarine, and poultry, for example), the generation of public tax revenues, and the development of infrastructure. Malaysia has built a substantial national industrial base from palm oil, starting by refining CPO instead of exporting it, and moving on to oleochemical manufacturing and associated industries, such as the production of *vanaspati* (solidified oil for cooking), lubricants, cosmetics, soap, surfactants, glycerol, and, most recently, biodiesel. A higher tax on exports of CPO relative to its derivatives has been one of the major incentives to accomplish this transformation (Rasiah 2006).

Indonesia has been much slower to develop downstream industries, but in 2011 it introduced similar tax incentives. Nonetheless, in frontier regions such as Papua Province in Indonesia, with few prospects for food processing industries, forward linkages are likely to be weak (Obidzinski et al. 2014).

Value addition for soybeans begins with its initial processing into protein meal and oil; further opportunities to add value include refined oil and derived products (as for palm oil), and the use of meal to develop the livestock industry. Brazil has been slow to capture these opportunities. Unprocessed beans account for some two thirds of the value of Brazil's soybean exports, compared with only about one quarter in Argentina. Brazil's high value-added tax discourages local processing, whereas Argentina taxes exports of unprocessed soybeans heavily. Even so, Brazil is the world's largest exporter of chicken meat and a significant exporter of pig meat, so the equivalent of about 10% of its current exports of soybean products is exported in the form of livestock. Nonetheless, Goldsmith and Hirsch (2006) estimate the value added per dollar of soybeans produced is eight times higher in Illinois than in Brazil as a result of the much larger downstream conversion to meat and dairy products in Illinois.

The evidence also suggests the presence of significant economic growth linkages that stimulate local economic development in producing regions. One analysis found that 84% of the income generated from oil palm in Riau Province (Indonesia's largest producer of oil palm) was spent locally, and most inputs were sourced locally, providing significant indirect benefits to village people who were not producing oil palm (Budidarsono et al. 2013). At the same time, the fact that a major share of oil palm is produced on large plantations or by relatively better off smallholders using migrant labor may widen income disparities within a region and reduce local economic benefits (Mukti et al. 2014; Obidzinski et al. 2014). For Indonesia as a whole, Edwards (2015) using district-level data, estimated that a 10% increase in the land area devoted to oil palm in a district from 2001 to 2009 increased district GDP by 2.4% over the long run, and reduced the poverty rate by 10%. In the 10 districts with the largest increase in oil palm area, this expansion accounted for 70% of total poverty reduction.

Likewise, in Brazil, Weinhold et al. (2013) found strong increases in household incomes and rural GDP, and reduced poverty in the *municipios* (counties) where soybean production expanded most. Soybean-producing *municipios* also had higher income inequality, however, probably reflecting the concentration of production on large farms. The intensification of soybean production systems through double-cropping has similarly positive impacts on incomes and local development through employment, migration into the producing area, and investment (VanWey et al. 2013). These knock-on effects of soybeans on overall economic growth in the Cerrado appear to have become stronger over time (Richards et al. 2015).

Increased revenue from local and national taxes generates other economic benefits. In Indonesia, taxes provided more than US$2 billion annually at the national level alone (Organisation for Economic Co-operation and Development 2012b). In Papua New Guinea, corporate taxes paid by one company, New Britain Palm Oil Ltd. (NBPOL), account for 2.6% of all national tax revenue (ITS Global 2011). Likewise in Sarawak, Malaysia, a 7.5% tax on palm oil production is a major source of state revenue. One potential source of local government revenue—land tax revenue—is often small, however. In Paraguay, where soybeans are the major crop, the land tax averages only US$0.16/ha annually.

Finally, those companies with strong codes of corporate social responsibility provide additional benefits in the form of local physical and social infrastructures, especially roads, schools, and health facilities. NBPOL in Papua New Guinea provides comprehensive health care to 32,000 employees, pays school fees through secondary school for employees' children, and spends US$3 million annually on local infrastructures (ITS Global 2011). Similarly, in Papua Province of Indonesia, local communities identified improved roads and social services as a significant benefit from oil palm expansion (Andrianto et al. 2014).

Clearly, the economic benefits of these new industries based on oil crops are large and are transforming many rural areas. A visitor to towns in Riau Province, Indonesia, or Mato Grosso State, Brazil, will be impressed by their rapid expansion; the apparent prosperity generated by the demand for inputs, transport, and logistics from the new industries; and the knock-on effects from expenditures by employees of the new industries.

WAGE EMPLOYMENT ON LARGE FARMS AND PLANTATIONS

The potential for direct job creation differs greatly in the oil palm and soybean industries. Oil palm is a major source of wage employment in the big producing countries, because production of this perennial crop is not yet fully mechanized. Using standard figures of 6 to 8 ha per worker, we estimate oil palm production in Indonesia employs almost 2 million workers, not including jobs generated elsewhere in the value chain. This number far exceeds the estimated 1.3 million employed in textiles, another major export sector in Indonesia. Many of those employed in oil palm production are, of course, smallholders who are self-employed on their own plots. Even so, about half are wage earners in large plantations, and most smallholders also hire workers for some tasks. In West and Central Africa, the production and small-scale, semimechanized processing of palm oil are also major sources of employment, especially in the off-season, when food crops are not being produced.

In contrast, employment on the large soybean farms of the Cerrado in Brazil is estimated at one worker for every 167 to 200 ha (Bickel and Dros 2003). Busto et al. (2015) showed that for Brazil as a whole, soybean area planted per worker rose dramatically from around 12 ha in 1980 to nearly 100 ha in 2011, resulting from the move to the Cerrado, where farms are larger and highly mechanized. Total hired employment in the sector may not exceed 0.1 million (Bailis 2014), although as many as 1 million workers may be employed in the industry as a whole when self-employed family labor and those employed in postharvest transport, handling, and processing are added (Food and Agriculture Organization of the United Nations 2007). Employment in value-adding industries such as poultry and pig production is also significant (Hosono and Hongo 2015). In India, where smallholders produce soybeans using semimechanized practices, one worker is employed per 7 ha (similar to oil palm), indicating that well more than 1 million workers are employed (mostly self-employed).

Given the large number of jobs in oil palm production and the fact that land for oil palm production tends to be located in sparsely populated areas, migrant labor is the norm in the industry. As many as 2 million workers move long distances within countries (especially from Java and Sulawesi to Sumatra, Kalimantan, and Papua in Indonesia) or move across national borders (especially from Indonesia to Malaysia). A large share of the oil palm workers in Malaysian plantations, reportedly at least 80%, are attracted from neighboring countries.

Given the millions of migrants involved and the historical record of poor working conditions in plantations, surprisingly little field research has been carried out on labor conditions and relations in the industry. The story is likely to be quite varied. After analyzing migrant labor in Malaysia, Pye et al. (2012) presented a useful typology (Table 8.1). Workers in the large plantation companies generally have the best conditions, extending from compliance with official labor standards and migration regulations to the provision of training, subsidized housing, treated water, electricity, insurance, and other amenities (Norwana et al. 2011). Even so, these plantations outsource many tasks to specialized labor contractors, who set their own conditions that may be difficult for a company to monitor. The major problems are found when unscrupulous contractors recruit illegal immigrant workers. These migrants may enter a state of debt bondage in which they have little recourse to the law because of their illegal status. No statistics exist on the relative numbers in each category of worker shown in Table 8.1, but recent investigations by civil society and the media have drawn attention to labor abuses in oil palm (for example, see Skinner [2013]). Certifying sustainable supply chains is one way to improve standards (Chapter 9).

The industry employs male and female workers, usually with different responsibilities—for example, men for harvesting and women for weeding. Progressive companies will have an explicit gender policy to promote female workers into more skilled jobs such as vehicle operation and maintenance, but changes in the industry and in labor supply can often have negative effects on women in the workforce. For example, after plantations are established in an area and the supply of local labor has increased, some companies have moved from recruiting families for permanent positions with associated housing and schooling benefits to recruiting individual males for permanent positions and then contracting casual labor, often female, for low-paid and more seasonal jobs (Li 2014). Rising official minimum wages in Southeast Asia will likely accelerate efforts to reduce labor costs by outsourcing, with the attendant risks of labor abuses, especially of female laborers.

The circumstances of wage laborers on oil palm plantations are therefore quite heterogeneous, depending on company size and visibility, the use of contract labor, and the prevalence of illegal migrant labor. Many workers experience a "precarious regime" at best (Pye 2015). At the same time, growing labor scarcity and increasing wages in both Indonesia and Malaysia should improve incomes of wage employees who operate freely in the rural labor market. Wiggins and Keats (2014) provide some indication of these changes. Rural daily wages in real terms in South Kalimantan Province of Indonesia (a major oil palm-producing province) increased from US$3.29 in 2007 to US$4.76 in 2010 in 2010 dollars. Yet, real wages in Central Java (which supplies many

Table 8.1 A typology of migration experiences in the Malaysian oil palm industry

Typology	*Type of worker*	*Description*
1a	Estate and mill workers	Legal workers with work permits; semipermanent employment of families, direct contract with employer, competitive wages with health security, and decent housing and schools on the estate; debt not a major problem
1b	Smallholder workers	Working directly for smallholders or small private plantation, often legally and with work permit; relatively competitive wages, working on a longer term basis; housing and health coverage vary
2a	Contract workers by task	Working for contractors but with permit or semilegal status; short in/out contracts, relatively high wages, debt for permit fee that is deducted for passage; housing varies with some "living in the forest"; health coverage variable
2b	Superexploited contract workers	Illegal, working for subcontractors at extremely low wages, living in debt bondage
3	Outsourced	Shifting strategy, working directly but as freelancers for subcontractors; higher wages, freedom of choice; sometimes with initial debt, but using the outsourcing system to their own benefit

Source: Adapted from Pye, O., R. Daud, Y. Harmono, and Tatat. 2012. Precarious lives: Transnational biographies of migrant oil palm workers. *Asia Pacific Viewpoint* 53 (3): 330–342.

migrants) were only US$2.27 per day in 2010, less than 20% of the agricultural wage in Malaysia, which in that same year was US$11.75 per day. These differentials provide big incentives to use migrant labor, and as long as substantial numbers of migrant workers are illegal and their working conditions remain unregulated, the problems of labor abuse and debt bondage will persist.

IMPACTS ON FOOD SECURITY

The impacts of oil crops on food security can be examined at multiple levels—among households, across countries, and globally. In Chapters 4 and 6 we discussed the role of vegetable oils in global food security, particularly the provision of affordable calories and minimum fat requirements to the world's poorest, and the tradeoff in allocating resources to the production of food versus biofuels.

At the household level, cash crops such as oil palm and soybeans are often assumed to undermine food security by reallocating resources away from food production. For example, rice area has fallen in the Jambi and Riau Provinces of Indonesia by about 0.8

Mha since 1995 with the rapid expansion of oil palm. Subsistence food production is only one path to food security; another is for household members to generate income from cash crops to purchase food. In general, households learn to balance the production of cash crops and food crops in ways that enhance food security (von Braun and Kennedy 1994; Wiggins, Henley and Keats 2015). Oil crops are no exception. Overwhelming evidence shows that when smallholders plant oil palm, they increase their incomes substantially (Susila 2004; Rist et al. 2010; Cramb and Curry 2012; Cahyadi and Waibel 2013; Krishna et al. 2015). Oil palm is much more profitable than other crops, especially food crops such as rice. Feintrenie et al. (2010) estimate that in Jambi Province of Indonesia, oil palm provides about 10 times the returns to land as rice and 20 times the returns to labor. Not surprisingly in an area that has good access to markets, most households no longer produce rice and depend on the market for their food staple. Even so, Euler et al. (2015) found that in Jambi Province, adoption of oil palm leads to a 13% increase in per-capita calorie consumption and a 22% increase in calories from nutritious foods. Likewise, growers in Costa Rica who have given up food crop production to specialize in oil palm report positively on their improved food security (Beggs and Moore 2013).

In India, smallholders have made a conscious decision to diversify away from traditional food staples such as sorghum, motivated by the much higher incomes they generate from soybeans—their major cash crop. Where soybeans are grown for home consumption as well, as in Nigeria, they make an important nutritional contribution to protein-deficit households (Sanginga et al. 1999).

In more remote regions with poorly developed markets, oil crops are combined with food crop farming to ensure household food security. For example, in Papua New Guinea, women allocate 2.5 times more labor to food crops than to oil palm (Koczberski and Curry 2005). Involving women in oil palm production can also make important contributions to food security. The Mama Lus fruit scheme organized through NBPOL encourages women to pick up loose palm fruits, for which they are paid separately. Women receive 29% of household income through this scheme and spend the extra income on better foods, schooling, and health (Fisher et al. 2012).

Nonetheless, important caveats are attached to the role of smallholder oil crops in food security. The first is that much depends on the type of household that adopts the crop. Initially, under state-sponsored schemes in Malaysia and Indonesia, the poorest and landless households benefited; but, over time, with the shift to independent smallholders, oil palm was adopted by wealthier households with better access to land, labor, and capital (Krishna et al. 2015; Schwarze et al. 2015). The poorest, nonadopting households may not benefit directly from producing oil palm, although they benefit from wage employment whether hired by smallholders or by larger plantations. In contrast, soybean adoption in India, with its lower upfront capital costs, has been strongly pro-poor (Chand 2007).

A second caveat is the risk of depending on a single crop as the major source of income. Although prices have trended upward since about 2001, commodity prices are notoriously volatile and subject to extended periods of depressed prices. This risk is especially marked for tree crops, because it takes many years for supply to adjust to demand. Households that produce food crops as well as oil palm (as in Papua New Guinea) have a safety net when prices are low, as do households that have diversified to nonfarm jobs (as in Peninsular Malaysia). Household surveys in Indonesia's oil palm-producing regions indicate that 60% to 80% of household income is generated by oil palm, suggesting the risk of hardship is considerable when prices are low (Susila 2004; Cahyadi and Waibel 2013; Lee et al. 2014b). Because many smallholders are relatively recent adopters of oil palm, they have yet to experience a serious price downturn. The landing, when it comes (and it will!), could seriously undermine household food security.

Reliance on a few varieties of one crop with a relatively narrow genetic base also presents significant production risks from disease and pest outbreaks. The large oil palm-producing countries in Southeast Asia have, by and large, escaped such an occurrence to date, yet the outbreak of bud rot disease in Colombia is a cautionary tale for producers everywhere (Chapter 2). A similar risk prevails in the major soybean-producing regions of Brazil, which have already experienced a serious incursion of soybean rust, leading to a prohibition on sowing soybeans during certain months to break the disease cycle. Incentives to introduce more diversified soybean–livestock–forestry systems as part of the Brazilian "ABC plan" to promote low-carbon agriculture have met with limited adoption (Gil et al. 2015).

For national food security, potential concerns related to risk and diversification must be considered. World prices are relatively more volatile for tropical commodities than for other traded commodities and manufactured products (Food and Agriculture Organization of the United Nations 2004). Countries that depend on a single commodity for export earnings and import a significant share of their food staples may face serious foreign exchange shortages in periods of world price spikes.

For the major oil crop exporters, these concerns are minor. Few countries appear to have a major overdependence on exports of oil crops and their derivatives. Palm oil is a major export earner in Malaysia and Indonesia, but both countries have diversified their exports. The bulk of their export earnings now come from manufactured goods and services; palm oil makes up less than 10% of foreign exchange earnings. Brazilian exports are likewise diversified. Even in smaller countries, such as Paraguay, soybeans make up less than one third of export earnings. And because prices of oil crops, unlike prices of other tropical commodities (such as beverages and rubber), correlate closely with prices of cereals, demand for foreign exchange for importing cereals track export earnings from oil crops.

ENGAGING SMALLHOLDERS AND COMMUNITIES IN OIL PALM

A Simple Framework for a Complex Issue

A key driver of inclusive growth is the involvement of smallholders—a challenge for a crop such as oil palm, which must be processed immediately (generally in large mills) and requires considerable upfront investment. A countless variety of business models, often quite complex, can be used to engage smallholders and communities in oil palm production.

To explore the range of business models for oil palm, we use a framework based on the contribution of the various resources used in the value chain—capital, labor, land, and management—and the distribution of the ownership of those resources among various stakeholder groups: namely, smallholders and their communities, migrants, private investors, and the state (Figure 8.1). The assets owned by different stakeholder groups are often complementary, giving rise to opportunities for partnerships. For example, smallholders and their communities have access to family labor, retain customary land rights, and possess local knowledge; private agribusiness companies have access to capital markets, specialized technology and management, and links to global commodity markets; and the state may contribute land, specialized technical services, and financing.

Within this simple framework, a fundamental driver of the distribution of benefits is the extent that smallholders participate directly in production. The level of smallholders' participation is fundamental because, as producers, they capture returns to all factors of production, particularly returns to land and labor. If production is organized in large plantations, smallholders gain only to the extent they receive wages from working on plantations, and those wages more than compensate them for the

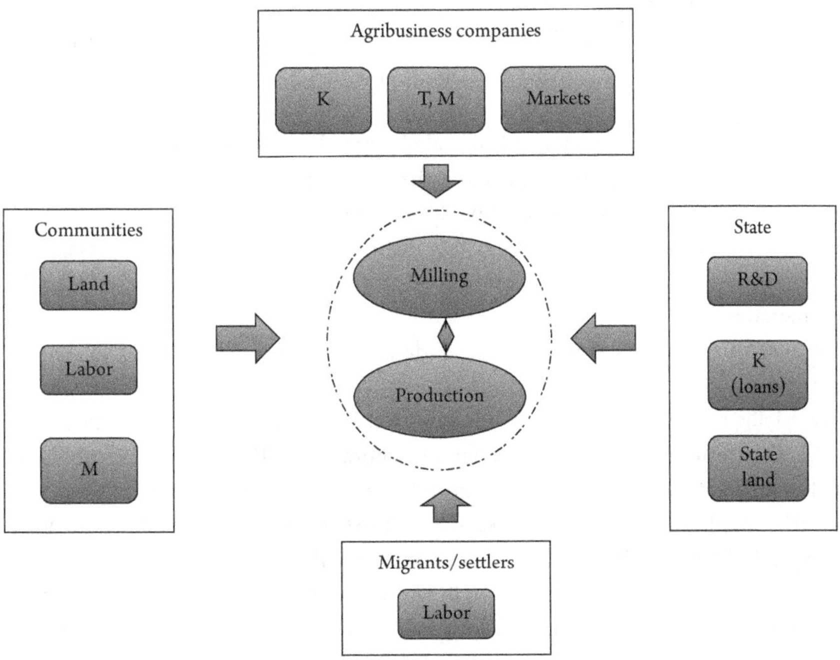

Figure 8.1 Key stakeholders contributing resources to oil palm production. Four major stakeholder groups—in situ communities, companies, migrants, and the state—may each contribute resources according to their asset endowment to provide the land, labor, management (M), capital (K), technology (T), and access to markets needed for production, which must be linked closely to milling, which is mostly in the hands of companies. R&D, research and development.

opportunity cost of their time in other activities. If they have secure land rights, they may also be able to capture returns through renting or selling land to oil palm companies, especially land that is currently uncultivated. In the frontier regions where oil crops are expanding, this option tends to be limited by poorly functioning land markets and the lack of secure land rights.

There is ample evidence that smallholders find oil palm profitable. In Malaysia, annual income from a 3.8-ha plantation was estimated at US$8160 in 2010 (Simeh 2011). Feintrenie et al. (2010) estimated that the annual income from a 2-ha plot was US$7000; they also estimated the returns to labor at US$50/day, which is more than 10 times the daily wage rate, and higher than for any other crop. Not surprisingly, many smallholders are keen to produce oil palm.

If oil palm is so profitable to smallholders, why do they not dominate the production stage of the value chain as they do for most crops in low-wage economies? First, as discussed in Chapter 2, the need to process palm fruit soon after it is harvested incurs high transaction costs; planting and harvesting by many smallholders must be coordinated with the available capacity of a large mill. Second, oil palm is a new crop to independent smallholders in Southeast Asia (but not in Africa), so they face considerable learning and information requirements for successful production. Finally, capital costs are high for perennial crops relative to annual crops. To establish a plantation, a producer must sustain the costs of developing land, acquire quality planting materials, and care for the crop until the first harvest in the third or fourth year—an amount that often exceeds US$3000/ha and cannot be repaid for at least 10 years after planting (Rist, Feintrenie, and Levang 2010). When a plantation is established, significant requirements for short-term working capital arise, especially for purchasing fertilizer, the largest component of oil palm's annual production costs. To be sure, smallholders in Africa use few inputs, but the resulting income is usually inadequate to move producing households out of poverty unless they engage in processing as well. Well-functioning financial markets or other institutional mechanisms to access inputs are therefore critical for smallholders' participation in the industry to be sufficiently remunerative.

Consistently, surveys show that independent smallholders—lacking capital, know-how, and experience, and using poor-quality seedlings and insufficient inputs—obtain lower yields than large estates (see, for example, Vermeulen and Goad [2006], Zen et al. [2006], and the International Finance Corporation [2013]). Independent smallholders also obtain yields that are significantly lower (by 25%–38%) than yields obtained by smallholder outgrowers (Lee et al. 2014b). If yields and profitability are very low, smallholder households may be better off selling their labor to a large estate, provided wage standards are observed. On the other hand, well-capitalized smallholders earn incomes at least 50% more than the smallholder average (Norwana et al. 2011).

Much experimentation has occurred with models linking smallholders to state agencies and/or private agribusiness companies. The varied motivations for this experimentation have included finding ways to overcome the asset deficits of smallholders (specifically, capital and technology) as well as companies (specifically, access to land and labor). Building on various typologies of these arrangements (such as those by Cramb and Curry [2012], Vermeulen and Goad [2006], and Oxfam International [2014]) and asset complementarities, we constructed Table 8.2, which shows the contributions of major stakeholders in some of the common arrangements found in the oil palm sector.

These relationships can be analyzed further with respect to how the partners share (Vermeulen and Cotula 2010):

- *Ownership*: Who owns the assets?
- *Voice*: Who has the decision-making and negotiating power?
- *Risk*: How are production and price risks shared?
- *Reward*: How are costs and benefits shared?

The resulting business models may range from spatially dispersed and independent smallholders who exercise full decision-making powers and appropriate all benefits, to large, contiguous, and centrally managed blocks in which smallholders have an equity share in the final output but possess little decision-making power. In either case, smallholders may be organized into cooperatives to obtain credit, inputs, and advisory services; contract with mills; or negotiate with companies for land deals and joint ventures.

The design and outcomes of the various business models depend on the extent to which the ownership of assets is well defined and secure. Secure land ownership is particularly important, given that poorly defined or overlapping rights may lead to conflicts between companies and communities. The state or private companies may claim large tracts under the law, but often communities use that same land under customary tenure arrangements the state does not recognize. Power relations are also a fundamental determinant of how complementary assets are combined through various types of contractual and partnership arrangements, and how the benefits are shared. If smallholders and their communities are unorganized and uninformed, they are likely to get the short end of the stick in negotiations with companies.

State-Managed Schemes for Smallholders

State-organized settlement schemes for smallholders have a long record, and most have now been phased out. As mentioned in Chapter 2, in Malaysia FELDA initiated a program in 1960 to resettle poor and landless households by developing large contiguous blocks in state forest land into oil palm and rubber plantations. The settlers contributed labor during the early years and then took over some management tasks as the trees matured. After repaying the land development debt, they received full title to their individual plots. FELDA has involved 113,000 settler households since its inception. Plantation management has been restructured a number of times but was eventually recentralized, and in lieu of title to a specific plot, settlers received shares in the company. FELDA has generally been rated a success as an antipoverty program (Pletcher 1991), although it was costly in terms of state resources, and its approach to smallholder development was quite paternalistic (Sutton 1989). Rapidly increasing

Table 8.2 Examples of business models used in oil palm production based on who contributes different production factors

Business model	*Contributions by*				
	Local communities	*Migrants*[a]	*Private companies*	*State agencies*	*Example*
Settlement smallholders	Labor, management	Labor, management	—	Processing, capital, land, technology, management	FELDA (Malaysia)
In situ managed smallholders	Land, labor	Labor	—	Processing, capital, land, technology, management	SALCRA (Malaysia)
Independent smallholders	Land, labor, capital, management	Labor	Processing	Technology, capital	Thailand
Contract farmers	Land, labor, management	Labor	Processing, capital, management	—	Palma Tica (Costa Rica)
Vertically integrated cooperatives	Land, labor, capital, management, processing	Labor	—	—	HonduPalma (Honduras)
Nucleus–outgrower scheme	Land, labor, management	Labor, management	Processing, technology, management	Capital, land, facilitation	Many companies, (Indonesia) GOPC (Ghana)
Vertically integrated company in joint venture with communities	Land, labor	Labor	Processing, capital, technology, management	Facilitation, capital	Many companies (Sarawak, Malaysia). NBPOL (Papua New Guinea)
Vertically integrated company	Labor	Labor	Processing, capital, technology, management	Land (if based on a concession)	Most companies (Indonesia)

[a] Migrants may become smallholders in their new communities.

FELDA, Federal Land Development Authority. GOPC, Ghana Oil Palm Company. SALCRA, Sarawak Land Consolidation and Rehabilitation Authority. NBPOL, New Britain Palm Oil Ltd.

wages in Malaysia eventually meant that FELDA had difficulty attracting labor from settler households and even attracting settlers themselves. Accordingly, one arm of FELDA embarked on direct plantation development and management, and has subsequently transformed these plantations into one of the world's largest palm oil producers, FELDA Global Ventures, in which settler households hold the largest block of shares (Chapter 2).

A somewhat different model is used in situ by the Sarawak Land Consolidation and Rehabilitation Authority (SALCRA), a state-owned authority in Sarawak, Malaysia. SALCRA enters into agreements with local communities to establish oil palm plantations that SALCRA manages. After deducting all costs and a management fee, the returns are distributed to the individual landowners. SALCRA is essentially a state-managed plantation that is unlikely to be optimal, given the inefficiency and rent-seeking that tend to affect such operations. Yields and returns on SALCRA plantations appear to be significantly below the average for private companies. Even so, returns to communities are higher than for some of the joint venture schemes discussed later, and farmers do receive full title to their land.

Private Partnerships of Nucleus Estates with Smallholders

An important business model to involve smallholders has been the association of a nucleus plantation estate with outgrowers (NES). As we saw in Chapter 2, this model has been used extensively in Indonesia in a number of formats, as well as in Africa. The state or donors have financed the establishment costs (the World Bank and Asian Development Bank were enthusiastic supporters during the early years). Sometimes the mill may provide annual working capital for inputs with the costs deducted from the mill payment—a practice followed, for example, by Ned Oil in Sierra Leone (Small 2014).

Most evaluations have graded the NES as reasonably successful (for example, Zen et al. 2006) or in some instances highly successful (Jelsma et al. 2009). Many experienced serious problems, however, usually involving three sets of issues (summarized here and discussed in relation to supply issues in Chapter 2).

The first set of issues includes frequent complaints about state appropriation of community land without compensation, a lack of transparency in contractual agreements, a lack of consultation with local communities, and unequal benefit sharing (Jurgens et al. 2010). In many cases, no written contract substantiated the arrangements concluded with smallholders, who did not understand their loan repayment obligation had to be met to gain title to their plot, nor did they understand their long-term rights to land after trees were no longer productive (Rist et al. 2010).

The second set of issues involved the widely varying degree to which individual farmers in NESs were able to take their own decisions versus executing orders from the company. In the most successful cases, units of about 50 ha are managed by a cooperative work unit (Box 8.1). In other cases, smallholders were little more than laborers on a centrally managed plantation, although they received a share of profits in addition to wages. Furthermore, after many government functions were decentralized in

Box 8.1 Hindoli: A Successful Smallholder Outgrower Scheme

Hindoli is a nucleus estate with 10,000 ha (under expansion) owned by Cargill Tropical with an associated outgrower scheme of 20,000 ha. The 10,000 outgrowers are organized into 20 cooperatives. Establishment of the outgrowers was financed by long-term loans from a state-owned bank, and annual working capital is provided through the cooperatives' own savings and loan associations. Cargill supports the outgrowers through a professional staff of 31 people who provide technical assistance in agronomy, business management, organization, and human resources. The oil produced is sold under contract to one of two Cargill–owned mills. Smallholders in the scheme earned an average US$5500 in 2012, and their yields were higher than on the estate. Assisted by Cargill, the outgrowers were the first smallholders in Indonesia to be certified by the RSPO. Oil palm has been the route out of poverty for these smallholders; many families now send children to university.

Because the trees are nearly 20 years old, the looming challenge is to finance both the replanting costs as well as living costs for the 3 to 4 years until the young trees can start being harvested. An amortization fund has been established for this purpose, but additional financing will be required through commercial bank loans.

Source: Authors' personal interview with a Hindoli Cooperative, May 2014.

Indonesia, local governments came to play the facilitating role in agreements between communities and companies. Some local governments with resources have pursued policies favoring smallholders, whereas others have exhibited rent-seeking behavior that favored companies (McCarthy et al. 2012).

The third set of issues concerns the sharp decline in smallholders' participation following the withdrawal of state financing and after decentralization. Smallholder participation commonly corresponds currently to only 20% of the area of an NES (the minimum required for a plantation license), versus up to 70% to 80% of the area during the early years (McCarthy et al. 2012). Many estates are now managed centrally, with smallholders receiving equity shares but not retaining individual plots (Vermeulen and Goad 2006). The success of these new arrangements is still uncertain, although many see them as a step backward in the long struggle to achieve a balance between large-scale models and smallholders (McCarthy et al. 2012).

Outgrower schemes are also used extensively for oil palm in West Africa (Chapter 2), where donors have usually funded the cost of establishing outgrower plantations. Although these schemes have enjoyed some success, an important difference in Africa is the presence of an alternative market for selling FFBs to small-scale processors. As a result, many outgrowers may engage in side-selling and do not fulfill their contractual obligations to mills, which means mills often cannot recover the cost of the loans they extend to outgrowers (Fold and Whitfield 2012). Even so, the o-verall result has been positive for outgrowers as well as the many independent growers who step in to sell to mills with excess capacity (as a result of outgrowers' side-selling) (Väth and Gobien 2014).

Finally, in Latin America, contract farming is common, and mills depend entirely on purchased fruit. Small and medium growers enter into a contract with a mill to supply their FFBs, usually based on a price formula, in return for support from the milling company. In some cases, mills may provide long-term capital under contract to small and medium producers. For example, Palma Tica in Costa Rica works with medium-scale farmers averaging 33 ha to convert low-productivity pasture to oil palm. Farmers sign a contract of 12 to 14 years underwritten by mortgages on the land (Beggs and Moore 2013). Obviously, a secure land title and a strong legal system are needed for such long-term financing to work.

Joint Ventures with Communities

The joint venture model has been promoted in both Malaysia and Indonesia as a way for communities that control uncultivated land to partner with private investors. In Malaysia, the Sarawak Land Custody and Development Authority had brokered 33 such joint ventures by 2009, enabling companies to access land for large-scale production in return for providing communities with an equity share in company profits. By 2009, only one joint venture had issued dividends, however, leading to a number of protests by communities. The joint ventures were structured in a way that provided incentives for companies to use creative company structures to avoid dividend distribution (Cramb 2012). Nor did communities have a voice on company boards. Given the general unhappiness with these arrangements, recent joint ventures are moving toward a fixed annual payment for land (Cramb and Sujang 2011).

Joint ventures have had much more success in Papua New Guinea. There, so-called Incorporated Land Groups lease land to companies (through an arrangement with the state) that pay a modest annual rental of US$19/yr plus a percentage of the value of the harvested product. The largest and most successful company, NBPOL (now part of Sime Darby), also provides company shares to communities, giving them a further sense of ownership. One drawback of the land groups was that a change to the Forestry Act in 2007 caused a rash of companies to lease land for timber extraction in the guise of developing it for plantations that never materialized.

Independent Smallholders

Independent smallholders account for an increasing proportion of palm oil production in most countries. Thailand, the world's third-largest producer, has quietly built its palm oil industry on the basis of independent smallholders. Some 70% to 80% of its output of 1.6 Mt is produced by 120,000 smallholders averaging 3.9 ha (Dallinger 2011). Most smallholders work independently of mills, and many mills depend entirely on purchased fruit (Beall 2011).

One reason that smallholders dominate palm oil production in Thailand is that national agricultural policy has consistently favored a smallholder model. A second reason is that land for large-scale plantations is difficult to acquire. Thai farmers have secure land tenure, a ceiling has been placed on the size of private land holdings, and foreign ownership of farmland is prohibited. A ban on forest clearing has closed the land frontier since 1989, and large concessions are simply not available. Smallholders have shifted land from rice and rubber production to oil palm, along with previously unused (degraded) land.

Third, the state-owned Bank of Agriculture and Agricultural Cooperatives has a long history of providing credit effectively to agriculture in Thailand (it is one of the few state-owned banks to do so). With secure land titles, farmers can obtain the long-term financing needed to establish their oil palm plantations, and acquire the short-term financing required for working capital. Finally, on the milling side, the Thai Oil Palm and Palm Oil Industries Development Plan has provided significant incentives for investing in processing capacity (Chavananand 2013). Milling capacity is double the size of production, so farmers face a competitive market to sell fresh fruit.

Despite these achievements, the industry faces two major problems: the quality of fruit delivered to mills is low, resulting from the lack of an effective grading system and competition among mills to buy fruit, and the average capacity use of mills is low (Beall 2011; Dallinger 2011). The overall average oil yield per hectare in Thailand is only about half that in neighboring Malaysia, in part because of a less favorable climate for oil palm and in part because of the low quality of fruit delivered to mills. Some mills have moved to contract farming, and some farmers have moved to cooperative milling to upgrade quality and make better use of mill capacity (Preechajarn 2010; Thongrak and Kiatpahtomchai 2012).

Independent smallholders are also a fast-growing sector in Indonesia, where an estimated 1.5 million smallholders plant some 3.5 Mha or one third of the total area (Obidzinski 2015). Independent growers not only obtain significantly lower yields, but also generally sell through an intermediary at a 20% price discount relative to the price for direct delivery to a mill (International Finance Corporation 2013; Lee et al. 2014b). Surveys have identified a lack of access to technical assistance, financing for replanting, and labor for timely harvesting as other major constraints facing independent smallholders (Lee et al. 2014b).

Malaysia is addressing these constraints through a special fund of US$300 million under the National Key Economic Areas Program administered by the MPOB. The program, which targets independent smallholders in areas where yields are low, provides replanting grants of US$2500 to US$3000/ha, planting materials, technical advice through MPOB extension officers, and support to organize cooperatives that sell directly to the mill. According to the MPOB, the program has reached 135,000 ha to date and has achieved a 15% to 20% increase in prices by selling directly to mills. Private companies are initiating similar programs. For example, Cargill Tropical has started a scheme to improve productivity for independent smallholders located near its plantations in Indonesia.

Cooperatives offer important advantages to independent growers. They can provide advisory services to smallholders and help them to purchase inputs in bulk, obtain financing, and link to or own cooperative mills. The cooperative model is widely used in Thailand; for instance, the Krabi Oil Palm Cooperative Federation has 15,000 farmers with about 3 ha each and its own large mill (Preechajarn 2010).

Medium-size farmers with 10 to 100 ha in Latin America largely work in cooperatives such as HONDUPALMA in Honduras (SNV et al. 2009) and Copalcol in Colombia (Rugeles et al. 2011). Along with its own mills, HONDUPALMA owns several downstream industries, including a biodiesel plant that has obtained RSPO certification. Copalcol is sufficiently established to run its own savings and loan and advisory services programs, as well as to forge a close working relationship with a mill that is also a member of the cooperative.

Empowering Smallholders: Examples from Other Tree Crops

The overview of business models that engage smallholders and communities indicates their considerable variation across space and time in design and performance (McCarthy et al. 2012). There is evidence, however, that involving smallholders may create efficiency–equity tradeoffs. This point is highlighted by the detailed analysis of Cramb and Ferraro (2012) in Table 8.3, which suggests that the state-run SALCRA scheme generates only half the benefits per hectare of private estates but provides higher benefits to local people. Many of the business models have resulted in worsening income distribution, because only better-off smallholders are able to capture

Table 8.3 Simulated distribution of stakeholder benefits from a 15,000-ha plantation by type of business model, Sarawak (US$ million discounted net present value), 2007

Stakeholder	*State-run smallholder scheme*	*Joint venture*	*Private company with concession*
Local people			
Dividends to land and capital	119.8	68.1	—
Wages	9.9	9.9	9.9
Total	129.7	77.9	9.9
Foreign labor	11.2	17.9	17.9
Estate management	3.3	6.3	6.3
Private investor	—	123.3	209.9
Government			
Dividend to capital	—	21.1	—
Land rent	—	—	2.8
Company tax	—	77.6	77.4
Total	—	98.7	80.2
Grand total	144.1	324.1	324.1

Yields on joint ventures are assumed to be the same as for private companies with concessions, but in practice they have been lower.

Source: Data are from Cramb, R. A., and D. Ferraro. 2012. Custom and capital: A financial appraisal of alternative arrangements for large-scale oil palm development on customary land in Sarawak, Malaysia. *Malaysian Journal of Economic Studies* 49 (1): 49–69.

opportunities offered by oil palm, and/or large plantations assume a major share of the benefits.

The major reasons for these tradeoffs are poorly functioning financial markets, insecure land rights, weak advisory services, and lack of strong farmer and community organizations to negotiate a fair deal with investors or millers. In the long run, the state can foster a better environment for smallholder development through concerted actions to build sustainable rural financial institutions, secure property rights, provide strong research and advisory services, and facilitate the emergence of local organizations. In the short run, a well-governed state might overcome some of smallholders' asset deficits by providing improved planting materials and extension programs (Barlow and Tomich 1991), although the state itself is often captured by the economically powerful, who act against the interests of smallholders (McCarthy 2010).

Institutional models used for other tree crops could be adapted to oil palm. One option is to establish an autonomous smallholder development authority, financed through a percentage levy on exports, to provide critical services to smallholders (replanting grants, advisory services, and processing, for example). The Smallholder Tea Development Authority of Sri Lanka and the Rubber Industry Smallholder Development Authority of Malaysia are two parastatals that have used this model successfully to transform their respective industries from a production base of large plantations to smallholders (Byerlee 2014). When the industry is sufficiently well organized, these parastatals can be privatized under majority smallholder ownership. For example, the Kenyan Tea Development Authority provides a range of services to growers, including processing. It is owned by more than half a million smallholders, who produce and process 70% of Kenya's tea (Byerlee 2014). Another example is the Federation of Coffee Growers of Colombia, which manages a levy on coffee exports to provide a wide variety of services to its half a million members, made up overwhelmingly by smallholders.

In the push for private-sector solutions to development, the potential role of these types of collective action in supporting the development of a smallholder oil palm sector has been overlooked. Some of the earlier business models, such as Indonesia's NES programs, showed that smallholders can produce oil palm successfully with little sacrifice in yields, but they lost momentum after state and donor financing dried up. The use of a levy is an alternative source of financing, but it requires a well-run parastatal or preferably a well-organized smallholder sector that can manage its own resources and collective action effectively. The levy recently implemented in Indonesia, in part to support smallholders, does not appear to have either of these attributes.

ENGAGING SMALLHOLDERS AND COMMUNITIES IN SOYBEAN PRODUCTION

Unlike oil palm, soybeans are an annual crop that usually requires little establishment capital and no close coordination with a mill. The range of business models for including smallholders is more limited in soybean production than in oil palm. In Brazil, relatively high labor costs of around US$50/day (including wages and benefits) for farm workers in the Cerrado encourage the development of large, fully mechanized farms. The significant pioneering risks, transaction costs, and land development costs of opening the Cerrado frontier also favor large operations. These considerations, coupled with economies in purchasing inputs and marketing output, seem to indicate that the role for smallholders is fairly circumscribed. In neighboring Paraguay, some companies with an explicit corporate social responsibility objective, notably the Desarrollo Agrícola del Paraguay, have attempted to establish outgrower schemes in adjacent communities using hired machinery, but they have had little success because soybean technology differs so greatly from the technology used to produce traditional food crops (Guereña 2013). The extremely high and growing inequality of farmland distribution in Paraguay (Gini coefficient of 0.91) does not bode well for an inclusive growth model based on its principal crop: soybeans.

These experiences contrast with the strong smallholder and pro-poor orientation of the soybean expansion in Central India. Because soybeans are entering farming systems through intensification or crop substitution and wages are low (less than US$3/day), soybeans have been adopted rapidly through manual or semimechanized operations. The development of cooperatives that undertake their own processing or contract with processors has also facilitated smallholders' expansion into soybean production.

These experiences have important implications for the nascent soybean sector in Africa, where rural wages are relatively low and the rural population continues to grow rapidly. The Indian model seems much more appropriate for this setting than the Brazilian model, especially in areas of medium to high population density, where the crop is being introduced into existing farming systems (Nigeria). In areas where population density is low, there may be a role for larger, more mechanized operations to pave the way during the early years, but the objective should be to foster an industry based on small and medium-size family farms. The fact that sub-Saharan Africa has to absorb 200 million additional people in rural areas during the next 20 years provides little choice, in contrast to Brazil where a vibrant nonfarm drew labor out of agriculture.

SUMMING UP

The bottom line for this chapter is that rapid expansion of oil crops in the tropics has provided major economic benefits to the handful of countries that have participated, but this expansion has often missed opportunities to reduce poverty. Smallholders still account for less than half the oil palm area, and their yields and economic benefits are lower than they could be. Still, even when oil palm is produced on large plantations, it has generated millions of jobs (in Indonesia, more than the textile industry), although wages and labor conditions are sometimes below standard, especially when large numbers of illegal immigrants are employed through contractors. Large-scale mechanized soybean production, on the other hand, has been much less conducive to job creation, and in tropical Latin America, soybean production based on smaller family farms accounts for a small percentage of production. Obviously, there are major exceptions to these general findings. Smallholders have led the oil palm expansion in Thailand (although with modest yields) and soybean expansion in India (again with low yields).

The challenge of linking smallholders to global value chains is especially daunting for oil palm, considering the high initial costs of establishing plantations and the requirements of coordinating with mills. We have seen, however, that the state has often tilted the playing field against smallholders through policies related to land concessions, milling licenses, and weak extension systems. We recognize the importance of private investment in downstream milling, in logistics, and in overcoming some of the pioneering costs and risks of establishing new crops in new areas. At the same time, we see no reason why oil palm in more established areas could not move toward a smallholder-based industry (as in other tree crops), either through strong, state-supported services or innovative and inclusive business models with companies. Likewise, as soybeans spread in Africa, it will be important to maximize opportunities for small and medium commercial producers.

Finally, the generation of employment and the conditions under which workers are employed are largely unexplored subjects. More focused field research would likely reveal a wide range of conditions from excellent to abysmal. Given the large and growing number of people employed in oil palm, field research on labor use and working conditions is a fertile area for producing a better understanding of the economic and social impacts of oil palm.

9

LAND USE AND THE SUSTAINABILITY CHALLENGE

OIL CROPS AND THE CHANGING PHYSICAL, SOCIAL, AND ENVIRONMENTAL LANDSCAPE

A primary motivation for this book is that oil crops have had a profound impact on global land use in recent years. Recall that, since 1970, the area planted to oil crops increased by an astonishing 150 Mha, 100 Mha of that during the two decades from 1993 to 2013, or double the area of Spain or Thailand. More than 50 Mha of that area was used to grow soybeans, mostly in South America and India, followed by 16 Mha for rapeseed (in Canada, Europe, and Australia), and 10 Mha for oil palm (in Southeast Asia). The expansion of rapeseed was uncontroversial, because it occurred mostly on land that had been used for other crops. In contrast, soybean and oil palm production expanded mostly on previously uncultivated land. Not surprisingly, oil crops rank high in land-use change and GHG emissions in the tropics, as seen in Table 9.1, with soybean in second place in land-use changes and oil palm in third place in emissions. Note including emissions from conversion of peatland in Table 9.1 would likely put oil palm in second place.

The dramatic expansion of oil crops in the tropics from the early 1990s occurred just as a host of environmental concerns captured the world's attention, especially climate change and species loss—two issues encapsulated in the rapid clearing of biodiverse tropical forests and woodlands. Media campaigns shone a bright spotlight on tropical deforestation and often equated it with the planting of oil palm in Southeast Asia and soybeans in the Amazon. Under this scrutiny, evidence is emerging that the impact of oil crops on forests varies by country and over time, as well as by the type of forest converted—primary tropical forest, secondary logged and degraded forest, or woodland and scrubland.

Aside from their environmental dimensions, changes in land use on the agricultural frontier have important human dimensions, especially with respect to the rights of local communities to access and use land and forest resources. During the early years of frontier settlement in Brazil, and until the present day in Southeast Asia, local communities (many of them indigenous) have been engaged in a struggle—associated closely with the expansion of tropical oil crops—to own and control land and forestall the encroachment of large agribusiness companies and immigrant settlers. This struggle is also playing out across the new frontiers for oil crops, such as Paraguay (soybeans) and Myanmar (oil palm).

Table 9.1 Summary of land-use changes and greenhouse gas emissions for four commodities in the pan-tropics

Product	*Land-use change (Mha/yr), 2000–2011*	*Emissions GtCO$_2$/yr, 2000–2011*
Total	9.4	3.7
Beef	2.7	1.0
Soybean	0.5	0.1
Oil palm	0.3	0.2
Wood products	0.4	0.4

The total includes the four commodities and all others commodities. Emissions for oil palm are underestimated because emissions from conversion of peatland are not included. GtCO$_2$, Gigatons of carbon dioxide.

Source: Data are from Henders, S., U. M. Persson, and T. Kastner. 2015. Trading forests: Land-use change and carbon emissions embodied in production and exports of forest-risk commodities. *Environmental Research Letters* 10 (12): 125012.

In this changing landscape of tropical oil crops, with its broad and complex social and environmental dimensions, the experiences of Brazil and Indonesia provide sharp contrasts. Up until around 2005, Brazil was the largest single source of tropical deforestation. At the peak of deforestation in 2004, Brazil lost 2.78 Mha of forest in the Amazon and converted 0.89 Mha of natural vegetation to other uses in the Cerrado (Figure 9.1). Brazil was also the world's largest emitter of GHGs associated with land-use change. Deforestation declined sharply in the Brazilian Amazon after 2004, and now Indonesia has superseded Brazil as the nation with the largest annual loss of primary tropical forests (Figure 9.2). Indonesia also adds large quantities of GHGs from the conversion and cultivation of peatlands. In total, Indonesia emitted 1.22 Gg (gigagrams) of equivalent carbon dioxide in 2012 from land-use changes alone, 50% more than Brazil, the second-largest emitter of GHGs caused by land-use changes. A reduction in deforestation (and draining of peatlands) in Indonesia comparable with that achieved by Brazil would go a long way to reducing global GHG emissions from land-use change, not to speak of conserving valuable tropical biodiversity and providing other environmental benefits.

In this chapter we open the discussion of the sustainability and land-use challenges surrounding tropical oil crops with detailed studies of how Brazil and Indonesia have managed their forest resources as oil crops became increasingly important to their economies. As described in Chapters 2 and 3, Indonesia and to a lesser extent Brazil have also experienced serious conflicts over land rights on the agricultural frontier. To illustrate further the human costs of land-use changes, we turn to case studies of large land acquisitions for oil crops in Sarawak in Malaysia, Mozambique, and Liberia. The final part of this chapter lays out a broad policy agenda to promote sustainable agricultural development that improves both social and environmental outcomes. We do not espouse win–win approaches but recognize that tradeoffs are needed—and that the hard question is how to minimize those tradeoffs.

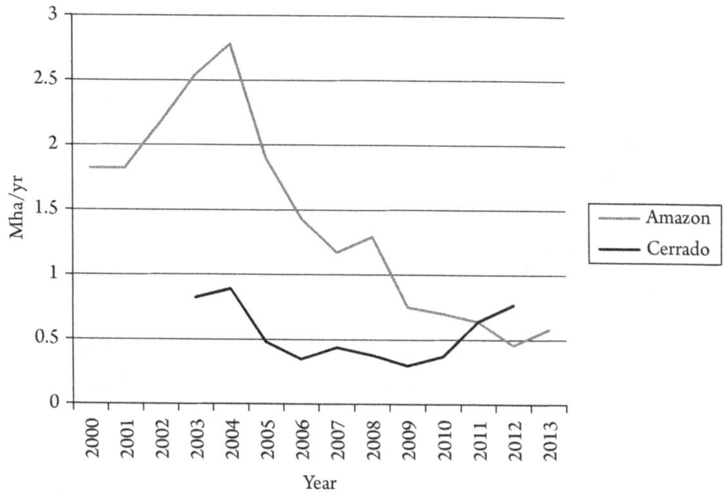

Figure 9.1 Deforestation in the Brazilian Amazon and Cerrado, 2000 to 2013, based on primary forest loss.
Source: Soares-Filho, B., R. Rajão, M. Macedo, A. Carneiro, W. Costa, M. Coe, H. Rodrigues, and A. Alencar. 2014. Cracking Brazil's forest code. *Science* 344 (6182): 363–364.

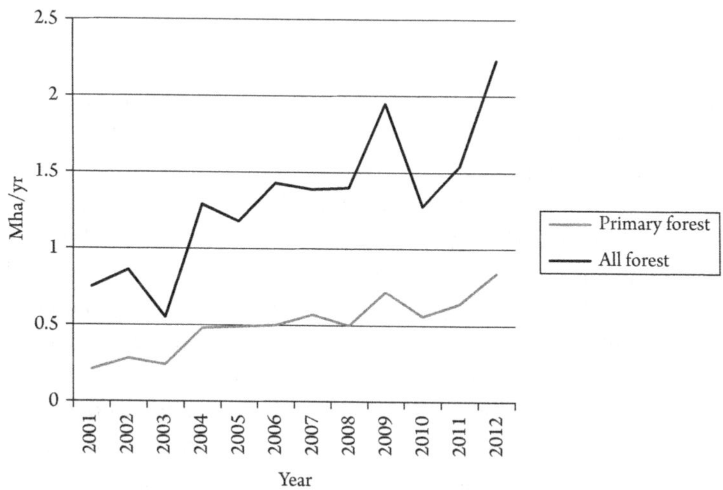

Figure 9.2 Deforestation in Indonesia, 2000 to 2013. Estimates are based on satellite imagery. Official government statistics show a decline in deforestation in Indonesia during the past decade.
Source: Data are from Margono, B. A., P. V. Potapov, S. Turubanova, F. Stolle, and M. C. Hansen. 2014. Primary forest cover loss in Indonesia over 2000–2012. *Nature Climate Change* 4: 730–735 for primary forest loss; and Global Forest Watch. 2015. Country profiles. http://www.globalforestwatch.org/countries (accessed September 15, 2015) for all forests with tree cover of more than 30%.

BRAZIL: EXPANDING SOYBEAN PRODUCTION WITHOUT CUTTING FORESTS

From 1990 to 2012, Brazil added 20 Mha for crop agriculture, added another 35 Mha devoted to pasture and other land uses, and lost 57 Mha of forest (FAOSTAT). About 80% of this agricultural expansion occurred in two biomes—the Cerrado and the Amazon (Lapola et al. 2014). The soybean expansion occurred largely in the Cerrado (Chapter 3), but until 2006, a serious, persistent concern was that soybeans would move into the Amazon at the expense of its forests. Soybean area exploded from 2001 to 2006 with the addition of some 9 Mha, and the expansion correlated strongly with deforestation (Macedo et al. 2012). During that period, soybean area in the Amazon biome expanded by about 15% per year as new infrastructures opened access to river ports to the north, new European and (later) Chinese markets emerged, and a sharp currency devaluation made Brazilian soybean exports highly competitive (Nepstad et al. 2006). At the time, some 100 Mha in the Amazon were considered suitable for profitable soybean production (Vera-Diaz et al. 2008).

Using satellite imagery, Morton et al. (2006) estimate that about one third of the new soybean area appearing in 2001 to 2004 in the major soy-producing state of Mato Grosso replaced Cerrado savannah and woodland, one third replaced forests, and one third replaced pasture (Figure 9.3). By far the largest immediate cause of deforestation was the conversion of forests to pasture for cattle, which accounted for 72% of all forest losses, compared with 17% for soybeans. Arvor et al. (2012) attribute only 13% of the forest losses from 2001 to 2006 to soybeans.

Since 2004, the conversion of natural vegetation to cropland and pasture has fallen sharply in the Amazon, but to a much smaller extent in the Cerrado. Macedo et al. (2012) estimate that forest losses in Mato Grosso dropped from an annual average of 0.76 Mha in 1996 to 2005 to 0.08 Mha in 2012. After 2006, less than 1% of the soybean crop was sown on newly cleared forest in the Amazon biome, and soybean area in the Amazon biome has peaked at about 9% of total soybean area (Rudorff et al. 2011; Garrett et al. 2013). Although soybean area has expanded by another 5 Mha since 2006, most of this increase came from intensifying production on land that was already cropped, and from converting pasture to cropland. One estimate is that the conversion of Cerrado woodland and savannah has decreased from 1.6 Mha per year during the early 2000s to 0.65 Mha in 2010 (Lapola et al. 2014), with much of the recent conversion occurring in the frontier areas of Mapitoba (Chapter 3).

The evidence points to success in slowing deforestation in the Amazon, and some progress in slowing the clearing of the Cerrado, but in neither case is victory ensured. Much discussion focuses on whether the conversion of pasture to soybean farms in the Cerrado displaced cattle production farther to the north, onto new pastures created by clearing forests. Statistical models support the displacement hypothesis, especially before 2006, when forest losses were high (Arima et al. 2011; Richards et al. 2014). We are cautious about interpreting statistical relationships as causality. Displacement is hypothesized to occur via higher beef prices brought about by reduced pasture area, via specialized but moveable assets in beef production, or via speculation that

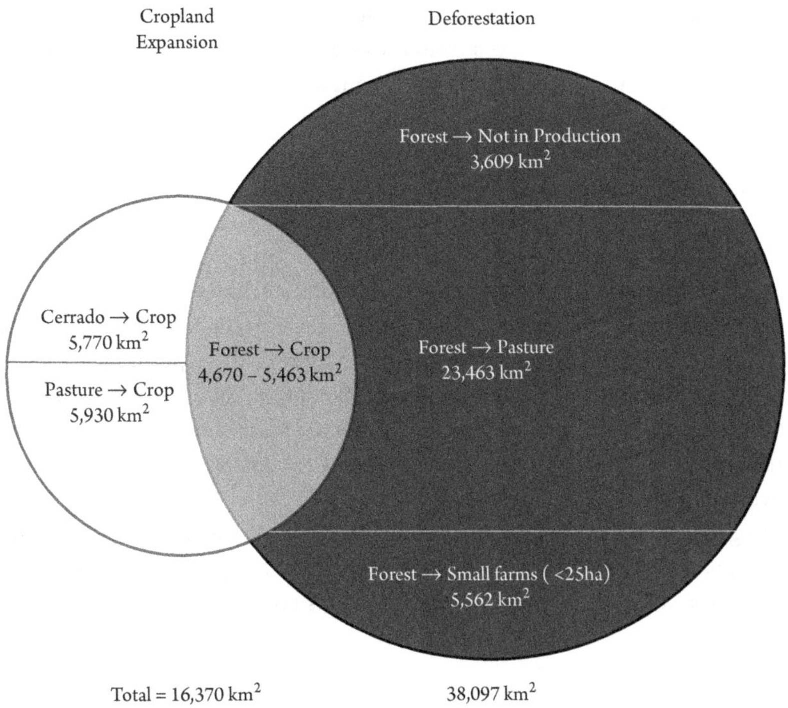

Figure 9.3 Relationship between cropland expansion and deforestation, Mato Grosso, 2001 to 2004. Except where noted all conversions are for commercial farms with over 25 ha.
Source: Morton, D. C., R. S. DeFries, Y. E. Shimabukuro, L. O. Anderson, E. Arai, F. del Bon Espirito-Santo, R. Freitas, and J. Morisette. 2006. Cropland expansion changes deforestation dynamics in the southern Brazilian Amazon. *Proceedings of the National Academy of Sciences United States of America* 103 (39): 14637–14641. Copyright (2006) National Academy of Sciences, U.S.A.

land values will increase when pasture is converted eventually to soybean production (Richards et al. 2014). The notion that beef prices are an underlying cause of displacement seems implausible, given that Brazil, for much of the period under consideration, has been a major beef exporter and therefore beef prices are largely set by world prices. The specialized assets of the beef industry, in the form of the animals and the skills of cattle producers, have some plausibility as causes of displacement, but we see no convincing field evidence that traces the trajectory of displaced cattle farmers and their herds. The most plausible driver of displacement is that cattle ranchers clear land without expecting to profit from raising cattle but anticipating that soybean infrastructure and markets will arrive eventually and allow them to profit from selling the land. Perhaps more fundamentally, clearing land for pasture is a means of gaining title to land that will appreciate in value. Land prices have indeed increased throughout the region as the soybean frontier has advanced. Our overall assessment—in the absence of good field data to back statistical correlations—is that the indirect effects on forest loss of soybeans displacing cattle farming northward into the Amazon contribute only modestly to forest losses, largely through raising land values by converting to soy production.

The continued clearing of the Cerrado, although much less controversial than the clearing of the Amazon forests, is a serious cause for concern. The carbon footprint generated by clearing Cerrado land is only about one quarter the size of the carbon footprint from clearing forest land, although this gap may narrow if soil carbon is included in the calculations (Macedo and Davidson 2014). Even so, the Cerrado harbors considerable biodiversity; 40% of its flora, for example, is endemic (Klink and Machado 2005), and the region performs crucial hydrological functions for Brazil's river systems. Biodiversity and other environmental services could be preserved by establishing sufficiently large protected areas and natural corridors in strategic locations, but such precautions seem to have been ignored in the Cerrado, where the spatial pattern of development has been driven by the suitability of land for specific activities and by access to markets. By law, farmers and ranchers must set aside 35% of their land for conservation, but this is likely to fragment conserved areas severely. Only 8% of the Cerrado is protected and some 40 Mha of the Cerrado could still be converted legally to agriculture.

Setting aside these caveats, it is important to ask what drove the sharp reduction in deforestation in Brazil, and to examine how oil crops are likely to coexist with Brazil's natural areas in the years to come. The next sections provide some answers to these questions.

A Complex of State and Private Actions to Reduce Deforestation

Longtime observers of deforestation in Brazil have identified at least five major actions that helped to reduce forest losses significantly in the Amazon biome (Boucher et al. 2013; Nepstad et al. 2014). First, the Brazilian government adopted a national climate change plan that specified targets to reduce GHG emissions from the Amazon biome by 80% by 2020. This plan had teeth, provided by the enactment of the Climate Change Law of 2009 (Boucher et al. 2013). The president of Brazil at that time, Luis Inácio Lula da Silva, and his minister for the environment, Marina Silva, provided top-level leadership for these initiatives. Together, the Climate Change Law and plan provided the vision and tools for reducing deforestation (Nepstad et al. 2014).

Second, civil society campaigned hard against the expansion of soybean farming into forested areas. A broad coalition of national and international nongovernmental organizations, their action spearheaded by a 2006 Greenpeace report, *Eating Up the Amazon,* was instrumental in establishing a moratorium on purchases of soybeans from land deforested during or after 2006. The big five multinational soybean processors and exporters, as well as other members of the Associação Brasileira das Indústrias de Óleos Vegetais (Brazilian Association of Vegetable Oil Industries), supported this moratorium. A similar moratorium for beef was put into place in 2009. Critically, these moratoriums are enforced by real-time satellite monitoring of forest clearing—a tool pioneered by Brazilian environmental agencies.

Third, the new law and the technical capacity to monitor forest losses were backed by greatly increased will and capacity to implement land-use regulations on the ground. The government developed a list of 36 *municipios* (counties) with the highest level of deforestation and withdrew financing and subsidies from agricultural producers in those locations (Nepstad et al. 2014). A new system of property registration (through the Cadastro Ambiental Rural) and environmental licensing (Licenciamento Único Ambiental) is being introduced as a prerequisite for receiving financing from banks (Rudorff et al. 2011). These new systems enable the owners of newly cleared areas to be identified immediately and makes it possible to assess landowners' compliance with requirements for on-farm conservation of natural areas (set at 35% of farm area in the Cerrado and 80% in the Amazon).

Fourth, in addition to the command and control system just described, payments for environmental services have acted as incentives. At the national level, the government of Norway under REDD+ committed more than US$1 billion to support Brazil's deforestation and GHG targets, which was paid out fully in 2015 based on the progress realized. In addition, quotas for environmental reserves (such as the 35% quota in the Cerrado) were made tradable, so that one producer who wished to clear more land could purchase rights to do so from producers who wished to conserve more land than legally required. A fund for low carbon agriculture also provided cheap loans for landowners to adopt sustainable practices, although its use has been limited.

Finally, soybean area was able to expand through the intensification of crop and livestock production. The area that was double-cropped (mostly soybean with maize) increased from 1.5 Mha in 2000 to 4.3 Mha in 2010 in Mato Grosso (VanWey et al. 2013), and the pasture stocking rate improved by 45% between 1990 and 2011 (Lapola et al. 2014). This intensification was the result of better governance of forest resources, which reduced the effective supply of land and provided the incentive to intensify.

Other factors have also proved to be important. The area of protected natural woodland and forest grew from about 3% to 8% in the Cerrado (Soares-Fihlo et al. 2014). More impressively, 43% of the Amazon is now protected. Indigenous peoples have been granted title to large areas in the Amazon, and land tenure administration has improved greatly. Previously, the cheapest route to a secure the title to land was to convert forest to pasture (Nepstad et al. 2006). As many as 40% of farmers in the Amazon had no land title (Lapola et al. 2014) until a massive titling program was initiated in 2009. Farther south in the main soybean-producing areas of the Cerrado, land titling was regularized earlier. Garrett et al. (2013) report almost all soybean farmers there had land titles.

It is impossible to disentangle the effects of these various policy measures, but together they may explain as much as half of the reduction in forest losses (Assunção, Gandour, and Rocha 2012). The new policies clearly have broken the strong link between soybean prices and cropland conversion. For example, Hausman (2012), looking at the period from 1973 to 2005, estimated elasticities for soybean area of 0.9% and for cropland area of 0.27% with respect to soybean prices. Others, such as Arima et al. (2011), found equally strong relationships between soybean area and prices. In the most recent period from 2006 to 2012, deforestation declined by 75% whereas soybean prices more than doubled—making the reductions in deforestation and land conversion all the more remarkable.

Can Intensification without Land Expansion Be Sustained?

With such measures in place to slow deforestation, has the expansion of soybeans at the expense of natural vegetation run its course? Brazilian experts believe average yields of 3.9 t/ha are attainable; if they are, soybean area should stabilize at around current levels to meet future demand (Santana et al. 2011). Even if soybean yields increase slowly because technology fails to advance or climate change imposes limitations, soybean area should not grow by more than 1.0% per year—a level of growth that can be achieved by converting degraded pasture area (estimated at 28–40 Mha) to soybean production (Strassburg et al. 2014) (Chapter 7).

The achievements in the forest zone remain to be consolidated (for example, by extending the soybean moratorium, which runs through 2016). At the same time, the focus on conservation in the Cerrado is likely to intensify. About 60% of the original vegetation and natural pasture remains in the Cerrado, and about 11% of the land is cropped. Natural vegetation continues to be converted to other uses at a moderate pace, however, so it is critical to manage future land conversion in ways that maximize the preservation of biodiversity hotspots and natural corridors. Some 40 Mha could still be converted legally to agriculture in the Cerrado (Soares-Filho et al. 2014), although the remaining land (in lower and drier environments) may be less desirable for soybeans. When land quality is taken into account, less land may be available than commonly believed (Spera et al. 2014).

The environmentally responsible coexistence of soybean production and forests may have progressed in Brazil, but what about potential "leakages" of soybean production to neighboring countries where the governance of forest resources is weak? From 2001 to 2012, soybean area in Paraguay continued to expand at about double the rate of soybean area in Brazil, and it experienced one of the fastest rates of deforestation in the world. Paraguay has a zero-deforestation policy but its capacity to monitor and implement them is notably weaker than Brazil's. Likewise, Bolivia's soybean area grew during the 2000s partly at the expanse of forests (Hecht 2005), and Argentina saw a substantial loss of forests to soybeans (Gasparri et al. 2013). These "leakages," arising from differences in land and forest governance from one country to the next, undermine net progress toward reduced deforestation globally (Gasparri et al. 2013).

As with soybeans, considerable concern surrounds the potential negative impact of oil palm on forests in Brazil. Brazil's nascent oil palm industry is centered in the State of Pará, which has an abundance of degraded pasture. The largest company, AgroPalma, has 40,000 ha under production, and big players linked to the energy sector are stepping up their oil palm investments. Oil palm production could expand rapidly, without further deforestation, on the estimated 32 Mha of degraded pasture lying within the Amazon biome and considered suitable for oil palm (Yui and Yeh 2013; Vilella et al. 2014). Strict measures to direct the expansion of the industry to these areas are in place and may provide a model for other countries (Box 9.1).

Box 9.1 Can Brazil Become a Sustainable Palm Oil Producer?

The Brazilian government developed the Sustainable Palm Oil Production Program to promote oil palm in areas deforested before 2007, based on results of a detailed zoning exercise presented in the Agro-Ecological Zoning of Oil Palm in Deforested Areas of the Amazon (ZAE-Palma). The zoning exercise relied on agroclimatic data combined with information on land cover. As part of the national biofuel program, investors receive a bundle of incentives from tax breaks, lines of credit, and R&D and advisory services to implement the palm oil program. The intention is to plant 329,000 ha by 2015, with an ultimate target of 4.3 Mha. If this plan is realized, Brazil will eventually become one of the world's leading palm oil producers.

Through strict application of the ZAE-Palma zoning results, satellite monitoring, property registration, and the rural environment registry (the Cadastro Ambiental Rural, which already applies to the soybean and cattle industries), the development of the palm oil industry has a strong sustainability focus. The AgroPalma company became fully certified by the RSPO in 2011. In conformity with Brazilian environmental law, AgroPalma maintains 60% of its land bank or 64,000 ha under natural forests, and carefully safeguards and monitors the biodiversity within its boundaries. It recently received the highest score of any oil palm company under the Sustainable Palm Oil Transparency Toolkit of the Zoological Society of London.

Source: www.Agropalma.com.br; Villela, A. A., B. J. D'Alembert, L. P. Rosa, and M. V. Freitas. 2014. Status and prospects of oil palm in the Brazilian Amazon. *Biomass and Bioenergy* 67: 270–278; Yui, S., and S. Yeh. 2013. Land-use change emissions from oil palm expansion in Pará, Brazil depend on proper policy enforcement on deforested lands. *Environmental Research Letters* 8 (4): 044031.

INDONESIA: STOPPING OIL PALM DEVELOPMENT WILL NOT STOP DEFORESTATION

How Much Forest Has Been Cleared for Oil Palm?

Indonesia has the largest area of tropical forest in Southeast Asia—91 Mha, covering about half the country. It also has one of the fastest rates of tropical forest loss, and in recent years it has cleared more primary forest than any other tropical country (Margono et al. 2014). In the eyes of many in civil society and the media, the rapid depletion of Indonesia's tropical forests is synonymous with the rapid development of its oil palm industry (for example, see Gillis [2014]). Oil palm is an important driver of deforestation in Indonesia, but there are many other drivers as well.

Of necessity, estimates of forest loss during earlier periods are based on secondary data reported by the government. From 1975 to 2005, Indonesia lost 40 Mha of forest, and oil palm area increased by 5.5 Mha. Even if all that oil palm was planted by clearing forests, it would cover only 14% of the forest area that was cleared. Oil palm accounted for about half of the expansion in agricultural area, and rice accounted for much of the

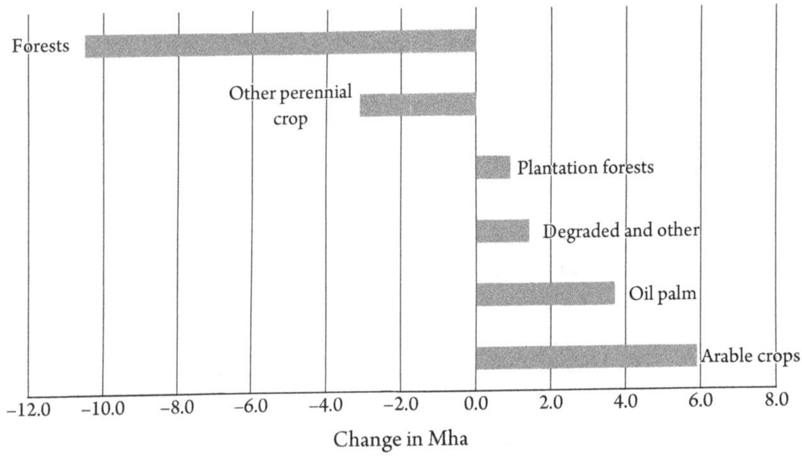

Figure 9.4 Changes in land use in Indonesia based on secondary statistics, 1994 to 2005.
Source: Calculated from data in Wicke, B., R. Sikkema, V. Dornburg, and A. Faaij. 2011. Exploring land-use changes and the role of palm oil production in Indonesia and Malaysia. *Land Use Policy* 28 (1): 193–206.

rest (Wicke et al. 2011). Considering that some oil palm was planted in place of other crops, especially rubber, and also that it was planted on land converted from natural areas other than forests, perhaps 1.7–3.0 Mha of oil palm replaced forests between 1990 and 2005 (Fitzherbert et al. 2008). Figure 9.4 illustrates these land-use changes for 1994 to 2005, which is when oil palm took off in Indonesia.

More recent estimates of forest loss rely on satellite imagery. In one of the most comprehensive exercises, the RSPO commissioned a scientific panel to quantify the role of oil palm in forest loss (Gunarso et al. 2013). The panel concluded that 33% of the oil palm planted from 1990 to 2010 was planted on forest land, although only 6% of that land was undisturbed forest (Table 9.2). The share of forest land converted to oil palm production has increased over time, rising during the most recent period (2005–2010) to 37% of new oil palm area; in Kalimantan, it was 44%. Likewise, Abood et al. (2014), analyzing forest losses only in the concession areas, estimated that of the 6.6 Mha of forest lost from 2000 to 2010, 1.9 Mha was replaced by forest plantations, 1.8 Mha was logged, and 1.6 Mha was planted to oil palm. Lee et al. (2014a), using satellite information, analyzed only forest loss in Sumatra and attributed 19% directly to clearing forests for oil palm.

From 2005 to 2010, the expansion of oil palm on converted forest land was equivalent to 18% of the total loss of forest area in Indonesia (Gunarso et al. 2013). Other losses in forest area came about through logging, pulp plantations, other types of agriculture, and forest fires. Likewise, Busch et al. (2014) estimate that oil palm expansion in forest land accounted for 10% of the total loss of forest in Indonesia from 2000 to 2010. Gunarso et al. (2013) observed a trajectory in land use, starting with undisturbed forest and moving on to logged forest, often followed by fire, before land is converted to palm plantation. Depending on the time lapse and land ownership in this

Table 9.2 Prior use of land planted to oil palm in Indonesia, 1990 to 2010

Prior use	*1990–2000 (%)*	*2001–2005 (%)*	*2005–2010 (%)*	*1990–2010 (%)*
Undisturbed forest	16.1	0.2	0.7	6.2
Disturbed forest	27.7	11.3	36.3	27.3
Shrubs and grassland	18.4	16.9	38.9	26.3
Agroforest and plantation	35.2	66.2	14.6	34.1
Intensive agriculture	0.7	3.8	5.4	3.3
Bare soil and other	2.0	1.5	4.1	2.7
Total (%)	100	100	100	100
Total oil palm planted (Mha)	2.34	1.48	2.57	6.39

Source: Data are from Gunarso, P., M. E. Hartoyo, F. Agus, and T. J. Killeen. 2013. Oil palm and land-use change in Indonesia, Malaysia and Papua New Guinea. Paper presented at the technical panels of the 2nd Greenhouse Gas Working Group Roundtable on Sustainable Palm Oil, Kuala Lumpur, Malaysia, 2013.

sequence, the logging operations in undisturbed forests in some cases could be construed as part of the development of the oil palm plantation. In Indonesian Borneo, for example, one quarter of the area in timber concessions in 2000 was converted to plantations by 2010 (Gaveau et al. 2013). In other cases, concessions allocated for oil palm plantations were logged but never planted, indicating that logging was the main motivation for seeking the concession.

What emerges from these studies is that oil palm is a major driver of deforestation, yet even without the expansion of oil palm, Indonesia would have suffered a high rate of deforestation from other drivers such as logging, pulp plantations, and the production of other crops.

The Environmental Costs of Land-Use Changes

Indonesia's unabated pace of deforestation raises global and local concerns. Its forests provide environmental services on a global scale: they mitigate climate change by sequestering GHGs and they conserve valuable biodiversity. Locally, Indonesia's forests provide equally important services. For example, they protect watersheds and secure the livelihoods of local communities that depend on forest resources—estimated at 7 million people (Srinavas et al. 2015).

The conversion of forested areas to oil palm plantations and other uses has major implications for the richness of biodiversity. Oil palm plantations are not suitable habitats for most forest species (Savilaakso et al. 2014); they support half as many vertebrate species as primary forest (Fitzherbert et al. 2008). The orangutan is the iconic face of the global protest against oil palm, but the problem extends far beyond the plight of a single species. Oil palm plantations, for example, conserve only about 20% of the bird species found in tropical forests. Even a highly degraded forest contains more biodiversity than an oil palm plantation. Smallholders' oil palm plots that usually include remnants of forests are likely to conserve more biodiversity than the large plantations (Sayer et al. 2012; Savilaakso et al. 2014).

The importance of tropical forests for reducing GHG emissions is well known. An undisturbed tropical upland forest conserves 189 t/ha of carbon compared with 104 t/ha for disturbed upland forests and 36 t/ha for an oil palm plantation (Agus et al. 2013). In Indonesia, the conversion of tropical peatlands, forested or not, to oil palm plantations and other uses makes an even greater contribution to GHG emissions than the conversion of forest alone. Indonesia has more than half the world's tropical peatlands (Ramdani and Hino 2013). Because they are poorly drained, peatlands are not often used for agriculture; but, for companies that can invest in drainage systems, peatlands present an appealing target for conversion to plantations. In 2010, an estimated 1.7 Mha of oil palm in Indonesia was growing on peatlands (Gunarso et al. 2013). These plantings represent 18% of total oil palm area, but in 2006 to 2010 they accounted for 64% of the GHG emissions from changes in land use caused by oil palm (Agus et al. 2013). Of all emissions occasioned by changes in land use from 2006 to 2010, emissions brought about by oil palm production accounted for 18% (which seems consistent with the land-use changes discussed previously). The remaining emissions were the result of logging, degradation, fires, and clearing for forest plantations and other agriculture (Agus et al. 2013).

The most visible evidence of GHG emissions and the high environmental cost of poor land management in Indonesia is the thick haze that blankets Southeast Asia during many dry seasons. In the El Niño year 2015, 120,000 fires on 2 Mha were reported in Indonesia, causing daily emissions that exceeded the average daily emissions of the entire US economy (Harris et al. 2015; *The Economist* 2015). The risk to the environment from burning is especially high on carbon-rich peatlands, where fires can persist for months on end. More than half the fires in 2006 occurred outside the concessions mostly in smallholder systems, and oil palm accounted for one third of the fires within concessions in Sumatra and two thirds in Kalimantan (Marlier et al. 2015).

The haze, emanating from uncontrolled burning of forest and peatland in Indonesia, imposes a huge economic and health burden across the entire region (Marlier et al. 2013). In 2013 and 2015, Singapore and many areas of Indonesia had to close schools and restrict outdoor activities during the peak of the smoke haze, and Indonesia was forced to issue an apology to Singapore. More than 100,000 deaths from acute respiratory problems are attributed to the fires in Southeast Asia and many more in El Nino years (Johnston et al. 2012). Economic losses are reported in the tens of billions of dollars and, ironically, the haze was so dense in 2015 that it affected yields of oil palm negatively. The regional union of Southeast Asian countries, Association of South East Asian Nations (ASEAN), has drawn up protocols[1] to monitor and prosecute

[1] The ASEAN Agreement on Transboundary Haze Pollution (Association of South East Asian Nations 2002) and the ASEAN policy on zero burning (Association of South East Asian Nations 1999).

landowners if burning is identified on their land. Progress has been slow on the ground because Indonesia has not ratified the agreement and does not disclose concession ownership as required by the protocols.

Deforestation as a Governance Problem

On paper, Indonesia has numerous laws and regulations to promote a sustainable palm oil industry—reportedly at least 150 laws and regulations for oil palm production alone (Wakker 2014), some 580 laws pertaining to nonforest land, and 2000 legal instruments governing forest land (Srinavas et al. 2015). The implementation of these laws and regulations is highly flawed as a result of weak capacity, multiple jurisdictions, decentralized governance, and outright corruption (Brockhaus et al. 2012).

The public sector has little capacity to implement and monitor laws and regulations. The Forestry Research and Development Agency has only 32 researchers, and 76% of public-sector forestry personnel have no qualification beyond high school (World Bank Group 2008). An Environmental Impact Assessment is a prerequisite for establishing a plantation, but the quality of the assessments is generally poor, and local governments in particular are ill prepared to evaluate their quality.

Many aspects of the governance problem originate with the national concession program. Two thirds of all land in Indonesia belongs to the Forest Estate, and 70% of the oil palm has been planted on Forest Estate land, mainly through state concessions. The Forest Estate nominally lies under the jurisdiction of the Ministry of Forestry, whereas local governments are responsible for allocating concessions. This arrangement produces considerable tension between the ministry and local governments, which often do not follow national laws and guidelines. For example, local governments have issued permits for 6.9 Mha of plantations on forest land without approval of the Ministry of Forestry (Wakker 2014). Land concessions awarded by local governments sometimes encroach on protected areas and national parks. In other cases, companies obtained concessions for oil palm but had no intention of planting it; their major interest was to extract timber (Sheil et al. 2009). In short, failure to comply with laws and regulations supporting sustainability is widespread in the industry (Wakker 2014).

The lack of transparency in land concessions favors noncompliance and corruption. Treated as a state secret, data on concessions and their owners are difficult to access. A digitized map of concessions has only just become available through the World Resources Institute. Stakeholders have no input into the process of awarding concessions because the availability of concessions is not publicized. Reports on corruption in local government awards of land concessions are rife. In Riau Province, the major oil palm-producing province, a location permit or principal permit may be worth US$86,000 to US$258,000 for a 5,000-ha concession, many times the official price (Hadinaryanto 2014). The Corruption Eradication Commission recently charged the governor of Riau Province with accepting a bribe from a small oil palm company. He is the third consecutive governor of Riau to be indicted for corruption.

Glimpses of Progress

Although oil palm has undoubtedly been a major boost to the economies of the producing areas in Sumatra and Kalimantan (Chapter 8), the cost in terms of environmental sustainability has often been high. Yet there is reason for cautious optimism that Indonesia may be turning the corner on deforestation and land use policies. Several recent developments underpin this assessment.

Opportunities to Expand Oil Palm on Degraded Land

We see oil palm expanding in Indonesia for the foreseeable future, albeit at a much slower pace (Chapter 7). The government of Indonesia has set a target of 9 to 13 Mha of additional planting during the next decade. The good news is that sufficient land is available for conversion to oil palm without further deforestation or use of peatlands (perhaps as much as 32 Mha of degraded land) (Wakker 2014). One type of degraded land where oil palm can be grown profitably is the extensive *alang alang* grasslands (Fairhurst and McLaughlin 2009). However, use of degraded land has its own challenges of securing land rights for plantation investors while respecting rights of existing land users.

Payments for Environmental Services

The government of Norway, under REDD+, negotiated a moratorium by Indonesia on forest clearing in return for payment of up to US$1 billion, based on the achievement of agreed performance indicators. The agreement is not perfect; it protects only about one quarter of currently forested land. Half the forest area can still be logged or converted to plantations (Edwards et al. 2012). Land already allocated in concessions for plantations is exempted—as much as 6 to 7 Mha (US Department of Agriculture 2013). Even so, the agreement reinforces the protection of deep peatlands and requires the development of a detailed geo-referenced database of degraded land. Overall, Busch et al. (2014) estimated that if the moratorium had been implemented in 2000, deforested area would have declined by a very modest 1.3% to 3.5%.

National and Provincial Leadership

National and local leadership for a sustainable palm oil industry appears to be emerging. The moratorium, despite its flaws, was a personal cause of former President Yudhoyono, who set a target of reducing GHG emissions by 26% to 41% by 2020. The new president, Joko Widodo, has promoted better coordination by merging the environment and forestry ministries, has streamlined land administration agencies, and has extended the moratorium, but the jury is still out on his commitment to reducing deforestation. Both national leadership and court decisions are also gradually strengthening recognition of customary land rights.

At the local level, the governor of Central Kalimantan Province has outlined a specific plan to reduce deforestation by 80% and to double smallholder participation in the oil palm sector. An electronic Plantation Monitoring System is a central element of this plan. The monitoring system will provide public access to information on plantations to monitor their compliance with agreed sustainability targets. It will also identify and map degraded land that can be used for plantations (Scherr et al. 2015). An inventory of smallholders has been prepared to provide support and technical advice to them. If it succeeds, this initiative could provide a model for other provinces and districts for building a sustainable industry.

Strong Corporate Commitments to Sustainability

In preparation for the 2015 Paris Climate Conference, several of the major oil palm plantation and trading companies made strong commitments to building a sustainable supply chain. Wilmar International and several other leading companies committed to securing their supply chains to comply with "no deforestation, no development on peatlands, and no exploitation of people and local communities" by 2015. Given that Wilmar accounts for a reported 45% of traded palm oil, this step is a major advance in enforcing sustainability standards, although it is clear that broader reforms are needed for Wilmar to implement its commitments (as we discuss later).

"LAND GRABS" FOR OIL CROPS

The global discourse on tropical oil crops emphasizes overwhelmingly the environmental dimensions of sustainability, including the conservation of biodiversity and sequestration of carbon in tropical forests. At the local level, the social dimensions of sustainability—especially the social risks of expanding oil crop production—take precedence. Social risks include infringements on local land rights that ignite outright conflict between local communities and companies or immigrant farmers.

Land rights are a social issue for many reasons. In the tropics, oil crops moved for the most part into the land frontier—the critical area that local people use for extensive agriculture or that has been converted to agriculture only recently, where land rights are notoriously and poorly defined. Often, the customary rights of people living on the land frontier are not reflected in national land laws. Surveys of land use and ownership in these recently and still sparsely settled areas are scarce.

In established regions with secure tenure and functional land administration systems, such as southern Brazil or Peninsular Malaysia, land for oil crops is obtained through a market transaction between private parties. Where land markets work well, land rarely becomes a major source of tension, unless ownership is highly unequal.

In contrast, the frontier regions in many countries are enmeshed in a web of conflicting statutory land law, generally based on Western legal codes, and customary law, based on local communities' long-standing traditions of land use and allocation. Colonial governments commonly introduced statutory systems that vested ownership of all land in the state—a practice that most countries continued after independence. Such systems tend to give only cursory recognition to customary rights. As a result, considerable ambiguity has surrounded customary tenure of forest land used for extensive agricultural systems relying on long fallows. Local rights to use forests for hunting, grazing, and the extraction of timber and nontimber products are rarely recognized unequivocally. Even where the law recognizes customary rights explicitly, how the law translates into practice depends very much on land information and land administration systems, among other processes, which in turn strongly influence social outcomes (Smith and Naylor 2014).

National states have been recognized historically as the sovereign power controlling land and natural resources within their jurisdiction, but recent international conventions and agreements are exerting a strong influence on land rights. The United Nations Declaration on the Rights of Indigenous Peoples of 2007 has been especially important in securing explicit recognition of indigenous land rights. Land rights more generally were addressed in the FAO's *Voluntary Guidelines on the Responsible Governance of Tenure of Land, Fisheries, and Forests in the Context of National Food Security*, approved by the Committee on World Food Security in 2012 (Food and Agriculture Organization of the United Nations 2012). Through these initiatives, the recognition and application of customary law within national land laws gained impetus.

We now turn to three case studies to show how tensions surrounding land play out in practice for oil crops. These cases were selected to cover countries with high, medium, and low scores for land rights and access—Malaysia (scoring 4.5 points of 5.0 points), Mozambique (4.0 points), and Liberia (2.9 points)—based on the International Fund for Agricultural Development (2011) scorecard. They illustrate problems in land rights that are fairly typical for countries at these stages of development.

Sarawak, Malaysia

Sarawak is a state of Malaysia populated in the interior by diverse indigenous groups collectively known as the Dayaks. The Dayaks traditionally practiced long-fallow agricultural systems and exploited forest resources. During the colonial era, when Sarawak was administered autonomously under British protection, customary law, known as *native customary rights* (NCRs), was largely respected (Cramb 2007). Because the local administration discouraged a plantation mode of production based on large land concessions, smallholders integrated rubber trees successfully into their traditional agricultural systems, and rubber became an important cash crop. When the British assumed full administration of Sarawak after 1946, a 1958 land law recognized NCRs but froze their extension.

After independence, and especially with the palm oil boom from 1990, the state eroded NCRs incrementally (Colchester et al. 2009). A 1994 amendment allowed NCRs to be extinguished with minimal compensation. A 1996 amendment placed the burden of proof on communities to claim NCRs (Cramb 2007). State policy shifted from smallholder development to commercial exploitation of natural resources for timber and large plantations. Many concessions were awarded for plantations on land that indigenous groups regarded as theirs under NCRs. The Land Custody and Development Authority has promoted joint ventures with local communities as a way for companies to access land for oil palm, but communities have seen few benefits to date, as we discussed in Chapter 8.

In brief, a patron–client state has emerged in which the state favors commercial interests consistently over the rights of indigenous groups (Cramb 2013). Cozy relations between investors and government officials and their families, and alleged corruption in land deals are documented extensively by Global Witness (2013) and the Bruno Manser Fund (Straumann 2014). Government policies that fail to record NCRs on the ground reinforce investors' interests. The state has made very little progress in surveying NCRs to allow formal registration of title, and land administration authorities do not accept surveys commissioned by indigenous communities.

The Dayaks have fought to preserve their remaining land—perhaps less than 20% of the area of Sarawak. One of their major strategies is to contest company encroachment on their lands through the courts. Malaysia has a relatively capable and independent judiciary system (Colchester et al. 2012). Many from the Dayak elite studied in Australia and New Zealand during the period when courts handed down major decisions on indigenous land rights in those countries (the Mabo case in Australia in 1992 and the Maori Council case upholding the Treaty of Waitangi in New Zealand in 1987). The enactment of the United Nations Declaration on the Rights of Indigenous Peoples in 2007 also provided impetus to legal recourse as a means of resolving land conflicts. Public interest lawyers, mostly Dayaks, have assisted the communities in contesting land claims by investors.

A number of important court cases succeeded, although some decisions are under appeal (Cramb and Sujang 2011; Colchester et al. 2012). Legal recourse is a long and expensive undertaking, yet it has raised awareness and given new life to NCRs in Sarawak. Indigenous groups are also using their land proactively to plant oil palm both as a lucrative investment and as a means of strengthening their land claims (Cramb and Sujang 2013).

Mozambique

Our second example is Mozambique, where some companies and the multigovernment ProSAVANA project target large-scale commercial farming, especially of soybeans (Chapter 3). Mozambique is a postconflict country in which the government gave high priority to land policy after peace was restored. It quickly established a national land commission, which led to enactment of the Land Law of 1997. This law is widely regarded as progressive because the customary usufructuary rights of existing users (Direito de Uso e Aproveitamento dos Terras [DUATs]) are fully recognized (Deininger and Byerlee 2011; Smith and Naylor 2014). DUATs are held by about 90% of the rural population and can be registered, although registration is not required and only 12% of communities have done so, mainly because of the cost.

Mozambique is a country with a liberal foreign investment policy and abundant land. Only 6 Mha of an estimated 36 Mha suitable for crops is currently cultivated. On paper, Mozambican law balances community rights (especially the rights of smallholders) with incentives for investors (Deininger and Byerlee 2011). The law requires investors seeking land to consult with communities and obtain their agreement before requesting government approval.

Intense competition to obtain land for large-scale farming projects ensued when commodity prices rose during the 2000s. A legal process that seemed adequate on paper was marred in practice by little actual consultation with the affected communities, a failure to survey the land in question, and an absence of safeguards and mechanisms to resolve disputes. Even communities that registered DUATs sometimes found the land they were entitled to use overlapped with land granted to investors (Deininger and Byerlee 2011). Or, communities that were farming abandoned state farms found themselves in conflict with investors (as we described in Chapter 3 with the Quifel soybean project) (Norfolk and Hanlon 2012). The strong demand from investors created numerous opportunities for rent-seeking by government officials and politicians (Fairbairn 2013) and speculation by investors. As problems became ever more apparent, the government put a moratorium on new land transfers to investors in 2010 and canceled some of the leases that had not been developed. The case of Mozambique shows that even where progressive laws are on the books, building an efficient land administration and information system is critical to the success of proposed large investment programs such as ProSAVANA (Chapter 3).

Liberia

Our final case relates to the high transaction costs for companies attempting to access land in Liberia and the concomitant high costs to the livelihoods of communities that lose their land to investors. These high costs reflect Liberia's extremely poor ranking in land governance indicators (International Fund for Agricultural Development 2011).

Liberia has a long history of providing land concessions to foreign plantation companies, beginning with Firestone Rubber Company, which received a concession of 400,000 ha in 1926 to develop what became the world's largest rubber plantation. Decades later, Dalton (1965) noted the very limited development impacts of that investment, which marginalized indigenous people in the interior relative to the elite Americo-Liberians, descendants of the immigrant settlers who were originally slaves in the Americas.

Ethnic tensions and clashes over access to land and other natural resources lay at the heart of the devastating civil wars that raged from 1989 to 2003. Afterward, the government of Liberia renegotiated existing concessions and awarded new ones, largely for oil palm and rubber. Although the concessions are transparent—the contracts may be examined freely on the Internet—they were made between the government and the companies, with little consultation with local communities. In addition, the contracts were awarded in an environment of extremely poor land governance, characterized by the World Bank (2008b) as highly unequal and chaotic.

The land laws (consisting of the Hinterland Law of 1949, the Aborigines Law of 1956, and the Public Land Law of 1973) refer to both customary and statutory systems, but contradictory and overlapping elements in these legal texts create "massive confusion" (Unruh 2009). Statutory law always prevails over customary law, and the marginalization of local communities is apparent in antiquated references to "civilized" and "uncivilized natives." The distinction between public lands and customary lands remains vague, even for communities that registered their customary lands under the 1956 Aborigines Law. Aside from the substance of the law, the land administration is described as being in complete disarray, and whatever cadastral surveys have been conducted are in a "deplorable state" (World Bank 2008b). Capacity to administer land laws and regulations in Liberia is extremely weak and fragmented across ministries.

Against this background, in 2009 a large concession of 311,000 ha (44,000 ha for outgrowers) was provided to Sime Darby, the Malaysian oil palm giant, building on a concession that Sime Darby had acquired from the merger with Guthries Plantations in 2007. On paper, such a contract presents many opportunities. Only 5% of land in Liberia is cultivated. Sime Darby is a well-established and respected company that anticipated investing US$3 billion in the venture (three times Liberia's GDP at the time) and creating 35,000 jobs.

The contract states the "concession area is free and clear of all encumbrances" yet alludes vaguely to the possibility of existing users (Government of Liberia 2009). Naively accepting the contract at face value, Sime Darby recognized the problem of customary users only when it started operations. Not surprisingly, conflicts with local communities erupted and became a lightning rod for civil society campaigns (Basta! and Les Amis de la Terre 2012; Siakor 2012). Independent studies by Colombia University and Warwick University support communities' claims that they have been displaced (Lanier et al. 2012; Evans and Giffiths 2013).

Sime Darby realized belatedly that it was following neither its own company standards nor those of the RSPO (although it was one of its first members) by failing to consult with local communities, mapping existing land users, and securing free prior and informed consent to access land (Lomax et al. 2012; Oxford Economics 2014). Sime Darby has made some progress by signing an agreement with one community. The government of Liberia, in deciding to act as the arbiter of the land issues, has not helped, because it clearly lacks the capacity to do so. Messy disputes continue over the appropriate compensation for crop losses. The 44,000-ha outgrower scheme never eventuated because the government was charged in the contract with raising the needed investment funds. Although modest in size relative to the planned plantation of 220,000 ha, the outgrower scheme was a critical component of the contract.

In light of the prevailing land policy and land administration system, the contract was risky from the start, and the government and the company have to share responsibility for the ensuing problems. The option of working in situ with smallholders who were already producing and processing oil palm using local methods does not seem to have been considered and is still not discussed either by the company or the government. As of this writing, the project is way behind schedule, and the tragic Ebola outbreak put it on hold in 2015. Meanwhile, a multistakeholder national land commission is working to reform the antiquated land laws, and a new national land policy has been approved that gives full recognition to customary rights.

Poor Land Governance Is Pervasive on the Forest Frontier

The three case studies highlight four pervasive weaknesses of land governance that forestall the development of a sustainable oil crop sector. First, land laws generally do not accord sufficient tenure security to holders of customary rights, especially the rights to forests and to long bush–fallows that are typically held by communities in sparsely populated areas. Second, even where land laws do give equal rights to customary users, as in Mozambique, land administration systems and surveys are weak and fragmented, undermining the spirit of the law. Third, companies, aided by strong commercial pressures, exploit weaknesses in land laws and land administration systems to gain access to land, often through close relationships with state officials that invite rent-seeking and corruption. Finally, the countries with the weakest land governance are often the ones with the most land available to expand cultivation (Arezki et al. 2015).

For all of these reasons, actions to strengthen land and forest governance must be high on the agenda for the development of a sustainable oil crop sector. The following section outlines the most important policies to support this goal.

TOWARD A SUSTAINABLE OIL CROP SECTOR

The five case studies introduced here illustrate the enormous challenge of moving tropical oil crops toward a more sustainable future as well as the very heterogeneous local conditions in which that task will be pursued, which will require a correspondingly diverse set of approaches. In all cases, however, a clear priority for long-term progress is to develop policies and institutions that support better land and forest governance. A specific agenda to achieve this goal would include unequivocal recognition of customary tenure, protection of forests with high biodiversity and carbon stocks (and devolution of other forest land to communities), recording of tenure rights, establishment of an efficient and transparent land administration system, and development of a strong judicial system (for example, see Deininger [2014] and Deininger et al. 2012).

This agenda addresses long-term needs. Promising options for moving forward in the short to medium term start with efforts to reduce the demand for cropland by increasing yields, improving the state's management of large land concessions, and adopting spatial planning and landscape approaches to reduce conflicts among protecting environmentally sensitive areas, respecting the rights of existing land users, and expanding the area planted to oil crops. The use of incentives is a another practical option to promote sustainable land-use changes. Examples include voluntary private certification systems for agricultural products, and payments for environmental services provided by forests—notably, global services under REDD+.

Improving Land Governance

Better land governance—the capacity to protect the rights of communities and vulnerable groups while providing incentives and reducing transaction costs to private companies and smallholders alike—is critical for sustainable oil crop production. Deininger and Feder (2009) define three key elements of better governance: an appropriate legal and institutional framework, a comprehensive information system, and impartial institutions to enforce rights (the bureaucracy and the courts). We would add to this list the capacity to screen and monitor large land-related investments. These tasks are challenging; they will require sustained effort by states, the private sector, and civil society for many years.

In strengthening the legal and institutional framework, the first priority is for national land law to recognize customary law at a level at least equal to statutory law. Fortunately, a growing number of countries (including Mozambique, as we have seen) have reformed their land laws to elevate the status of customary law. Part of any such reform will be to rationalize state ownership of land—specifically, to identify state land that generates important national public goods, such as forest protection and conservation of biodiversity, and to devolve remaining state land to private actors, especially the communities using that land.

When they have recognized customary law fully, countries must initiate a formal process to demarcate and register customary rights, record land transfers, and build up a land cadaster. Low-cost methods for delineating community boundaries on the ground, combined with the declining cost and increasing accuracy of satellite imagery, mean land rights can be registered relatively inexpensively at the community level in the medium term. Community rights, established with safeguards to prevent elites from capturing these rights and low-cost administrative systems to record land transfers, will give households and communities the security to invest in their land and engage with investors. Poorly defined community land rights are one reason for the failure of joint oil palm ventures between communities and commercial interests in Indonesia and Malaysia. Civil society has an important role in promoting customary rights and monitoring processes on the ground to enforce those rights.

Improving the Management of Large Land Concessions

An urgent priority is to reform the widely used system for granting land concessions, especially concessions for oil palm plantations in Southeast Asia and West and Central Africa. In an era of increasing land scarcity, the state's power to allocate large tracts of land at nominal prices invites rent-seeking, inefficiencies, deforestation, and (not least) abuse of the rights of local land users. In the long run, the state should withdraw from direct land transactions by relinquishing state ownership and building community rights and capacity to enter into land transfers or joint ventures with companies.

In the medium term, the priority is to make land concessions fully transparent by publicizing requests for land, conducting participatory mapping of existing land users, auctioning land to establish market prices, and publicizing the concession contracts that are in force. Requiring grantees of land concessions to adhere to private standards, such as the RSPO, would also help to introduce processes such as free, prior, and informed consent to land concession transactions and provide recourse to a dispute resolution mechanism if needed (as discussed later). Land concession contracts could also include clauses stipulating the minimum area that must be allocated to smallholders and the minimum area to be set aside for conservation.

Saving Land by Increasing Crop Yields

A common proposal to reduce the pressure for oil crops to replace tropical forests is to raise the productivity of land already under oil palm, chiefly by investing more in R&D and diffusing best management practices to raise yields. This "land-sparing approach" is often attributed to Norman Borlaug, "father of the green revolution" and Nobel Laureate, who claimed that the green revolution saved more than 1 billion ha from being converted to crop agriculture (Borlaug 2007). Following similar logic, many analysts support stepped-up efforts to increase oil palm and soybean yields to save the world's remaining tropical forests. In particular, industry officials frequently note that the expansion of oil palm with its very high yield of oil per hectare relative to all other oil crops is land saving in itself. Our analysis in Chapters 2 and 3 indicates substantial potential exists to increase oil palm yields, especially for smallholders, but the potential to increase soybean yields is lower.

As we have also noted, some argue that increasing the yields and productivity of crops grown on the forest frontier will not save land; by raising the return to land, they will provide incentives to expand crop area at the cost of forests and other natural areas. This proposition is sometimes known as Jevon's Paradox after the 19th-century economist who observed that increasing the efficiency of resource use may increase the use of that resource. Angelsen and Kaimowitz (2001) first articulated this argument for land use by building on a series of case studies. One case study contends convincingly that the movement of soybeans into Brazil's Cerrado and Amazon biomes was supported strongly by advances in technology that raised soybean yields and productivity sharply (Kaimowitz and Smith 2001). Recent statistical analysis at the *municipio* level also supports a strong relationship between soybean yields and area expansion in the Cerrado (Barretto et al. 2013; Garrett et al. 2013).

Even so, increases in yields in major exporters such as Brazil (soybeans) and Indonesia (oil palm) will put downward pressure on world prices that could negate the incentives to expand area. The relationship between yields and land use is thus more complex than at first sight and ultimately depends on balancing local and global effects through markets. We have made a start in analyzing those effects using a global economic model (Global Trade Analysis Project) that includes vegetable oils as a single sector (that is, it assumes perfect substitution among oils), and that models the conversion of land among forests, pasture, and crops within a global market framework. The results suggest that increasing soybean and oil palm yields within realistic expectations during the next 25 years would have a negative impact on forest area in the exporting countries (Brazil for soybeans, and Malaysia and Indonesia for oil palm) (Stevenson et al. 2011; Villoria et al. 2013). This analysis provides support for Jevon's Paradox that increased yields may offer incentives to expand cropping at the expense of forests (about 0.1% increase in cropland in each case). At the same time, a small increase in area in the exporting countries is counterbalanced by land savings in other countries arising from lower prices for vegetable oils in world markets. Although the increase in yields for oil palm does provide a small net savings in land and GHGs globally, the analysis does not account for the greater biodiversity value of the tropical forests that are cleared in the exporting countries.

The bottom line is that increasing the yield of tropical oil crops by itself will probably have little impact on saving tropical forests. Higher yields will need to be combined with stronger governance of forests. Poor forest governance makes it is easier and more economic for producers of oil crops to increase production through expanded area (extensification), but more rigorous governance will give producers incentives to intensify. The recent experience with the cattle industry in Brazil demonstrates this point; only after forest governance was greatly improved did cattle producers invest decisively in more productive pastures (Lapola et al. 2014). For smallholders where yield gaps are greatest, additional financing and technical support will also be needed to intensify production.

Adopting Land-Use Planning and Landscape Approaches

Given that some expansion of oil and other crops in the tropics is inevitable, an obvious priority is to avoid expansion into land that has high conservation value and high carbon stocks, or that provides other valuable environmental services. In reality, large areas of degraded land could be converted profitably to crop agriculture. The so-called Bonn Challenge of 2011 calls for 150 Mha of such land to be restored to natural areas or sustainable agriculture by 2020. Plans are underway to restore 20 Mha of such land in Latin America.

A growing number of spatial databases identify and estimate the area of land suitable for crop agriculture, including degraded land or land with natural vegetation of low environmental value. Most such analyses in the tropics start by excluding areas of natural vegetation with a carbon stock greater than 40 t/ha, along with protected areas, peatlands, and areas with settled agriculture. Dinerstein et al. (2014) estimated that 125 Mha of degraded land is available in the tropics and neotropics; 40% is in Brazil, 14% in India, 6% in Mozambique, 5% in Myanmar, and 4% in Indonesia. This area includes 55 Mha in units of 5000 ha or larger suited for large-scale farming. Fischer and Shah (2010) use a similar approach to identify land suited for producing particular crops (land that is not forested, is not protected, and has a population density of less than 25 persons/km^2). Their global estimate for soybeans is 138 Mha; for oil palm, it is 22 Mha. These figures drop sharply—to 83 Mha for soybeans and 3.9 Mha for oil palm—if remote areas more than 6 hours away from a market town are excluded. Again, the available land is concentrated in a few countries of Latin America and Africa.

Similar studies have been carried out at the country level. At least 7 Mha of degraded land are thought to be suitable for conversion to oil palm production in Indonesia (some estimates are much higher), much of it *Imperata* grassland (Gingold 2010). In Brazil, EMBRAPA estimates that at least 32 Mha of mostly low-quality pasture is suited for oil palm (Stokes et al. 2014), and 40 Mha is suited to soybeans (Gibbs et al. 2015). In the Republic of the Congo, Feintrenie et al. (2014) estimated that 1 Mha of degraded land or other land of low environmental value, in areas with a very low density of about 2 persons/km^2, is suited to large-scale oil palm production.

These estimates are only a starting point for considering the other dimensions of a strategy to convert degraded land to oil palm production. First, much of this land is fragmented and not amenable to large-scale farming,[2] although it may serve well for smallholder farming within a landscape mosaic. Second, most uncultivated land is still used in some way or claimed by local people (Lambin et al. 2013). Participatory mapping is a difficult but essential exercise for identifying users, their legitimate claims, and instances where their rights are poorly defined in relation to others' rights or where they overlap with others' rights. Attempts to use contested land will be subject to the weaknesses inherent in land law and land administration systems in countries such as Indonesia.

Many countries have zoned land for a specific use. Brazil has an advanced zoning system that ties financing and mill licenses to specific zones for specific crops. In Indonesia, estimates suggest that by directing new oil palm plantings away from peatlands and high-carbon stock forests, oil palm expansion targets in Kalimantan could be met with a 35% to 59% reduction in emissions over a "business as usual" policy and at only modest cost to industry profitability (Austin et al. 2015). However, zoning based on agroclimatic suitability and reduced risks to forests and peatlands may require a change in legal status. In Indonesia, attempts to rezone cleared Forest Estate for agriculture have run into a bureaucratic brick wall despite the potential to reduce deforestation. One pilot effort for one plantation remains unresolved after 5 years and an expenditure of US$200,000 (Rosenbarger et al. 2013).

Top-down approaches to mapping land use and users are a partial answer to the question of how to encourage more sustainable land use. In recent years, participatory landscape approaches have come into favor because they combine the social and environmental dimensions of sustainability. Landscape approaches look beyond individual land parcels to consider the whole landscape in ways that enhance environmental services such as biodiversity and hydrological functions while integrating social and economic opportunities. They ostensibly put people at center stage by emphasizing sustainable livelihoods (Koh et al. 2009; Sayer et al. 2014). A process that includes participatory modeling of scenarios, in conjunction with tools to support negotiations among multiple stakeholders, has proved useful in addressing the several functions of land use and complex tradeoffs within different scales (Sandker et al. 2007; Dewi et al. 2014). These approaches are especially promising in a transparent setting that builds trust (the best experiences have been in Costa Rica and some parts of Brazil). Their success is undermined easily by unequal powers among stakeholders and by poor governance, but they can be aided by strong environmental laws, such

[2] As mentioned, Dinerstein et al. (2014) estimated that less than half of degraded land in the tropics and neotropics could be used for large-scale farming.

as requirements for commercial farms and plantations to set aside a minimum area for conservation.

Paying for Environmental Services and Realizing the Potential of REDD+

Payments for environmental services can provide incentives to forest owners to conserve forests. The REDD+ program[3] was conceived by the United Nations Framework Convention on Climate Change with much fanfare in 2007 as a way to save forests and mitigate climate change. The idea was to set up a global mechanism to provide compensation for environmental services, especially carbon sequestration and conservation of biodiversity. Over time, the emphasis has shifted from the development of private markets for carbon toward performance-based payments by governments to countries that commit to reduce deforestation. Norway has been the largest donor. Many international agencies and some countries have mounted pilot activities to test and refine the concept, but the overall level of funding is still low.

A review of 23 pilot initiatives by the Center for International Forestry Research "reveals huge challenges with implementing REDD on the ground" (Sills et al. 2013, p. xxiii). These initiatives highlight the critical role of secure tenure, free prior and informed consent, equitable benefit sharing, transparency, good administrative capacity, and elimination of corruption in land and forest governance. Accordingly, the focus is on helping countries, local governments, and communities to be "ready for REDD+" through activities that build the wide range of capacities needed to strengthen governance of forest resources.

Even when countries or communities meet the conditions for REDD+, the opportunity costs of conserving forests may be much higher than payments for carbon sequestration, given likely carbon prices. This dilemma is especially relevant for oil palm because of the high payoffs from converting forested land to oil palm production (Butler et al. 2009; Venter et al. 2009; Fisher et al. 2011; Terauchi et al. 2014). The exceptions are peatlands—carbon emissions are especially high when they are cleared (World Bank Group 2008)—and areas that are relatively marginal for oil palm (Abram et al. 2014). In addition, REDD+ needs to include compensation for local governments that depend on taxes (or even bribes) from land concessions and commercial agriculture (Irawan et al. 2013; Nantha 2014).

The large anticipated flow of international funding for REDD+ has not materialized, yet strategic contributions such as the government of Norway's commitment of US$1 billion each to Brazil and Indonesia, based on agreed performance indicators, appear to have heightened national commitments to improve the governance of forest resources (although in Indonesia that commitment has yet to be expressed in reduced deforestation rates) (Birdsall et al. 2014; Seymour et al. 2015).

[3] The "+" was added to REDD to include activities aimed at conservation of forest carbon stocks, sustainable management of forests, and enhancement of forest carbon stocks.

Certifying Sustainability: Will Private Standards Work?

Considering that funding for environmental services has been small to date in relation to the market value of oil palm and the other products driving deforestation, the adoption of voluntary certification programs by private producers will be critical to encourage sustainable land use (Boucher 2015). The growing controversies over the social and environmental costs of oil crops and the failure of states to reduce those costs have led the private sector to develop and adopt voluntary incentives for improving sustainability. The most important approach has been voluntary certification under the RSPO, but many other certification schemes have been initiated, especially for biofuels.

The Roundtable on Sustainable Palm Oil

The RSPO originated in 2004 on the initiative of the largest trader and consumer of palm oil (Unilever) and a large environmental NGO (the World Wildlife Fund for Nature), together with the Malaysian Palm Oil Association and others. The RSPO has a multistakeholder membership and governance. In 2015, the organization had 2600 members. The largest subset of members came from the consumer goods manufacturing sector (41%) (Roundtable on Sustainable Palm Oil 2014); only 8% were growers and only about 3% were civil society organizations. Neither the governments of Malaysia nor Indonesia are members (Nesadurai 2013; CSR Asia 2014).

RSPO certification was initially implemented in 2007 through a set of principles and criteria (P&C) with accompanying guidelines, which by 2013 had been revised and approved to include eight principles and 42 criteria (Roundtable on Sustainable Palm Oil 2013). At first, the certification standards emphasized environmental concerns. A key criterion was that any new plantings since 2005 could not occupy land classified as primary forest, and a prior assessment was required to identify areas of high conservation value. As of 2013, resulting from the pervasive land conflicts between oil palm plantations and local farmers and communities, at least 10 RSPO criteria addressed land rights and relations with local communities (Table 9.3). RSPO members at the country level are encouraged to develop their own national interpretations of the P&C, adapted to their laws and institutions, and subject to approval by the global governing body. The RSPO also set up a mechanism to receive and act on complaints against its members with regard to noncompliance with its Code of Conduct or the P&C in the case of certified operations, and it has a grievance mechanism to address issues in its own performance. By 2015, some 20% of the palm oil supply was certified through a number of supply chain mechanisms (Box 9.2).

The RSPO is one of many voluntary certification schemes that have been set up by industry or by third parties to meet growing consumer demands for environmental sustainability, especially for conservation of tropical forests (examples include the Rainforest Alliance Network) and for social justice (such as Fair Trade certification). Within this growing family of standards, the RSPO follows common norms for environment standards but is relatively more comprehensive on criteria for land rights and community relations (Nesadurai 2013; Byerlee and Rueda 2015). Even its critics

Table 9.3 Major social and environmental sustainability criteria, Roundtable on Sustainable Palm Oil

Land rights, smallholders, and community relations	*Environmental conservation of land and forest resources*
• Proof of legitimate ownership or rights to use the land by company, and rights to land are not legitimately contested • Participatory mapping to show recognized users (legal and customary) • Free prior and informed consent of previous users • Fair compensation to previous users (even if land allocated as government concession) • Social Impact Assessment to assess impacts on livelihoods and subsistence • Open and transparent means to communicate with local communities, including transparency in dealing with smallholders in purchasing FFBs • Contributions to local development • Grievance procedure to address conflicts	• New plantings since 2005 may not replace primary forest • Avoid use of fragile soils, including peatlands • High Conservation Value assessment to identify and maintain biodiverse areas • Plans to minimize net GHG emissions (but no specific targets) • Ban use of fire for land preparation

An additional requirement is that an Environmental and Social Impact Assessment must be carried out with full consultation and made public, and a protocol must be put in place to monitor agreed measures to mitigate land conflicts and environmental impacts. In early 2016, the RSPO issued a second higher level certification (RSPO Next) to comply with the commitment made by the major trading companies in late 2014 to zero deforestation, zero peatland, and no exploitation of communities. FFBs, fresh fruit bunches; GHG, greenhouse gas.

Source: Based on Roundtable on Sustainable Palm Oil. 2013. *Principles and criteria for the production of sustainable palm oil.* Kuala Lumpur: Roundtable on Sustainable Palm Oil.

agree that the RSPO is a major step forward on land rights relative to the state-led systems (Pichler 2013).

The RSPO P&C represent a struggle between global and national governance (Hospes and Kentin 2014). Consumer demands in rich countries do not necessarily translate into local priorities. Already, some local communities have protested strict environmental requirements on conservation that may infringe on their rights and livelihoods (CSR Asia 2014). In other cases, standards for global markets may conflict with national laws and regulations. RSPO requirements to set aside land of high conservation value may conflict, for example, with state regulations that demand full development of land allocated through concessions.

Palm oil is one of the first "bulk commodities" (commodities of standard and undifferentiated quality and relatively low unit value) to attempt large-scale certification,

Box 9.2 Supply Chain Mechanisms for Certification by the Roundtable on Sustainable Palm Oil

Supplies can be certified in several ways. In the GreenPalm market, certified producers sell GreenPalm certificates to buyers who want to support sustainable palm oil (Green Palm Sustainability 2015), but the palm oil from certified growers is not segregated in trade. Certified growers can also sell into a market where their palm oil is bulked with other certified oil, or certified oil is kept fully traceable along the supply chain to the final retailer or user. A "mass balance" approach allows certified oil to be mixed with noncertified oil, as long as the amount certified at the end point of the supply chain does not exceed the total certified oil entering the supply chain.

Source: Roundtable on Sustainable Palm Oil. *2014. An international multilevel stakeholder initiative transforming markets to make sustainable palm oil the norm.* Kuala Lumpur: Roundtable on Sustainable Palm Oil.

and as such the RSPO is a work in progress that continues to reflect learning and adaptation over time. Inevitably within a multistakeholder governance structure, tensions arise between industry actors and others not directly involved in the supply chain, which leads to compromise rather than consensus (McCarthy 2012; Schouten and Glasbergen 2012). Many compromises reflect tradeoffs between economic development and strict sustainability standards such as zero deforestation. A subgroup, the Palm Oil Innovations Group of companies and NGOs, is testing and demonstrating systems to tighten even further standards that might be incorporated in future revisions of the P&C. Tighter standards are likely to be needed to help companies comply with recent commitments by trading companies to zero deforestation, zero peatland, and no exploitation of local communities.

Some accuse the RSPO of making decisions that reflect the unequal power relations between industry and other stakeholders, such as civil society, smallholders, and labor (Pichler 2013; Marin-Burgos et al. 2014). This claim may have some justification, yet the RSPO does provide "discursive power and leverage to nongovernmental organizations and social movements," which is absent in many state-led schemes (McCarthy 2012, p. 1872).

The RSPO should not been seen as a panacea for the industry's poor sustainability image. Companies have major economic incentives to certify their production or supply chain under only two conditions. The first is if they benefit from market premiums for sustainably produced palm oil. However, sustainability standards are largely demanded by rich consumers that account for less than one quarter of global palm oil consumption. The big markets in Asia (China, India, and Pakistan) seek cheap vegetable oil, and so far their consumers show little inclination to pay a premium for certified oil. As a result, the price premium is very small (0.3%–2.5% and a bit higher for segregated oil) and insufficient to cover certification costs (Potts et al. 2014).

A second incentive for companies to certify production is to access finance through equity markets, institutional investors, or bank loans that increasingly use sustainability standards as a condition for funding. About 50 palm oil companies are publicly listed and must respond to shareholders' concerns, especially the concerns of institutional investors at the global level. About half of all palm oil companies are owned by private insider interests, however— typically families (Grayson and Stampe 2012) that will embrace sustainability objectives only to the extent they are financed through banks that impose sustainability criteria. Bank loans provide about 24% of the financing for oil palm companies (Shibao 2015).

That said, companies have noted that certification has other benefits. A series of small improvements in management gained through the certification process can lead to yield gains that more than cover the costs of certification (Levin 2012). In the longer run, the benefits of certification are likely to take the form of better labor relations and reduced conflicts with communities.

Perhaps the biggest challenge of certification is to extend it to smallholders, who supply more than one third of the market. The economies of scale in certification are well known, although special provisions have been developed for smallholders (Deininger and Byerlee, 2012; Mausch et al. 2009). Some smallholders working in groups have been certified, especially plasma associated with certified mills and nucleus estates in Indonesia. The certification challenge is particularly acute for independent smallholders. Donor-funded efforts to reach this group have encountered high certification costs per farmer, and reviews note the difficulty of scaling up (Beall 2011; Nagiah and Azmi 2013). In its P&C, the RSPO does not encourage partnerships with smallholders explicitly as a way to share benefits.

Despite these challenges, palm oil certification generally imposes stricter standards and already covers a larger share of the market than certification efforts for any other bulk commodity. The RSPO system has progressed much faster than its sister Roundtable on Responsible Soy (Box 9.3). The palm oil certification system is far from perfect—particularly in its monitoring and enforcement practices—but it is a much more transparent and consultative process than state-based approaches or even those of some of its critics. Standards are becoming tighter over time. Some unmet demands, such as full traceability through the supply chain, seem unrealistic (and unnecessary) for all but the most demanding markets that are willing to pay a substantial premium. Some regions, notably Sabah State in Malaysia and Central Kalimantan State in Indonesia, are moving toward "jurisdictional" certification of all palm oil produced in their territory to reduce transactions costs. Even if they succeed, market-driven standards are unlikely to substitute completely for fair and transparent state laws and regulations, especially in an environment where governance of land and forest resources is poor (McCarthy 2012; Lambin et al. 2014).

Other Certification Initiatives for Oil Palm

The RSPO is a voluntary certification scheme. Mandatory schemes are also being introduced on both the user and producer ends of the supply chain. On the user end,

Box 9.3 Certifying Sustainable Soybean Production

Unlike efforts with palm oil, efforts to certify that soybeans are produced sustainably have met with the additional complication that most producers plant GM varieties that civil society and many consumers regard as incompatible with sustainable production. ProTerra, a Brazil-based group initiated in 2006, certifies non-GM soybeans for sustainable practices. It had reached more than 5% of Brazilian production by 2008, but since then its share of production has fallen whereas the share of GM seed has increased. Even the substantial price premium that ProTerra enjoyed for non-GM soybeans seems to have evaporated.

For soybeans, the equivalent to the RSPO is the Roundtable on Responsible Soy (RTRS). With its global mandate, the RTRS is structured quite similarly to the RSPO as an industry initiative with an opportunity for other stakeholders to participate. The RTRS did not get off the ground until 2011, in part because of the squabble over GM soybeans. Certification is picking up quickly, although it still covered only about 1 Mha, or less than 1% of the industry in 2014.

The RTRS faces three challenges to scaling up. The first is a low price premium for certified soybeans. Because of the moratorium imposed by traders on purchases of soybeans grown on recently deforested land in Brazil, the deforestation attributed to soybeans is now negligible, although the expansion of soybean production in the Cerrado is still associated with significant conversion of natural areas. Consequently, demand for certified soybeans is low, and indeed the price premium has been running at a meager 0.3% (Potts et al. 2014). Second, RTRS certification of GM soybeans undermines demand in some quarters. Third, the big Brazilian players—Associacao dos Produtores de Soja e Milho do Estado de Mato Grosso (the producer association) and the Associação Brasileira das Indústrias de Óleos Vegetais (Brazilian Association of Vegetable Oil Industries; the association of processors and traders), both of which supported the moratorium—have withdrawn from the RTRS because of its restrictions on land-use changes; they feel that Brazilian law, if enforced, ensures sustainable land use (Hospes et al. 2012). They have initiated another scheme, Soja Plus, based on implementation of existing laws, although it has yet to move to full certification.

the European Union, through its Renewable Energy Directive, or RED, developed standards for biofuels based on a 35% savings in GHG emissions, and biofuel feedstocks such as palm oil and soybean oil must comply with one of the European Union's approved certification systems. The EU standards are stricter and more specific with respect to GHG emissions than the RSPO standards, so the RSPO developed RSPO–RED, which includes the additional criteria that must be met under the EU guidelines. In addition there are standalone biofuel certification systems, such as the International Sustainability and Carbon Certification and the Roundtable on Sustainable Biofuels. Not surprisingly, given the focus on GHGs in the European Union, the approved certification systems for biofuels are often quite weak in their treatment of land rights and other social issues (German and Schoneveldt 2012). Under one of the weaker systems, the social impacts of certified vegetable oils used in biodiesel may well be negative.

On the producer end, the government of Indonesia required all companies to become certified under its Indonesian Sustainable Palm Oil (ISPO) system by the end of 2014 although full implementation has been delayed. The ISPO focuses on compliance with Indonesian laws and regulations. ISPO certification is likely to lack credibility to global customers resulting from the poor record of state regulation and enforcement in Indonesia. A similar Malaysian Sustainable Palm Oil certification scheme initiated by the government in 2015 shows promise, but consumers will find the multiplication of certification schemes confusing.

Sustainability Commitments by Palm Oil Traders

The growth of voluntary certification schemes was overshadowed in late 2014 when the major palm oil traders committed to zero deforestation, zero peatland, and no exploitation of local communities. The big breakthrough came in December 2013, when Wilmar, which accounts for about 45% of global trade in palm oil, made these commitments (Wilmar International 2013). Other major traders that account collectively for anywhere from 60% to 90% of the global trade in palm oil (estimates vary widely) have since joined. These commitments could make a huge difference if and when they are implemented, because most exports—including exports to the big markets of China and South Asia, where demand for certified oil is low—will have to meet standards higher than RSPO standards. The new commitments appear to result from a protracted campaign by some multinational buyers and civil society, the increasing access to real-time information on deforestation that can be mapped to companies, and pressure from the government of Singapore, where several palm oil giants (including Wilmar) are headquartered and there is a loud annual outcry about the smoke haze from Sumatra (caused in part by burning to clear land for new plantations).

Details such as the definition of high carbon stocks and the time frame are still being worked out. A study commissioned by the companies is proposing a zero net emissions metric that would allow some deforestation as long as there are appropriate emission offsets. With agreed definitions, it should be relatively straightforward to monitor aggregate deforestation and planting on peatlands through satellite imagery, but it will be much less straightforward for a large trader such as Wilmar, with more than 800 suppliers to implement full traceability from the field to the mill and along the supply chain. The commitments resemble Brazil's moratoriums on unsustainable soybeans and beef, but they will be much more difficult to implement in Indonesia, with its weaker state capacity, and poor land and forest governance.

It will also not be easy to define and monitor the commitment to "no exploitation" of local communities. Smallholders run the greatest risk of being sidelined in the market by the new commitments, which make no mention of supporting a balanced development model that gives equal weight to smallholders. Monitoring the social commitment will be quite tricky unless it is done under an enhanced RSPO model or land governance is reformed dramatically. The RSPO is already working on a set of standards to certify mills for compliance with the traders' new and more stringent standards on deforestation, peatlands, and land rights.

Finally, how might the major consuming countries of Asia (India, China, Indonesia, and Pakistan) respond if they conclude that prices of an essential food staple are rising to cover the costs of implementing the commitments? Asian-based trading companies conceivably could expand their operations or set up new operations to circumvent the commitments. COFCO, a large, state-owned Chinese food trading company, has already entered the soybean market in Brazil (Chapter 3). A further consideration is that about one third of Indonesian palm oil is consumed domestically, and an unknown share is traded through local companies that have not signed on to the commitments.

Will 2014 prove to have been the turning point in a long struggle to move oil palm to a more sustainable footing? As always, the devil is in the details. Short of full and credible certification of all suppliers against tightly specified standards, the controversy over oil palm is unlikely to fade away. A particular vulnerability—shared with the voluntary certification schemes—is that without support from governments, especially in Indonesia, companies will find it difficult to implement their commitments fully. The government of Indonesia in particular has sometimes been critical rather than supportive of the new commitments by the multinational trading companies.

CONCLUSION

Oil crops have driven economic growth and helped to increase rural incomes and reduce rural poverty (Chapter 8), but at the same time they have been associated with serious negative social and environmental outcomes. The media and civil society often equate tropical deforestation with the expansion of oil crops, yet the reality belies this simple formula. The causes of deforestation and social injustice on the forest frontier are many and complex. They include logging, operating timber plantations, raising cattle, and growing crops other than oil crops, which means that sustainability outcomes can be improved only by working at multiple levels.

Zero deforestation is a worthy, achievable target. It is also a very tangible target, in the sense that throughout the world it can be measured and monitored in real time though satellite imagery. A zero deforestation target should not be an impediment to expanding the industry. An ample supply of degraded land or pasture is available for growing oil crops (and other crops) to meet future market demand, provided that rights to such land can be negotiated with the existing users, or that those users become producers themselves.

Compared with a zero deforestation policy, implementing a "no exploitation" policy, or even better a smallholder-friendly policy, is far more challenging. Insecure land rights and a host of strong commercial interests often undermine the livelihoods of local communities. Some flexibility should be built into deforestation targets for smallholders, because, for them, moving to more profitable systems involving oil crops can be an important pathway out of poverty. This is especially so in Africa, which is poised to be the next frontier for oil palm expansion.

The environmental and social sustainability stakes are high, and the global community and industry have put much faith in incentive systems such as REDD+ and voluntary certification to win the sustainability game. Yet, the inescapable conclusion in this chapter is that those efforts are not enough. We believe incentive systems will have to be coupled with long-term efforts to resolve the deep, underlying problems of weak land and forest governance. Here, the state is the major player and, even if governance improves in one major producing country, production may well shift to another, unless governments and multinational companies commit to coordinated international action.

The reassuring news is that a great deal of progress has been made in the past few years, and the story of oil palm is evolving as we write. The moratorium imposed by major global traders arrested deforestation from soybean production decisively, and the similar stance taken recently by palm oil traders is a huge advance, even if big traders still face the daunting task of ensuring that all their suppliers meet sustainability standards.

We do not espouse win–win solutions on sustainability and recognize that judicious tradeoffs are needed. In particular, and in line with our conclusions in Chapter 8, we believe the full social benefits of the industry can be achieved only through greater inclusion of smallholders. Achieving this goal, especially where forest governance has been devolved to communities, may require tradeoffs between social and environmental outcomes. Although the global community has raised awareness of the threats to sustainability and offers incentives to reward more sustainable action, it will be "the people who live in the tropics who will forge the future fate of tropical forests" (Putz and Romero 2014, p. 502).

10

CONCLUSIONS
THE FUTURE WILL NOT BE LIKE THE PAST

TROPICAL OIL CROPS IN A GLOBALIZING WORLD

We began this book by describing the changes in tropical oil crops in recent decades as an agricultural revolution. Our in-depth review affirms the transformative nature of these changes. In a remarkably short space of about two decades, two oil crops—soybeans and oil palm—have become among the most important crops in world agriculture because of their contributions to food, feed, and fuel supplies. Oil palm is produced entirely in the humid tropics, and soybean production has moved decisively to the tropical zones in recent years.

The rise of these tropical oil crops has taken place within a globalizing world where the products of one region are consumed largely in another, so tropical oil crops and their products now rank first (soybeans and its derivatives) and third (palm and palm kernel oil) in world agricultural trade (wheat and wheat flour occupy second place). These two leading oil crops attracted massive investments of private capital to open up sparsely populated regions and created a host of dynamic industries in emerging economies, such as oil milling, intensive livestock, biodiesel, and oleochemical industries. Finally, these crops have been among the most important drivers of changes in land use in the tropics, where their production converted huge tracts of forest, savannah, and pasture into highly productive farm land.

In this final chapter, we recapitulate briefly findings on the major drivers of the tropical revolution in oil crops and its likely future trajectory. We conclude by reviewing options to make the future expansion of oil crop production more sustainable, especially in Africa.

THE MAJOR DRIVERS

In hindsight, we see that a host of factors aligned on the demand and supply sides to produce a "perfect storm" of change in tropical oil crops. Coming together during the 1990s, these factors drove a revolution in oil crops that endured into the first decade of this century.

First and foremost, demand expanded rapidly for both protein meal and vegetable oils. The rapid rise in consumption of livestock products in emerging economies that generated the demand for protein meal, and for soybean meal above all, is widely recognized. Less well known is the vibrant demand that emerged for vegetable oils for cooking and processed foods as these same consumers in emerging economies became richer. From 1991 to 2011, vegetable oils accounted for at least one quarter of the increase in calorie consumption globally. These increases in food demand for meat and edible oils were challenging in themselves, but they were made doubly challenging when the biodiesel industry—which depended heavily on vegetable oils as a feedstock—took off. This new industry, driven mostly by policies mandating the use of biofuels, has accounted for nearly half the increase in global vegetable oil demand since 2003.

These dramatic increases in demand could not have been satisfied without equivalent drivers on the supply side. In Brazil, cheap land became accessible with the construction of new highways, and new technologies allowed soybeans to grow well in the tropics. In Southeast Asia, even cheaper land became available for oil palm as governments made large concessions of forested and degraded land available to private companies to "develop." Timber exports were declining, and governments needed new sources of revenue and ways to use the land in their Forest Estate. The era of market liberalization and privatization starting during the late 1980s brought a surge of private capital, often foreign, to develop new supply chains, especially during the commodity boom of 2007 to 2014. State capital, often subsidized, was also available to the oil crop sector through national development banks. Easy access to land and capital, along with the availability of productive technologies, made oil crops very profitable and promoted their expansion.

The supply and demand sides were brought together by global trade liberalization and integration. As countries joined the newly created WTO, imports of vegetable oils and soybeans were liberalized almost everywhere, and trade exploded. Producers in a handful of exporting countries were linked to distant consumers and rapidly emerging new industries by investments in roads and ports and ever larger ships that greatly facilitated trade at relatively low costs. The most dramatic changes occurred in China and India. China's liberalization of soybean imports made it by far the world's largest importer of soybeans (mostly supplied by Brazil), and India's liberalization of vegetable oil imports made it the world's largest importer of vegetable oils (largely supplied by Indonesia).

Finally, the extraordinary success of the two dominant commodities—palm oil and soybeans—came from massive substitution of these products for traditional sources of vegetable oil (for example, palm oil replaced virtually all the coconut oil used in food in Indonesia) and livestock feed (soybean meal replaced waste materials and by-products in animal feed in China). We estimate these substitutions accounted for one half the global growth of palm and palm kernel oil and one quarter the growth of soybean meal consumption during 1991 to 2011. These substitutions reflected the intrinsic properties of the two commodities (such as the high protein content of soy meal), their low production costs, and the opening of world markets.

THE WINNERS AND LOSERS

As with all periods of rapid economic change, the revolution in oil crops has had its winners and losers. The revolution undoubtedly produced enormous economic benefits by creating new industries and transforming sparsely settled areas into dynamic growth poles with new towns and service providers along the value chain. Oil crops created millions of new jobs in Southeast and South Asia—regions with extensive poverty. Smallholders sometimes participated strongly in this growth, as in Indonesia (oil palm) and India (soybeans). The availability of cheap, plentiful vegetable oil has been important for improving food security in extremely poor populations by increasing their energy and fat intake.

Our review and many others also highlight the downsides of the revolution. Rapid growth presented opportunities for attaining more inclusive outcomes, but often they were missed. More could have been done in the leading producers to develop agrarian structures based on small and medium-size family farms and smallholders. Local communities without secure land tenure often lost livelihoods as large investors usurped their land rights. Although labor rights have received less attention than land rights, many laborers in oil palm cross international borders illegally and are sometimes subject to employment abuses and poor living conditions.

On the consumer side, consuming more than a certain amount of vegetable oil poses health risks that are possibly greater for palm oil, with its high level of saturated fats. Health risks are not the only risks. When oils are imported directly or indirectly as oilseeds, which is the case for 56% of world oil consumption, consumers are exposed to supply risks, because they are depending heavily on two commodities from a handful of exporting countries where plant disease epidemics, climatic shocks, and political instability could reduce supplies sharply. The same set of risks could undermine the livelihoods and food security of smallholder oil crop producers and workers in exporting countries if they depend largely on one cash crop.

Finally, the massive changes in land use associated with tropical oil crops have had high environmental costs in terms of GHG emissions and biodiversity losses, especially where oil crops have replaced tropical forest. All recent analyses conclude oil palm and soybeans, together with beef, and pulp and paper, are the four key commodities contributing to GHGs from land-use changes, which in turn account for about 15% of all GHG emissions.

Some signs point to convergence among global consumers, private actors, civil society, and local governments in finding ways to minimize the tradeoffs between economic benefits and social and environmental costs. Brazil, the major soybean producer, has made the most progress. Brazilian soybeans have contributed little to deforestation in recent years, although expansion into savannah and woodland continues, albeit more slowly, and is possibly displacing cattle farming to the Amazon frontier. As we write, a parallel story is unfolding in Indonesia, where major global traders are making strong commitments to achieve zero deforestation, zero use of peatlands, and no exploitation of local communities—but a difficult and uncertain path lies ahead when it comes to implementing these commitments on the ground.

THE FUTURE WILL NOT BE LIKE THE PAST

Alarmists predict oil crops will consume the remaining tropical forests. Our analysis sees growth in demand for tropical oil crops slowing sharply in the years to 2050 compared with recent experience. Our best-bet projection of demand for vegetable oils to 2050 is about 310 Mt, up from about 165 Mt in 2013 and representing an annual growth rate of 1.7%—a bit more than one third of the 4.8% growth rate from 2001 to 2013. We expect soy meal consumption to expand at an even more modest rate of 1.5% annually.

Such projections are fraught with uncertainty, but for a number of reasons we are confident growth in demand will be much slower toward 2050 than in the recent past. First, growth in demand for biofuel feedstocks cannot maintain the rapid pace of the past decade. This tapering off will be most apparent in the European Union—the major consumer of biodiesel today—especially because it is approaching the regulated maximum of renewable transport fuels that can be provided from foodstuffs. Some tropical countries, notably Brazil and Indonesia, are likely to compensate in part for the EU slowdown, but in our view neither India nor China (the two most populous countries) nor sub-Saharan Africa will become significant producers of biodiesel, given their high dependence on imported vegetable oils. Second, the use of vegetable oils for food will also grow more slowly than in the recent past; population growth will be slower, and the effects of increasing incomes will diminish as consumers in middle-income countries reach higher levels of vegetable oil consumption. The major exception will be sub-Saharan Africa, where both population and incomes are rising rapidly and vegetable oil consumption per capita is still very low. Third, the use of protein meal for feed is subject to a similar slowdown as China's meat consumption levels off (protein meal in feed concentrates has already hit optimal levels). Nonetheless, 2.5 to 3 billion people in South Asia and Africa still consume very little meat, and many will adopt diets with more meat as their incomes increase.

In our view, the area covered by oil crops does not have to expand greatly to meet future demand. This demand can be supplied largely through intensification of existing crop land and only a modest expansion in area. There will be no dramatic yield jumps, because yield gaps for oil crops in the major producers are already low relative to gaps in yields of other major staples, but steady progress is possible through genetic gains in yield. In oil palm, these gains will be realized slowly only because of the long replanting cycle, and attaining greater labor productivity will be even more important as wages in the main producers increase quickly. In addition, sufficient degraded land is available to accommodate area expansion, provided land governance and incentive systems can be developed to steer the expansion onto degraded lands and away from areas of natural vegetation with high carbon stocks and biodiversity values.

We expect the globalization of supply chains will ensure palm oil and soybeans remain highly competitive commodities, and their share in oil crops will continue to rise. China, which is investing heavily in logistics and infrastructure in Latin America, plans to link the Brazilian hinterlands by rail to Pacific ports. New infrastructure corridors emerging in Africa will facilitate imports for Africa's emerging consumers and poultry industries, and in some cases may even stimulate exports. The growth of palm oil and soybeans as substitutes for other oils and oil meals, respectively, will slow as they come to dominate their respective industries, however. Palm oil already accounts for 97% of the vegetable oil used in Indonesia, and soybean meal provides 72% of the protein meal in China (up from 14% in 1991).

We recognize that many unpredictable or unexpected developments will influence these projections. On the demand side, the wild cards include the future of African economies (can they maintain their recent accelerated growth?), the future of biodiesel policies in emerging countries such as Indonesia and Brazil (can they meet B10 or even higher mandates?), and meat consumption in India (will India's hundreds of millions of vegetarians join the ranks of major meat consumers as in China?). High dependency on imports—India's reliance on imported vegetable oils is a case in point—could cause a backlash against trade liberalization, with negative consequence for the growth of trade. On the supply side, the biggest wild card is how much land will be available for expanding oil palm in Asia and Africa, especially as environmental regulations tighten and current users gain more secure rights to their land. Whether this expansion is on forested or degraded land, the transaction costs of accessing large tracts of land surely will rise.

BETTER PROSPECTS FOR SUSTAINABILITY

We are cautiously optimistic that the future expansion of the sector can be managed in ways that minimize social and environmental costs. The much slower projected expansion will, in itself, make such costs more manageable. Recent experience in Brazil and very recent developments in Indonesia suggest that through pressure from civil society, consumers, enlightened companies, and local actors, the sector may have reached a turning point and is moving away from destroying forests and usurping local land rights to produce oil crops.

In the short to medium term, the best bets to improve sustainability involve full implementation of the commitment by palm oil traders to zero deforestation, zero peatlands, and no exploitation, combined with an upgraded certification systems to enforce standards based on those commitments (called RSPO Next) to help clean up supply chains. The performance-based contracts under REDD+ also provide important incentives to governments to support such commitments. On the social side, we see major new opportunities for smallholders to take advantage of the growing infrastructure of roads, mills, and logistics, as they have already done in Sumatra. Indeed, we see no reasons why smallholders, given the right sort of support, could not lead the future expansion of oil crops (Chapter 8). Still, a big challenge is how to support smallholders to adopt sustainable practices and to certify their production, given the high transaction costs and skill requirements they face.

In the long run, private standards, trader commitments, and moratoria cannot substitute for better land and forest governance. Better governance entails better geospatial information on land use and changes, reform of the land concession system, accelerated titling of land for communities and smallholders, an efficient and transparent land administration system, and a redress system. These same actions are also required to implement REDD+ during the long term. At the same time, we expect concerns about labor rights and conditions—overshadowed until now by concerns about land rights—to come to the fore as employers look further afield to recruit labor supplies. Reforms of biofuel mandates make economic and environmental sense but will face strongly entrenched political interests, especially during periods when vegetable oil prices decline.

AFRICA AS THE NEW FRONTIER

Africa is poised to take center stage in the tropical oil crop revolution in the coming decades. The region's demand for vegetable oils and protein meal already is expanding rapidly. On the supply side, Africa has the land and labor to meet this demand and perhaps become a significant exporter.

Oil palm, a traditional African crop, requires a judicious combination of improvements in the local supply chains of smallholders and small-scale processors with injections of outside capital, technology, and market expertise through private investors. If the focus is on increasing the productivity of existing producers, whose yields are only a fraction of potential yields, there is huge scope for meeting market demand, both regional and global, without expanding land area and by relying on smallholders. Approaches for joining outside capital and expertise with the assets of smallholders remain a work in progress, however. Learning from the Asian experience, we see that a critical role for the public sector is to provide incentives and support to the smallholder sector, especially when it comes to facilitating access to improved genetic stock, extension information, and long-term financing.

Oil palm strategies in Africa must also recognize the dual markets of unrefined red palm oil for local foods and refined palm oil for cooking oil and processed foods. With urbanization, the market for refined palm oil will expand faster. This expansion will shift the advantage gradually from small, semimechanized mills to larger industrial mills to meet quality standards, but with innovative business models, even the larger mills can be supplied by smallholders. The current emphasis on investment in large plantations is unlikely to be sustainable, given the complexities of African land markets and land rights.

Soybeans, a new crop in much of Africa, will inevitably expand with the attendant environmental costs of converting savannah and woodland to crop agriculture. Again, the focus should be on small and medium producers through contract farming and outgrower schemes to supply domestic and regional markets. Although the potential exists to develop large-scale soybean production for export in sparsely populated areas, especially along new infrastructure corridors, we see this option as a lower priority. Africa must absorb its growing labor force, and this massive challenge will not be met by developing fully mechanized, large-scale farms.

Even so, investors are exerting considerable pressure to gain access to land for tropical oil crops in Africa. Given the poor state of land markets and governance, expansion in this direction threatens the livelihoods of existing land users. The last large tract of tropical rainforest in the Congo Basin has been protected from deforestation and timber extraction by the lack of infrastructure, but the steady extension of infrastructure throughout Africa and better links to global markets pose a significant threat to these forests. Civil society, governments, and investors all have a role in strengthening governance of land and forest resources. The recent experience in Brazil and the unfolding story in Indonesia provide many lessons on how to accomplish this objective.

FINAL NOTE

Our review leaves us with a sense of optimism about the future benefits of growth in the oil crop sector. Slower growth will help to reduce the social and environmental costs of producing oil crops, and we see a new willingness to manage those costs. The commitments to sustainability by global consumers and supply chains will make it that much harder for states and companies to destroy forests and tread on the land rights of local communities. Finally, where there is a will, new tools allow stakeholders to monitor land users and changes in land use in real time, and they can identify and expose transgressors quickly. These tools, together with a growing concern about climate change and biodiversity loss, are shining a spotlight on tropical oil crops and global trade in their products—a spotlight that companies and states alike can no longer ignore.

REFERENCES

Aatola, H., M. Larmi, T. Sarjovaara, and S. Mikkonen. 2008. *Hydrotreated vegetable oil (HVO) as a renewable diesel fuel: Trade-Off between NO_x, particulate emission, and fuel consumption of a heavy duty engine*. Warrendale: SAE International.

Abbott, P. 2013. Biofuels, binding constraints, and agricultural commodity price volatility. NBER working paper no. 18873. http://www.nber.org/papers/w18873 (accessed 31 January 2016).

Abbott, P., C. Hurt, and W. E. Tyner. 2009. *What's driving food prices? March 2009 update*. Oak Brook: Farm Foundation.

Abood, S. A., J. S. H. Lee, Z. Burivalova, J. Garcia-Ulloa, and L. P. Koh. 2014. Relative contributions of the logging, fiber, oil palm, and mining industries to forest loss in Indonesia. *Conservation Letters* 8 (1): 58–67.

Abram, N. K., P. Xofis, J. Tzanopoulos, D. C. MacMillan, M. Ancrenaz, R. Chung, L. Peter, et al. 2014. Synergies for improving oil palm production and forest conservation in floodplain landscapes. *PLoS One* 9 (6): e95388.

Adjei-Nsiah, S., O. Sakyi-Dawson, and T. W. Kuyper. 2012. Exploring opportunities for enhancing innovation in agriculture: The case of oil palm production in Ghana. *Journal of Agricultural Science* 4 (10): 212–223.

Agarwal, D. K., S. D. Billore, A. N. Sharma, B. U. Dupare, and S. K. Srivastava. 2013. Soybean: Introduction, improvement, and utilization in India: Problems and prospects. *Agricultural Research* 2 (4): 293–300.

Agricultural Science and Technology Indicators and Malaysian Agricultural Research and Development Institute. 2012. *Agricultural science and technology indicators survey, 2012*. Unpublished. Washington, DC.

Agus, F., P. Gunarso, B. H. Sahardjo, N. Harris, M. van Noordwijk, and T. J. Killeen. 2013. Historical CO_2 emissions from land use and land-use change from the oil palm industry in Indonesia, Malaysia and Papua New Guinea. http://www.worldagroforestry.org/sea/Publications/files/report/RP0296-13.pdf (accessed 31 January 2016).

Ahmad, S. Z., and P. J. Kitchen. 2008. Transnational corporations from Asian developing countries: The internationalisation characteristics and business strategies of Sime Darby Berhad. *International Journal of Business Science and Applied Management* 3 (2): 21–36.

Alexandratos, N., and J. Bruinsma. 2012. World agriculture towards 2030/2050: The 2012 revision. ESA working paper no. 12-03. Rome, Italy: Food and Agriculture Organization of the United Nations.

Alltech. 2014. 2014 Alltech global feed survey summary. Nicholasville: Alltech. http://www.alltech.com/sites/default/files/alltechglobalfeedsummary2014.pdf (accessed 31 January 2016).

Alves, B. J. R., R. M. Boddey, and S. Urquiaga. 2003. The success of BNF in soybean in Brazil. *Plant and Soil* 252 (1): 1–9.

American Heart Association. 2015. Saturated fats. http://www.heart.org/HEARTORG/GettingHealthy/FatsAndOils/Fats101/Saturated-Fats_UCM_301110_Article.jsp (accessed 31 January 2016).

Andrianto, A., B. F. Sedik, H. Waridjo, H. Komarudin, and K. Obidzinski. 2014. The impacts of oil palm plantations on forests and people in Papua: A case study from Boven Digoel District. Working paper 163. Bogor, Indonesia: Center for International Forestry Research.

Angelsen, A., and D. Kaimowitz, eds. 2001. *Agricultural technologies and tropical deforestation.* Oxon: CABI Publishing.

Anggraeni, D., and Y. Zimmer. 2014. Palm oil: Economics of the driver of global vegetable oil markets. Paper presented at the Agri Benchmark Global Forum, Des Moines, Iowa, August 2014.

Arezki, R., K. Deininger, and H. Selod. 2015. What drives the global "land rush?" *The World Bank Economic Review* 29 (2): 207–233.

Arima, E. Y., P. Richards, R. Walker, and M. M. Caldas. 2011. Statistical confirmation of indirect land-use change in the Brazilian Amazon. *Environmental Research Letters* 6: 1–7.

Alsberg, C., and A. E. Taylor. 1928. *The fats and oils: a general view.* Food Research Institute, Stanford University.

Arvor, D., M. Meirelles, V. Dubreuil, A. Bégué, and Y. E. Shimabukuro. 2012. Analyzing the agricultural transition in Mato Grosso, Brazil, using satellite-derived indices. *Applied Geography* 32 (2): 702–713.

Association of South East Asian Nations. 1999. *Sixth ASEAN ministerial meeting on haze.* Bandar Seri Begawan: Association of South East Asian Nations.

Association of South East Asian Nations. 2002. *The ASEAN agreement on transboundary haze pollution.* Kuala Lumpur: Association of South East Asian Nations.

Assunção, J., C. C. Gandour, and R. Rocha. 2012. Deforestation slowdown in the legal Amazon: Prices or policies. Climate Policy Initiative working paper. San Francisco, CA: Climate Policy Initiative.

Athukorala, P.- C., and W.- H. Loke. 2009. Agricultural incentives in Malaysia: Trends, patterns and policy implications. *Malaysian Journal of Economic Studies* 46 (2): 151–173.

Aubrey, A. 2014. How tracing the oil in your Pop-Tarts may help save the rain forests. National Public Radio, February 21. http://www.npr.org/blogs/thesalt/2014/02/20/280257631/how-tracing-the-oil-in-your-pop-tarts-may-help-save-rainforests (accessed 31 January 2016).

Austin, K. G., P. S. Kasibhatla, D. L. Urban, F. Stolle, and J. Vincent. 2015. Reconciling oil palm expansion and climate change mitigation in Kalimantan, Indonesia. *PLoS One* 10 (5): e0127963.

Baffes, J., and A. Dennis. 2013. Long-term drivers of food prices. Policy Research working paper 6455. Washington, DC: The World Bank Development Prospects Group and the Poverty Reduction and Economic Management Network.

Bailis, R. 2014. Brazil: Biodiesel. In: *Sustainable development of biofuels in Latin America and the Caribbean*, ed. B. D. Solomon and R. Bailis, 103–126. New York, NY: Springer.

Barlow, C., and T. Tomich. 1991. Indonesian agricultural development: The awkward case of smallholder tree crops. *Bulletin of Indonesian Economic Studies* 27 (3): 29–53.

Barlow, C., Z. Zen, and R. Gondowarsito. 2003. The Indonesian oil palm industry. *Oil Palm Industry Economic Journal* 3: 8–15.

Barretto, A. G., G. Berndes, G. Sparovek, and S. Wirsenius. 2013. Agricultural intensification in Brazil and its effects on land-use patterns: An analysis of the 1975–2006 period. *Global Change Biology* 19 (6): 1804–1815.

Barros, S. 2013. *Brazil biofuels annual report 2013*. GAIN report no. BR13005. Global Agricultural Information Network. The Hague: US Department of Agriculture Foreign Agricultural Service.

Basta! and Les Amis de la Terre. 2012. Live or drive, a choice has to be made: A case study of Sime Darby operations in Liberia. Montreul: Basta! and Les Amis de la Terre. http://www.bastamag.net/IMG/pdf/Rap_LiberiaEN.pdf (accessed 31 January 2016).

Basu, S., K. S. Babiarz, S. Ebrahim, S. Vellakkal, D. Stuckler, and J. Goldhaber-Fiebert. 2013. Palm oil taxes and cardiovascular disease mortality in India: Economic–epidemiologic model. *British Medical Journal* 347 (3), 1–9.

Beall, E., ed. 2011. Smallholders in global value chains and certification: Evidence from three case studies. Environment and Natural Resources Management working paper no. 50. Rome, Italy: Food and Agriculture Organization of the United Nations.

Beggs, E., and E. Moore. 2013. *The social landscape of African oil palm production in the Osa and Golfito Region, Costa Rica*. San José: INOGO, Stanford Woods Institute for the Environment.

Bento Filho, J. B. de S. F., and C. E. de F. Vian. 2013. The Brazilian experience with the occupation of the Cerrados: The dynamics of large farms and small farms. Programa Cohesion Territorial para el Desarollo, documento de trabajo no 5. Santiago: RIMISP.

Berger, K. G., and S. M. Martin. 2000. Palm oil. In: *The Cambridge world history of food*, ed. K. F. Kiple and K. C. Ornelas, 397–410. Cambridge: Cambridge University Press.

Bhardwaj, M. 2015. Govt raises import taxes on vegetable oils. *LiveMint*. http://www.livemint.com/Politics/b04Ofy7DegwGtg3udzIC4I/Govt-raises-import-taxes-on-vegetable-oils.html (accessed 31 Jan 2016).

Bhatia, V. S., P. Singh, S. P. Wani, G. S. Chauhan, A. V. R. K. Rao, A. K. Mishra, and K. Srinivas. 2008. Analysis of potential yields and yield gaps of rainfed soybean in India using CROPGRO–soybean model. *Agricultural and Forest Meteorology* 148 (8–9): 1252–1265.

Bickel, U., and J. M. Dros. 2003. *The impacts of soybean cultivation on Brazilian ecosystems: Three case studies*. Amsterdam: World Wildlife Fund.

Bindraban, P. S., A. C. Franke, D. O. Ferrar, C. M. Ghersa, L. A. P. Lotz, A. Nepomuceno, M. J. M. Smulder, and C. C. M. van de Wiel. 2009. *GM-related sustainability: Agro-ecological impacts, risk and opportunities of soy production in Argentina and Brazil.* Wageningen: Plant Research International B.V.

Birdsall, N., W. Savedoff, and F. Seymour. 2014. The Brazil–Norway agreement with performance-based payments for forest conservation: Successes, challenges and lessons. CGD Climate and Forest Paper Series no. 4. Washington, DC: Center for Global Development.

Bisaliah, S. 1986. *Soybean development in India*. CGPRT no. 5. Bogor: Coarse Grains, Pulses, Roots and Tubers Centre.

Blasbalg, T. L., J. R. Hibbeln, C. E. Ramsden, S. F. Majchrzak, and R. R. Rawlings. 2011. Changes in consumption of omega-3 and omega-6 acids in the United States during the 20th century. *American Journal of Clinical Nutrition* 93 (5): 950–962.

Borlaug, N. 2007. Feeding a hungry world. *Science* 318 (5849): 359.

Boucher, D. H. 2015. The REDD/carbon market offsets debate: Big argument, small potatoes. *Journal of Sustainable Forestry* 34 (6–7): 547–558.

Boucher, D., S. Roquemore, and E. Fitzhugh. 2013. Brazil's success in reducing deforestation. *Tropical Conservation Science* 6 (3): 426–445.

Brinkley, J. 2013. Palm oil is proving unhealthy for people, environment. *SFGate*. December 13. http://www.sfgate.com/opinion/brinkley/article/Palm-oil-proving-unhealthy-for-people-environment-5063474.php (accessed 31 January 2016).

Brockhaus, M., K. Obidzinski, A. Dermawan, Y. Laumonier, and C. Luttrell. 2012. An overview of forest and land allocation policies in Indonesia: Is the current framework sufficient to meet the needs of REDD+? *Forest Policy and Economics* 18: 30–37.

Brown-Lima, C., M. Cooney, and D. Cleary. 2010. *An overview of the Brazil–China soybean trade and its strategic implications for conservation*. Arlington: The Nature Conservancy.

Bruinsma, J., ed. 2003. *World agriculture: Towards 2015/30: An FAO perspective*. London: Earthscan Publications Ltd.

Brümmer, B., O. Korn, K. Schlüssler, and T. J. Jaghdani. 2015. Volatility in oilseeds and vegetable oils markets: Drivers and spillovers. *Journal of Agricultural Economics*. (DOI: 10.1111/1477-9552.12141)

Budidarsono, S., A. Susanti, and A. Zoomers. 2013. Oil palm plantations in Indonesia: The implications for migration, settlement/resettlement and local economic development. In: *Biofuels: Economy, environment and sustainability*, ed. Z. Fang, 173–193. Rijeka: InTech.

Burgess, R., M. Hansen, B. A. Olken, P. Potapov, and S. Sieber. 2012. The political economy of deforestation in the tropics. *The Quarterly Journal of Economics* 127 (4): 1707–1754.

Busch, J., K. Ferretti-Gallon, J. Engleman, M. Wright, K. G. Austin, F. Stolle, S. Turubanova, et al. 2014. Reductions in emissions from deforestation from Indonesia's moratorium on new oil palm, timber and logging concessions. *Proceedings of the National Academy of Sciences of the United States of America* 112 (5): 1328–1333.

Business News. 2016. Malaysia, Indonesia protest France's progressive palm oil tax. *Business News*, February 5. http://www.thestar.com.my/business/business-news/2016/02/05/malaysia-and-indonesia-protest-frances-progressive-palm-oil-tax/ (accessed 31 Jan 2015).

Bustos, P., B. Caprettini, and J. Ponticelli. 2015. Agricultural productivity and structural transformation: Evidence from Brazil. Chicago Booth research paper no. 14-07, Fama-Miller working paper. Chicago, IL: University of Chicago.

Butler, R. A., L. P. Koh, and J. Ghazoul. 2009. REDD in the Red: Palm oil could undermine carbon payment schemes. *Conservation Letters* 2 (2): 67–73.

Byerlee, D. 2014. The fall and rise again of plantations in tropical Asia: History repeated? *Land* 3 (3): 547–597.

Byerlee, D., and X. Rueda. 2015. From public to private standards for tropical commodities: A century of global discourse on land governance on the forest frontier. *Forests* 6 (4): 1301–1324.

Cacho, M. M.,. 2016. Soybean agri-food systems dynamics and the diversity of farming styles on the agricultural frontier in Mato Grosso, Brazil. *The Journal of Peasant Studies*: 43: 419–441.

Cahyadi, E. R., and H. Waibel. 2013. Is contract farming in the Indonesian oil palm industry pro-poor? *Journal of Southeast Asian Economies* 30 (1): 62–76.

Carnegie Endowment for International Peace. 2010. The world order in 2010. Carnegie Endowment for International Peace: 1–29. http://carnegieendowment.org/files/World_Order_in_2050.pdf (accessed 31 January 2016).

Carrere, R. 2010. Oil palm in Africa: Past, present and future scenarios. World Rainforest Movement. WRM Series on Tree Plantations no. 15. http://wrm.org.uy/wp-content/uploads/2014/08/Oil_Palm_in_Africa_2013.pdf (accessed 31 January 2016).

Cassman, K. G. 2005. Soybean production and expansion in Mato Grosso, Brazil. *Nebraska Soybean Review* Fall, 10–15. Lincoln: Nebraska Soybean Association.

Castiblanco, C., A. Etter, and T. M. Aide. 2013. Oil palm plantations in Colombia: A model of future expansion. *Environmental Science and Policy* 27: 172–183.

Castiblanco, C., A. Moreno, and A. Etter. 2015. Impact of policies and subsidies in agribusiness: The case of oil palm and biofuels in Colombia. *EnergyEconomics* 49: 676–686.

Center for Disease Control and Prevention. 2012a. Dietary cholesterol. http://www.cdc.gov/nutrition/everyone/basics/fat/cholesterol.html (accessed August 5, 2014).

Center for Disease Control and Prevention. 2012b. Polyunsaturated fats and monounsaturated fats. http://www.cdc.gov/nutrition/everyone/basics/fat/unsaturatedfat.html (accessed August 5, 2014).

Center for Disease Control and Prevention. 2012c. Saturated fat. http://www.cdc.gov/nutrition/everyone/basics/fat/saturatedfat.html (accessed August 5, 2014).

Center for Disease Control and Prevention. 2014. Trans fat. http://www.cdc.gov/nutrition/everyone/basics/fat/transfat.html (accessed August 5, 2014).

Chaddad, F. 2014. BrasilAgro: Organizational architecture for a high-performance farming corporation. *American Journal of Agricultural Economics* 96 (2): 578–588.

Chand, R. 2007. Agro-industries characterization and appraisal: Soybeans in India. Agricultural Management, Marketing and Finance working document 20. Rome, Ital: Food and Agriculture Organization of the United Nations.

Chavananand, K. Latest challenges and opportunities in the Thai palm oil industry. Paper presented at Palmex Thailand, Surat Thani, Thailand, August 2013.

Chen, B. K., B. Seligman, J. W. Farquhar, and J. D. Goldhaber-Fiebert. 2011. Multi-country analysis of palm oil consumption and cardiovascular disease mortality for countries at different stages of economic development, 1980–1997. *Globalization and Health* 7 (45): 1–10.

Cheow, S. A. 2014. Five cooking oils: What's healthy and what isn't. *Straits Times*, July 14. http://www.straitstimes.com/news/singapore/health/story/5-cooking-oils-whats-healthy-and-what-isnt-20140714 (accessed 31 January 2016).

Cheyns, E., and S. Rafflegeau. 2005. Family agriculture and the sustainable development issue: Possible approaches from the African oil palm sector: The example of Ivory Coast and Cameroon. *Oléagineux, Corps Gras, Lipides* 12 (2): 111–120.

Chicago Mercantile Exchange. 2006. *An introduction to futures and options*. Chicago, IL: Chicago Mercantile Exchange.

Chicago Mercantile Exchange Group. 2010. *An overview of the edible oil markets: Crude palm oil vs soybean oil*. Chicago, IL: Chicago Mercantile Exchange Group.

Classen, S. F. 2013. Analysis of the discourse and background of the ProSAVANA Programme in Mozambique: Focusing on Japan's role. http://sodraafrikaidag.se/wp-content/uploads/2013/06/ProSavana-Analysis-Japanese-University.pdf (accessed 31 January 2016).

Colchester, M., T. Jalong, and W. M. Chuo. 2012. Sarawak: IOI-Pelita and the community of Long Teran Kanan. http://www.rightsandresources.org/documents/files/doc_5468.pdf (accessed 31 January 2016).

Colchester, M., N. Jiwan, Andiko, M. Sirait, A. Y. Firdaus, A. Surambo, and H. Pane. 2006. *Promised land: Palm oil and land acquisition in Indonesia: Implications for local communities and indigenous peoples*. Moreton-in-Marsh: Forest Peoples Programme.

Colchester, M., W. A. Pang, W. M. Chuo, and T. Jalong. 2009. *Land is life: Land rights and oil palm development in Sarawak*. Moreton-in-Marsh: Forest Peoples Programme and Perkumpulan Sawit Watch.

CONAB. 2014. Custo de Producao–Resumen. Agricultura Empresarial—Soja—Plantio Directo OGM – Alta Tecnologia. Safra de Verao -2013/14 – Primavera do Leste – MT. [Cost of Production—Summary. Commercial agriculture—Soybeans—Direct Planting and genetically modified—High technology. Summer season - 2013/14 = Primavera do Leste, Mato Grosso] http://www.conab.gov.br/OlalaCMS/uploads/arquivos/13_04_05_11_12_14_

Corley, R. H. V. 2009. How much palm oil do we need? *Environmental Science and Policy* 12 (2): 134–139.

Corley, R. H. V., and P. B. Tinker. 2003. *The oil palm*. Oxford: Blackwell Science.

Corley, R. H. V., and P. B. Tinker. 2016. *The oil palm*. 5th ed. Malden, MA: Wiley-Blackwell.

Cramb, R. A. 2007. *Land and longhouse: Agrarian transformation in the uplands of Sarawak*. Copenhagen: NIAS Press.

Cramb, R. A. 2012. Palmed off: Incentive Problems with joint-venture schemes for oil palm development on customary land. *World Development* 43: 84–99.

Cramb, R. 2013. A Malaysian land grab? The political economy of large-scale oil palm development in Sarawak. Land Policy Initiate working paper no. 50. Institute of Social Studies, Erasmus University. http://www.plaas.org.za/sites/default/files/publications-pdf/LDPI50Cramb.pdf (accessed 31 January 2016).

Cramb, R., and G. N. Curry. 2012. Oil palm and rural livelihoods in the Asia–Pacific region: An overview. *Asia Pacific Viewpoint* 53 (3): 223–239.

Cramb, R. A., and D. Ferraro. 2012. Custom and capital: A financial appraisal of alternative arrangements for large-scale oil palm development on customary land in Sarawak, Malaysia. *Malaysian Journal of Economic Studies* 49 (1): 49–69.

Cramb, R., and P. S. Sujang. 2011. "Shifting ground": Renegotiating land rights and rural livelihoods in Sarawak, Malaysia. *Asia Pacific Viewpoint* 52 (2): 136–147.

Cramb, R. A., and P. S. Sujang. 2013. The mouse deer and the crocodile: oil palm, smallholders, and livelihood strategies in Sarawak, Malaysia. *Journal of Peasant Studies* 40 (1): 129–154.

CSR Asia. 2014. Multi-stakeholder initiatives: Smallholders and inclusive business opportunities in palm oil. http://www.csr-asia.com/report/RIB_Asia_Palm_oil_report.pdf (accessed 31 January 2016).

Dallinger, J. 2011. Oil palm development in Thailand: Economic, social, and environmental considerations. In: *Oil palm expansion in South East Asia*, ed. M. Colchester and S. Chao, 24–51. Moreton-in-Marsh: Forest Peoples Programme and Perkumpulan Sawit Watch.

Dalton, G. 1965. History, politics, and economic development in Liberia. *Journal of Economic History* 25 (4): 569–591.

Daud, Z. A. M., D. Kaur, and P. Khosla. 2012. Health and nutritional properties of palm oil and its components. In: *Palm oil: Production, processing, characterization and uses*, ed. O.- M. Lai, C.- P. Tan, and C. C. Akoh, 545–560. Urbana, IL: AOCS Press.

Deaton, A. 1997. *The analysis of household surveys: A microeconomic approach to development policy*. Baltimore, MD: Johns Hopkins University Press.

Deese, W., and J. Reeder. 2007. Export taxes on agricultural products: Recent history and economic modeling of soybean export taxes in Argentina. *Journal of International Commerce and Economics* 1: 185–93.

de Gorter, H., D. Drabik, and D. R. Just. 2011. The economics of a blender's tax credit versus a tax exemption: The case of U.S. "splash and dash" biodiesel exports to the European Union. *Applied Economic Perspectives and Policy* 33 (4): 510–527.

de Gorter, H., D. Drabik, and D. R. Just. 2015. *The economics of biofuel policies: Impacts on price volatility in grain and oilseed markets*. London: Palgrave MacMillan.

de Gorter, H., D. Drabik, and G. R. Timilsina. 2013. The effect of biodiesel policies on world oilseed markets and developing countries. Policy Research working paper 6453. Washington, DC: World Bank, Development Research Group, Environment and Energy Team.

De Haan, C., P. Gerber, and C. Opio. 2010. Structural change in the livestock sector. In: *Livestock in a changing landscape, volume 1: Drivers, consequences, and responses*, ed. H. Steinfeld, H. A. Mooney, F. Schneider, and L. E. Neville, 35–50. Washington, DC: Island Press.

Deininger, K. 2014. Securing land rights for smallholder farmers. In: *New directions for smallholder* agriculture, ed. P. B. R. Hazell and A. Rahman, 401–429. London: Oxford University Press.

Deininger, K., and D. Byerlee. 2011. *Rising global interest in farmland: Can it yield sustainable and equitable benefits*? Washington, DC: World Bank Publications.

Deininger, K., and D. Byerlee. 2012. The rise of large farms in land abundant countries: Do they have a future? *World Development* 40 (4): 701–714.

Deininger, K., and G. Feder. 2009. Land registration, governance, and development: Evidence and implications for policy. *World Bank Research Observer* 24 (2): 233–266.

Deininger, K., H. Selod, and A. Burns. 2012. *The land governance assessment framework: Identifying and monitoring good practice in the land sector*. Washington, DC: World Bank.

Delgado, C., C. Narrod, and M. Tiongco. 2008. *Determinants and implications of the growing scale of livestock farms in four fast-growing developing countries*. Washington, DC: International Food Policy Research Institute.

Delgado, C. L., M. W. Rosegrant, H. Steinfeld, S. K. Ehui, and C. Courbois. 1999. Livestock to 2020: The next food revolution. Food, Agriculture, and the Environment discussion paper 28. Washington, DC: International Food Policy Research Institute.

Department of Agriculture and Cooperation. 2012. *All India report on agricultural census report 2005–06*. New Delhi: Ministry of Agriculture, Government of India.

de Rezende, G. C. 2006. *Labor, land and agricultural credit policies and their adverse impacts on poverty in Brazil*. Sao Paulo: Institute of Applied Economic Research.

Dewi, S., A. Ekadinata, D. Indiarto, A. Nugraha, and M. van Noordwijk. 2014. Negotiation support tools to enhance multifunctioning landscapes. In: *Climate-Smart landscapes: Multifunctionality in practice*, ed. P. A. Minang, M. van Noordwijk, O. E. Freeman, C. Mbow, J. de Leeuw, and D. Catacutan, 243–256. Nairobi: World Agroforestry Center (ICRAF).

Dinerstein, E., A. Baccini, M. Anderson, G. Fiske, E. Wikramanayake, D. McLaughlin, G. Powell, D. Olson, and A. Joshi. 2014. Guiding agricultural expansion to spare tropical forests. *Conservation Letters* 8 (4): 262–271.

Directorate of Economics and Statistics. 2015. Department of Agriculture and Cooperation, Government of India. http://eands.dacnet.nic.in/Default.htm (accessed August 1, 2015).

Dong, W., X. Wang, and J. Yang. 2015. Future perspectives of China's feed demand and supply during its fast transition period of food consumption. *Journal of Integrative Agriculture* 14 (6): 1092–1110.

Dube, R. 2015. Argentina to reduce export taxes on some agricultural products. *Wall Street Journal*, December 14. http://www.wsj.com/articles/argentina-to-remove-export-taxes-on-some-agricultural-products-1450100855 (accessed December 20, 2015).

Durand-Gasselin T., H. de Franqueville, Z. Hayun, J. C. Jacquemard, P. Amblard, F. Potier, and B. Nouy. 2003. *Is it possible and reasonable to produce 10 tonnes of oil per hectare of oil palms?* Medan: ISPOB.

Dutta, M., and A. Gulati. 2009. Analysis of supply and demand for agricultural products in India: Literature review. Working paper. Washington, DC: International Food Policy Research Institute.

Duty Calculator. 2015. Duty calculator. Pitney Bowes. http://www.dutycalculator.com (accessed 31 January 2016).

Dwyer, M. B. 2015. Trying to follow the money: Possibilities and limits of investor transparency in southeast Asia's rush for "available" land. CIFOR working paper no. 177. Bogor, Indonesia: Center for International Forestry Research.

Edwards, R. 2015. Is plantation agriculture good for the poor? Evidence from Indonesia's palm oil expansion. Job Market paper no. 2015-12. Canberra: Arndt-Corden Department of Economics, Australian National University.

Edwards, D. P., L. P. Koh, and W. F. Laurance. 2012. Indonesia's REDD+ pact: Saving imperilled forests or business as usual? *Biological Conservation* 151 (1): 41–44.

Ekboir, J. M. 2003. Research and technology policies in innovation systems: Zero tillage in Brazil. *Research Policy* 32 (4): 573–586.

Ekman, S.- M. Stensrud, and C. S. Macamo. 2014. Brazilian Development Cooperation in agriculture: A scoping study on ProSAVANA in Mozambique, with implications for forests. CIFOR working paper no. 138. Bogor, Indonesia: Center for International Forestry Research.

Engel, E. 1857. Die Productions- und Consumtionsverhältnisse des Königreichs Sachsen. *Zeitschrift des Statistischen Bureaus des Königlich Sächsischen Ministerium des Innere* 8 (9): 28–29.

Euler, M., V. Krishna, S. Schwarze, H. Siregar, and M. Qaim. 2015. *Oil palm adoption, household welfare and nutrition among smallholder farmers in Indonesia*. Goettingen: University of Goettingen.

European Commission. 2015. Renewable Energy Directive. https://ec.europa.eu/energy/en/topics/renewable-energy/renewable-energy-directive (accessed March 30, 2016.).

European Parliament. 2014. European Parliament Legislative Resolution of 11 September 2013 on the Proposal for a Directive of the European Parliament and of the Council Amending Directive 98/70/EC Relating to the Quality of Petrol and Diesel Fuels and Amending Directive 009/28/EC on the Promotion of the Use of Energy from Renewable Sources (COM(2012)0595—C7-033/2012—2012/0288(COD)). Procedure 2012/0288(COD). (accessed March 1, 2015).

European Union Delegation to Malaysia. 2012. The Malaysian palm oil sector: Overview. Agenzia per la Promozione all'estero e l'internazionalizzazione delle Imprese italiane. http://www.ice.gov.it/paesi/asia/malaysia/upload/173/Palm%20Oil_overview_2012.pdf (accessed 31 January 2016).

Evans, R., and G. Griffiths. 2013. *Palm oil, land rights and ecosystem services in Gbarpolu County, Liberia*. Research note 3. Reading, UK: Walker Institute for Climate System Research, University of Reading.

Exxon Mobil. 2014. *The outlook for energy: A view to 2040*. San Jose: Exxon Mobil.

Fairbairn, M. 2013. Indirect dispossession: Domestic power imbalances and foreign access to land in Mozambique. *Development and Change* 44 (2): 335–356.

Fairhurst, T., and D. McLaughlin. 2009. *Sustainable oil palm development on degraded land in Kalimantan*. Washington, DC: World Wildlife Fund.

Fane, G., and P. G. Warr. 2007. Distortions to agricultural incentives in Indonesia. Agricultural Distortions working paper 24. Washington, DC World Bank.

Fattore, E., C. Bosetti, F. Brighenti, C. Agostoni, and G. Fattore. 2014. Palm oil and blood lipid-related markers of cardiovascular disease: A systematic review and meta-analysis of dietary intervention trials. *American Journal of Clinical Nutrition* 99 (6): 1331–1350.

Fattore, E., and R. Fanelli. 2013. Palm oil and palmitic acid: A review on cardiovascular effects and carcinogenicity. *International Journal of Food Sciences and Nutrition* 64 (5): 648–659.

Feintrenie, L., W. K. Chong, and P. Levang. 2010. Why do farmers prefer oil palm? Lessons learnt from Bungo District, Indonesia. *Small-Scale Forestry* 9: 379–396.

Feintrenie, L., M. Nkoua, S. Saidi, and L. Gazull. 2014. Modelling as a tool for spatial planning of commodity production: The example of certified oil palm plantations in Central Africa. Paper presented at the 2014 World Bank Conference on Land and Poverty, Washington, DC, March 2014.

Ferdman, R. 2014. The generational battle of butter versus margarine. *The Washington Post*, June 17. http://www.washingtonpost.com/blogs/wonkblog/wp/2014/06/17/the-generational-battle-of-butter-vs-margarine/ (accessed 31 January 2016).

Fieldhouse, D. K. 1978. *Unilever overseas: The anatomy of a multinational 1895–1965*. Stanford, CA: Hoover Institution Press.

Fischer, T., D. Byerlee, and G. Edmeades. 2014. *Crop yields and global food security: Will yield increases continue to feed the world?* Canberra: Australian Centre for International Agricultural Research.

Fischer, G., and M. Shah. 2010. *Farmland investments and food security*. Laxenburg: International Institute for Applied Systems Analysis.

Fisher, B., D. P. Edwards, X. Giam, and D. S. Wilcove. 2011. The high costs of conserving Southeast Asia's lowland rainforests. *Frontiers in Ecology and the Environment* 9 (6): 329–334.

Fisher, H., C. Winzenried, and L. Sar. 2012. *Oil palm pathways: An analysis of ACIAR's oil palm projects in Papua New Guinea*. Canberra: Australian Centre for International Agricultural Research.

Fitzherbert, E. B., M. J. Struebig, A. Morel, F. Danielsen, C. A. Brühl, P. F. Donald, and B. Phalan. 2008. How will oil palm expansion affect biodiversity? *Trends in Ecology and Evolution* 23 (10): 538–545.

Flach, B., K. Bendz, R. Krautgartner, and S. Lieberz. 2013. EU biofuels annual 2013. GAIN report no. NL3034. The Hague: US Department of Agriculture Foreign Agricultural Service.

Flexor, G., K. Y. M. Kato, M. do S. Lima, and B. N. Rocha. 2011. Dilemas institucionais na promoção dos biocombustíveis: O caso do programa nacional de produção e uso de biodiesel no Brasil. *Cadernos do Desenvolvimento* 6 (8): 329–354.

Fliehr, O. 2013. Analysis of transportation and logistics processes for soybeans in Brazil: A case study of selected production regions. Thünen working paper 4. Braunschweig, Germany: Thunen Institute of Farm Economics.

Fold, N., and L. Whitfield. 2012. Developing a palm oil sector: The experiences of Malaysia and Ghana compared. DIIS working paper 2012:08. Copenhagen, Denmark: Danish Institute for International Studies.

Food and Agriculture Organization of the United Nations. 1994. Expert's recommendations on fats and oils in human nutrition. Agriculture and Consumer Protection Department. http://www.fao.org/docrep/t4660t/t4660t02.htm (accessed 31 January 2016).

Food and Agriculture Organization of the United Nations. 2004. *State of agricultural commodity markets 2004*. Rome: Food and Agriculture Organization of the United Nations.

Food and Agriculture Organization of the United Nations. 2007. *Future expansion of soybean 2005–2014: Implications for food security, sustainable rural development and agricultural policies in the countries of Mercosur and Bolivia*. Santiago: Food and Agriculture Organization of the United Nations.

Food and Agriculture Organization of the United Nations. 2010. *Fats and fatty acids in human nutrition: Report of an expert consultation.* Geneva: Food and Agriculture Organization of the United Nations.

Food and Agriculture Organization of the United Nations. 2012. *Voluntary guidelines on the responsible governance of tenure of land, fisheries and forests in the context of national food security.* Rome: Food and Agriculture Organization of the United Nations.

Food and Agriculture Organization of the United Nations. n.d. Statistical database of the United Nations Food and Agriculture Organization (FAOSTAT). Various dates. http://faostat3.fao.org/home/E (accessed May 4, 2016).

Foreign Agricultural Service. 2013. *Indonesia: Palm oil expansion unaffected by forest moratorium.* Washington, DC: US Department of Agriculture.

Foundation for Partnership Initiatives in the Niger Delta. 2011. *A report on palm oil value chain analysis in the Niger Delta.* Abuja: Foundation for Partnership Initiatives in the Niger Delta.

Friedman, M. 1957. *A theory of the consumption function.* Princeton, NJ: Princeton University Press.

Fry, J. 2009. The challenges facing the palm oil industry in the 21st century. Paper presented at the SCI, LMC International, Oxford, UK.

Fry, J. 2011. A global challenge: Markets and oil palm expansion. http://static.zsl.org/files/session-1-2-james-fry-a-global-challenge-markets-oil-palm-expansion-1463.pdf. (Accessed 31 January 2016)

Fukuda, H., A. Kondo, and H. Noda. 2001. Biodiesel fuel production by transesterification of oils. *Journal of Bioscience and Bioengineering* 92 (5): 405–416.

Gaiha, R., N. Kaicker, K. S. Imai, and G. Thapa. 2012. Demand for nutrients in India: An analysis based on the 50th, 61st, and 66th rounds of the N.S.S. Discussion Paper Series, RIEB, Kobe University. http://www.rieb.kobe-u.ac.jp/academic/ra/dp/English/DP2012-14.pdf (accessed 31 January 2016).

Gale, F. 2015. *Development of China's feed industry and demand for imported commodities.* Washington, DC: Economic Research Service, US Department of Agriculture.

Gale, F., D. Marti, and D. Hu. 2012. *China's volatile pork industry.* Washington, DC: Economic Research Service, US Department of Agriculture.

Garcia-Ulloa, J., S. Sloan, P. Pacheco, J. Ghazoul, and L. P. Koh. 2012. Lowering environmental costs of oil-palm expansion in Colombia. *Conservation Letters* 5 (5): 366–375.

Garrett, R. D. 2013. Interactions between global supply chains, land use, and governance: The case of soybean production in South America. PhD diss., Stanford University. http://purl.stanford.edu/zc839jm3057, Accessed January 8, 2016.

Garrett, R. D., E. F. Lambin, and R. L. Naylor. 2013. Land institutions and supply chain configurations as determinants of soybean planted area and yields in Brazil. *Land Use Policy* 31: 385–396.

Gaskell, J. C. 2012. The palm oil revolution in Asia. PhD diss., Stanford University.

Gaskell, Joanne. 2015. The role of markets, technology and policy in generating palm oil demand in Indonesia. *Bulletin of Indonesian Economic Studies* 51 (1): 29–45.

Gasparri, N. I., H. R. Grau, and J. Gutiérrez Angonese. 2013. Linkages between soybean and neotropical deforestation: Coupling and transient decoupling dynamics in a multi-decadal analysis. *Global Environmental Change* 23 (6): 1605–1614.

Gasparri, N. I., T. Kuemmerle, P. Meyfroidt, P. de Waroux, and H. Kreft. 2016. The emerging soybean production frontier in southern Africa: Conservation challenges and the role of south–south telecouplings. *Conservation Letters*. 9, 21–31.

Gaveau, D. L., M. Kshatriya, D. Sheil, S. Sloan, E. Molidena, A. Wijaya, S. Wich, et al. 2013. Reconciling forest conservation and logging in Indonesian Borneo. *PLoS One* 8 (8): e69887.

Gerasimchuk, I., R. Bridle, C. Beaton, and C. Charles. 2012. *State of play on biofuel subsidies: Are policies ready to shift?* Manitoba: International Institute for Sustainable Development.

German, L., and G. Schoneveld. 2012. A review of social sustainability considerations among EU-approved voluntary schemes for biofuels, with implications for rural livelihoods. *Energy Policy* 51: 765–778.

German, L., G. C. Schoneveld, and P. Pacheco. 2011. Local, social, and environmental impacts of biofuels: Global comparative assessment and implications for governance. *Ecology and Society* 16 (4): 29–43.

Gibbs, H. K., L. Rausch, J. Munger, I. Schelly, D. C. Morton, P. Noojipady, B. Soares-Filho, P. Barreto, L. Micol, and N. F. Walker. 2015. Brazil's soy moratorium. *Science* 347 (6220): 377–378.

Gil, J., M. Siebold, and T. Berger. 2015. Adoption and development of integrated crop–livestock–forestry systems in Mato Grosso, Brazil. *Agriculture, Ecosystems & Environment* 199: 394–406.

Gilbert, A., and L. Pinzon. 2013. *Colombia biofuels annual: Blend mandates maintain status quo as a biofuel policy vision remains unclear*. Global Agricultural Information Network. The Hague: US Department of Agriculture Foreign Agricultural Service.

Giller, K. E., M. S. Murwira, D. K. C. Dhliwayo, P. L. Mafongoya, and S. Mpepereki. 2011. Soyabeans and sustainable agriculture in southern Africa. *International Journal of Agricultural Sustainability* 9 (1): 50–58.

Gillis, Justin. 2014. Restored forests breathe life into efforts against climate change. *New York Times*, December 23 http://www.nytimes.com/2014/12/24/science/earth/restored-forests-are-making-inroads-against-climate-change-.html?_r=0. (Accessed 31 Jan 2016).

Gingold, B. 2010. FAQ: Indonesia, degraded land and sustainable palm oil. Insights: WRI's blog. World Resources Institute, November 4. http://www.wri.org/blog/2010/11/faq-indonesia-degraded-land-and-sustainable-palm-oil (accessed 31 January 2016).

Global Forest Watch. 2015. Country profiles. http://www.globalforestwatch.org/countries (accessed September 15, 2015).

Global Renewable Fuels Alliance. n.d. Global biofuel mandates. http://globalrfa.org/biofuels-map/ (accessed June 23, 2015).

Global Witness. 2013. *Inside Malaysia's shadow state: Backroom deals driving the destruction of Sarawak*. London: Global Witness.

Godoy, C. V., A. d F. Bueno, and D. L. Gazziero. 2015. Brazilian soybean pest management and threats to its sustainability. *Outlooks on Pest Management* 26 (3): 113–117.

Goh, K. J., P. H. C. Ng, C. K. Wong, and S. Arif. 2014. Yield potential of oil palm and its attainment in Malaysia. *Planter* 90 (1060): 503–520.

Gold, I. L., C. E. Ikuenobe, O. Asemota, and D. A. Okiy. 2012. Palm and palm kernel oil production and processing in Nigeria. In: *Palm oil: Production, processing, characterization, and uses*, ed. O.- M. Lai, C.- P. Tan, and C. C. Akoh, 275–298 Urbana, IL: AOCS Press.

Goldsmith, P., and R. Hirsch. 2006. The Brazilian soybean complex. *Choices* 21 (2): 97–104.

Gonzalez, P. R. 2015. In Argentine tax battle, bags of soybeans get the knife. *Bloomberg Business*, February 24. http://www.bloomberg.com/news/articles/2015-02-25/in-argentine-tax-battle-bags-of-soybeans-get-the-knife (Accessed 31 Jan 2016).

GOPDC. 2015. GOPDC. http:// http://www.gopdc-ltd.com (accessed September 10, 2015).

Government of Liberia. 2009. *An act to ratify the amended and restated concession agreement between the Republic of Liberia and Sime Darby Plantation (Liberia) Inc.* Monrovia: Ministry of Foreign Affairs.

Granger, C. W. J. 1969. Investigating causal relations by econometric models and cross-spectral methods. *Econometrica* 37 (3): 424–438.

Grayson, J., and J. Stampe. 2012. *Palm oil investor review: Investor guidance on palm oil.* http://wwf.panda.org/what_we_do/footprint/agriculture/palm_oil/solutions/responsible_financing/investor_review/ (accessed 31 January 2016).

Green, R., L. Cornelsen, A. D. Dangour, R. Turner, B. Shankar, M. Mazzocchi, and R. D. Smith. 2013. The effect of rising food prices on food consumption: Systematic review with meta-regression. *British Medical Journal* 364: 1–9.

Green Car Congress. 2014. Navigant forecasts global road transportation energy consumption to increase 25% by 2035; 84% from conventional fuels. *Green Car Congress*, July 28. http://www.greencarcongress.com/2014/07/20140728-navigant.html (accessed June 25, 2015).

Green Palm Sustainability. 2015. Supporting RSPO certified sustainable palm oil. http://greenpalm.org/ (accessed August 19, 2015).

Greenpeace. 2013. *Herakles farms in Cameroon: A showcase in bad palm oil production.* Washington, DC: Greenpeace USA.

Greenpeace International, 2006. *Eating up the Amazon*. Greenpeace, Amsterdam.

Guereña, A. 2013. *The soy mirage: The limits of corporate social responsibility: The case of the company Desarollo Agrícola del Paraguay*. Oxfam, Oxford: Oxfam Research Reports.

Gulati, A., and T. Kelley. 1999. *Trade liberalization and Indian agriculture: Cropping pattern changes and efficiency gains in semi-arid tropics.* New York, NY: Oxford University Press.

Gulati, A., A. Sharma, and D. S. Kohli. 1996. Self-sufficiency and allocative efficiency: Case of edible oils. *Economic and Political Weekly* 31 (13): A15–A24.

Gunarso, P., M. E. Hartoyo, F. Agus, and T. J. Killeen. 2013. Oil palm and land-use change in Indonesia, Malaysia and Papua New Guinea. http://www.tropenbos.org/publications/oil+palm+and+land+use+change+in+indonesia,+malaysia+and+papua+new+guinea (Accessed 31 Jan 2016)

Guterman, L. 2008. Colombia. In: *Distortions to agricultural incentives in Latin America*, ed. K. Anderson and A. Valdés, 159–187. Washington, DC: World Bank Publications.

Hadinaryanto, S. Eka. 2014. Palm oil, political interests and land use in Indonesian Borneo. *Mongabay*, April 24. http://news.mongabay.com/2014/04/special-report-palm-oil-politics-and-land-use-in-indonesian-borneo-part-i/ (accessed 31 January 2016).

Hardman and Co. 2012. *Commercial viability and value creation in West African palm oil production.* London: Hardman and Co.

Harris, N., S. Minnemeyer, F. Stolle, and O. A. Payne. 2015. Indonesia's fire outbreaks produce more emissions than entire US economy. *World Resources Institute.* http://www.wri.org/blog/2015/10/indonesia%E2%80%99s-fire-outbreaks-producing-more-daily-emissions-entire-us-economy (accessed 31 January 2016).

Hartley, C. W. S. 1967. *The oil palm.* London: Longmans.

Hausman, C. 2012. Biofuels and land-use change: Sugarcane and soybean acreage response in Brazil. *Environmental and Resource Economics* 51 (2): 163–187.

Hecht, S. B. 2005. Soybeans, development and conservation on the Amazon frontier. *Development and Change* 36 (2): 375–404.

Helfand, S. M. 1999. The political economy of agricultural policy in Brazil: Decision making and influence from 1964 to 1992. *Latin American Research Review* 34 (2): 3–42.

Helfand, S. M., and E. S. Levine. 2004. Farm size and the determinants of productive efficiency in the Brazilian center–west. *Agricultural Economics* 31 (2–3): 241–249.

Helleiner, G. K. 1966. *Peasant agriculture, government, and economic growth in Nigeria.* Homewood, IL: Richard D. Irwin, Inc.

Henders, S., U. M. Persson, and T. Kastner. 2015. Trading forests: Land-use change and carbon emissions embodied in production and exports of forest-risk commodities. *Environmental Research Letters* 10 (12): 1–13.

Henderson, J., and D. J. Osborne. 2000. The oil palm in all our lives: How this came about. *Endeavour* 24 (2): 63–68.

Henson, I. E. 2012. A brief history of the oil palm. In: *Palm oil: Production, processing, characterization, and uses*, ed. O.- M. Lai, C.- P. Tan, and C. C. Akoh, 1–31. Urbana, IL: AOCS Press.

Herakles Farms. 2011. Palm oil FAQ's. http://www.heraklesfarms.com/palm-oil-faqs.html (accessed January 30, 2015).

Herrero, M., P. Havlík, H. Valin, A. Notenbaert, M. C. Rufino, P. K. Thornton, M. Blümmel, F. Weiss, D. Grace, and M. Obersteiner. 2013. Biomass use, production, feed efficiencies, and greenhouse gas emissions from global livestock systems. *Proceedings of the National Academy of Sciences of the United States of America* 110 (52): 20888–20893.

HighQuest Partners and Soyatech, LLC. 2011. *How the global oilseed and grain trade works.* St. Louis, MO: U.S. Soybean Export Council.

Hofstrand, D. 2014. Tracking biodiesel profitability. Iowa State University Extension and Outreach. https://www.extension.iastate.edu/agdm/energy/html/d1-15.html (accessed 31 January 2016).

Hosono, A., and Y. Hongo. 2015. The impact of Cerrado development: Stable food supply and socially inclusive development with value chains. In: *Development for Sustainable Agriculture: The Brazilian Cerrado*, ed. A. Hosono, C. M. C. da Rocha, and Y. Hongo, 91–113. Palgrave Macmillan.

Hospes, O., and A. Kentin. 2014. Tensions between global scale and national scale governance: The strategic use of scale frames to promote sustainable palm oil production in Indonesia. In: *Scale-sensitive governance of the environment*, ed. F. Padt, P. Opdam, N. Polman, and C. Termeer, 203–219. West Sussex: Wiley.

Hospes, O., O. van der Valk, and J. van der Mheen-Sluijer. 2012. Parallel development of five partnerships to promote sustainable soy in Brazil: Solution or part of wicked problems. *International Food and Agribusiness Management Review* 15: 39–62.

Houthakker, H. S. 1959. The scope and limits of futures trading. In: M. Abromovitz (ed). *The allocation of economic resources: Essays in honor of Bernard Francis Haley*, Palo Alto: Stanford University Pres, 134–159.

Hoyle, D., and P. Levang. 2012. Oil palm development in Cameroon. Cameroon: World Wildlife Fund for Nature Report.

Huang, K., and B. H. Lin. 2000. Estimation of food demand and nutrient elasticities from household survey data. US Department of Agriculture TB-1887. Economic Research Service, USDA, Washington DC.

Huang, J., S. Rozelle, H. Yang, J. Bai, X. Deng, J. Wang, et al. 2014. *Urbanization and food security in China.* http://ageconsearch.umn.edu/bitstream/189685/2/Huang,%20J.pdf. (accessed 31 Jan 2016).

Huddleston, P., and M. Tonts. 2007. Agricultural development, contract farming and Ghana's oil palm industry. *Geography* 92 (3): 266–278.

Hungria, M., R. J. Campo, I. C. Mendes, and P. H. Graham. 2006. Contribution of biological nitrogen fixation to the n nutrition of grain crops in the tropics: The success of soybean (*Glycine max L. Merr.*) in South America. In: *Nitrogen nutrition and sustainable plant productivity*, ed. R. P. Singh, N. Shankar, and P. K. Jaiwal, 43–93. Houston, TX: Stadium Press.

Hunter, J. E., J. Zhang, and P. M. Kris-Etherton. 2010. Cardiovascular disease risk of dietary stearic acid compared with trans, other saturated, and unsaturated fatty acids: A systematic review. *American Journal of Clinical Nutrition* 91 (1): 46–63.

Huth, P. J., V. L. FulgoniIII, and B. T. Larson. 2015. A systematic review of high-oleic vegetable oil substitutions for other fats and oils on cardiovascular disease risk factors: Implications for novel high-oleic soybean oils. *Advances in Nutrition* 6: 674–693.

Hyman, E. L. 1990. An economic analysis of small-scale technologies for palm oil extraction in Central and West Africa. *World Development* 18 (3): 455–476.

Hymowitz, T. 2008. A history of the soybean. In: *Soybeans: Chemistry, production, processing, and utilization*, ed. L. A. Johnson, P. J. White, and R. Galloway, 1–31. Urbana, IL: AOCS Press.

Ibekwe, U. C. 2008. Role of women in oil palm fruit processing and marketing in Imo State, Nigeria. *The Social Sciences* 3 (1): 61–65.

Instituto Brasileiro de Geografía e Estatística. 2012. Área plantada/colhida.http://www.sidra.ibge.gov.br/bda/agric/default.asp?t=4&z=t&o=11&u1=1&u2=1 &u3=1&u4=1&u5=1&u6=1 (accessed November 9, 2012).

Independent Petroleum Association of America. 2015. Diesel fuel: The unsung hero. http://oilindependents.org/diesel-fuel-the-unsung-hero/ (accessed June 23, 2015).

Index Mundi. 2015. Indonesia palm oil exports by year. (accessed April 8, 2015). http://www.indexmundi.com/agriculture/?country=id&commodity=palm-oil&graph=exports (accessed April 8, 2015).

Indonesia-Investments. 2015. What you need to know about Indonesia's palm oil export levies. *Indonesia-Investments*, July 15. http://www.indonesia-investments.com/news/todays-headlines/what-you-need-to-know-about-indonesia-s-palm-oil-export-levies/item5744 (accessed 31 January 2016).

International Council on Clean Transportation. 2013. *Vegetable oil markets and the EU biofuel mandate*. San Francisco, CA: International Council on Clean Transportation.

International Energy Agency. 2011. *Technology roadmap: Biofuels for transport*. Paris: International Energy Agency.

International Finance Corporation. 2013. *Diagnostic study on Indonesian oil palm smallholders: Developing a better understanding of their performance and potential*. Washington, DC: International Finance Corporation.

International Fund for Agricultural Development. 2011. *Progress report on implementation of the performance-based allocation system*. Rome: International Fund for Agricultural Development.

International Monetary Fund. n.d. IMF primary commodity prices. http://www.imf.org/external/np/res/commod/index.aspx (accessed June 23, 2015).

International Trade Centre. 2012. *Palm products, global markets, and developments*. Geneva: International Trade Centre.

Iowa State University. 2015. Ag decision maker. File A1-21. Iowa State University of Science and Technology. http://www.extension.iastate.edu/agdm/crops/pdf/a1-21.pdf (accessed 31 January 2016).

Irawan, S., L. Tacconi, and I. Ring. 2013. Stakeholders' incentives for land-use change and REDD+: The case of Indonesia. *Ecological Economics* 87: 75–83.

Irwin, S., and D. Good. 2015. The relationship between stocks-to-use and soybean prices revisited. *FarmDoc Daily*. http://farmdocdaily.illinois.edu/2015/05/relationship-between-stock-to-use-and-soybean-prices.html (accessed 31 January 2016).

Islamic Invitation Turkey. 2014. Iran to totally ban palm oil use. *Islamic Invitation Turkey*, September 20. http://www.islamicinvitationturkey.com/2014/09/20/iran-to-totally-ban-palm-oil-use/ (accessed 31 January 2016).

ITS Global. 2011. *The economic benefits of palm oil in Papua New Guinea*. Melbourne: ITS Global.

James, C. 2014. Global status of commercialized Biotech/GM crops: 2013. ISAAA brief 46. Ithaca, NY: ISAAA.

Jeffries, E. 2015. Fat as an environmentalist issue. *Nature Climate Change* 5: 716–717.

Jelsma, I., K. Giller, and T. Fairhurst. 2009. *Smallholder oil palm production systems in Indonesia: Lessons from the NESP Ophir Project*. Wageningen: Wageningen UR.

Jepson, W. 2006. Producing a modern agricultural frontier: Firms and cooperatives in eastern Mato Grosso, Brazil. *Economic Geography* 82 (3): 289–316.

Jha, G. K., S. Pal, V. C. Mathur, G. Bisaria, P. Anbukkani, R. R. Burman, and S. K. Dubey. 2012. *Edible Oilseed supply and demand scenario in India: Implications for policy*. New Delhi: Indian Agricultural Research Institute, Division of Agricultural Economics.

Jiwan, N. 2012. The political ecology of the Indonesian oil palm industry. In: *The palm oil controversy in Southeast Asia: A transnational perspective*, ed. O. Pye and J. Bhattacharya, 48–76. Pasir Panjang: Institute of Southeast Asian Studies.

Johansson, O., and L. Schipper. 1997. Measuring the long-run fuel demand for cars: Separate estimations of vehicle stock, mean fuel intensity, and mean annual driving distance. *Journal of Transport Economics and Policy* 31 (3): 277–292.

Johnson, T. M. and J. Franco. 2009. Economic assessment of palm oil production for biodiesel in Colombia: Case study analysis for the central and eastern zones. Washington, DC: World Bank Group ESMAP.

Johnston, F., S. Henderson, Y. Chen, J. Randerson, M. Marlier, R. DeFries, P. Kinney, et al. 2012. Estimated global mortality attributable to smoke from landscape fires. *Environmental Health Perspectives* 120: 695–701.

Jurgens, E., T. Brown, J. Leitmann, P. van Hofwegen, S. Shetty, and D. Mitchell. 2010. *Environmental, economic, and social impacts of oil palm in Indonesia: A synthesis of opportunities and challenges*. Washington, DC: World Bank

Kaimowitz, D., and J. Smith. 2001. Soybean technology and the loss of natural vegetation in Brazil and Bolivia. In: *Agricultural technologies and tropical deforestation,* ed. A. Angelsen and D. Kaimowitz, 195–211. Wallingford: CABI in Association with Center for International Forestry Research.

Keats, S., and S. Wiggins. 2014. *Future diets: Implications for agriculture and food prices*. London: Overseas Development Institute.

Keys, A., J. T. Anderson, and F. Grande. 1965. Serum cholesterol response to changes in diet: IV. Particular saturated fatty acids in the diet. *Metabolism* 14 (7): 776–787.

Keyzer, M. A., M. D. Merbis, I. F. P. W. Pavel, and C. F. A. Van Wesenbeeck. 2005. Diet shifts towards meat and the effects on cereal use: Can we feed the animals in 2030? *Ecological Economics* 55 (2): 187–202.

Kilby, P. 1967. *The Nigerian palm oil industry*. Stanford, CA: Food Research Institute.

Klink, C. A., and R. B. Machado. 2005. Conservation of the Brazilian Cerrado. *Conservation Biology* 19 (3): 707–713.

Knoema. 2014. World Bank commodity price data (pink sheet), July 2014. http://knoema.com/WBCPD2014Jul/world-bank-commodity-price-data-pink-sheet-july-2014 (accessed May 5, 2015).

Koczberski, G., and G. N. Curry. 2005. Making a living: Land pressures and changing livelihood strategies among oil palm settlers in Papua New Guinea. *Agricultural Systems* 85 (3): 324–339.

Koh, L. P. 2007. Potential habitat and biodiversity losses from intensified biodiesel feedstock production. *Conservation Biology* 21 (5): 1373–1375.

Koh, L. P., P. Levang, and J. Ghazoul. 2009. Designer landscapes for sustainable biofuels. *Trends in Ecology & Evolution* 24 (8): 431–438.

Koh, L. P., and D. S. Wilcove. 2008. Is oil palm agriculture really destroying tropical biodiversity? *Conservation Letters* 1 (2): 60–64.

Kojima, Y., J. L. Parcell, and J. S. Cain. 2014. A demand model of the wholesale vegetable oil market in the U.S.A. Working paper, Department of Agricultural and Resource Economics, University of Missouri. http://ageconsearch.umn.edu/bitstream/162472/2/A%20Demand%20Model%20of%20the%20Wholesale%20Vegetable%20Oils%20Market%20in%20the%20U.S.A%20(Revised%20in%20March%202014)%20(1).pdf (accessed May 2, 2022).

Kotrba, R. 2015. Biodiesel blenders tax credit passes US House, Senate. *Biodiesel Magazine*, December 18. http://www.biodieselmagazine.com/articles/646582/biodiesel-blenders-tax-credit-passes-us-house-senate (accessed January 7, 2016).

Krapf, B. M., D. D. Raab, and B. L. Zwilling. 2014. Costs to produce corn and soybeans in Illinois—2013. *FarmDoc Daily*, March 21. http://farmdocdaily.illinois.edu/2014/03/cost-to-produce-corn-and-soybean-in-illinois-2013.html (accessed 31 January 2016).

Krishna, V., M. Euler, H. Siregar, Z. Fathoni, and M. Qaim. 2015. *Farmer heterogeneity and differential livelihood impacts of oil palm expansion among smallholders in Sumatra, Indonesia*. Goettingen: University of Goettingen.

Kruse, J. 2010. Estimating demand for agricultural commodities to 2050. Global Harvest Initiative, 1–26. http://globalharvestinitiative.org/Documents/Kruse%20-%20Demand%20for%20Agricultural%20Commoditites.pdf (accessed 31 January 2016).

Kumar, P., A. Kumar, S. Parappurathu, and S. S. Raju. 2011. Estimating demand elasticity for food commodities in India. *Agricultural Economics Research Review* 24: 1–14.

Kurup, S. A., G. K. Jha, and A. Singh. 2015. Technical and efficiency changes in oilseed sector in India: An analysis using Malmquist total factor productivity approach. Paper presented at the 29th International Conference of Agricultural Economists, Milan, Italy, August 12.

Lai, O.- M., C.- P. Tan, and C. C. Akoh, eds. 2012. *Palm oil: Production, processing, characterization, and uses*. Urbana, IL: AOCS Press.

Lambin, E. F., H. K. Gibbs, L. Ferreira, R. Grau, P. Mayaux, P. Meyfroidt, D. C. Morton, T. K. Rudel, I. Gasparri, and J. Munger. 2013. Estimating the world's potentially available cropland using a bottom-up approach. *Global Environmental Change* 23 (5): 892–901.

Lambin, E. F., P. Meyfroidt, X. Rueda, A. Blackman, J. Börner, P. O. Cerutti, T. Dietsch, et al. 2014. Effectiveness and synergies of policy instruments for land use governance in tropical regions. *Global Environmental Change* 28: 129–140.

Langeveld, H., J. Dixon, and H. van Keulen, eds. 2014. *Biofuel cropping systems: Carbon, land and food*. Oxon: Routledge.

Lanier, F., A. Mukpo, and F. Wilhelmsen. 2012. *Smell-no-taste: The social impact of foreign direct investment in Liberia*. New York, NY: Center for International Conflict Resolution, Columbia University.

Lapola, D. M., L. A. Martinelli, C. A. Peres, J. P. H. B. Ometto, M. E. Ferreira, C. A. Nobre, A. P. D. Aguiar, et al. 2014. Pervasive transition of the Brazilian land-use system. *Nature Climate Change* 4 (1): 27–35.

Larson, D. F. 1996. Indonesia's palm oil subsector. Policy Research working paper 1654. Washington, DC: World Bank.

Lee, J. S. H., S. Abood, J. Ghazoul, B. Barus, K. Obidzinski, and L. P. Koh. 2014a. Environmental impacts of large scale oil palm enterprises exceed that of smallholdings in Indonesia. *Conservation Letters* 7 (1): 25–33.

Lee, J. S. H., J. Ghazoul, K. Obidzinski, and L. P. Koh. 2014b. Oil palm smallholder yields and incomes constrained by harvesting practices and type of smallholder management in Indonesia. *Agronomy for Sustainable Development* 34: 501–513.

Levin, J. 2012. *Profitability and sustainability of palm oil production: Analysis of incremental financial costs and benefits of RSPO compliance*. Gland: World Wildlife Fund.

Li, T. M. 2014. The gendered dynamics of Indonesia's oil palm labour regime. Working paper no. 225. Singapore: Asia Research Institute, National University of Singapore.

Lin, J. Y. 2012. New structural economics: A framework for rethinking development. *The World Bank Research Observer* 26 (2): 193–221.

Ling, A. H. 2014. El Nino and palm oil production: Supply shock in the making? Pape presented at El Nino 2014, Effects and Implications, Malaysian Palm Oil Council, Kuala Lumpur, June 2014.

Lipow, G. W. 2008. *Price-elasticity of energy demand: A bibliography*. New York, NY: Carbon Tax Center.

Lobell, D. B., R. L. Naylor, and C. B. Field. 2014. Food, energy, and climate connections in a global economy. In: *The evolving sphere of food security*, ed. R. L. Naylor, 239–268. Oxford: Oxford University Press.

Lomax, T., J. Kenrick, and A. Brownell. 2012. *Sime Darby oil palm and rubber plantation in Grand Cape Mount County, Liberia*. Moreton-in-Marsh: Forest Peoples Program.

Lubungu, M., W. J. Burke, and N. J. Sitko. 2013. *Challenges of smallholder soybean production and commercialization in eastern province of Zambia*. IAPRI policy brief 62. Zambia: Indaba Agricultural Policy Research Institute.

Lynn, M. 1997. *Commerce and economic change in West Africa the palm oil trade in the nineteenth century*. Cambridge: Cambridge University Press.

MacDonald, J. M., P. Korb, and R. A. Hoppe. 2013. *Farm size and the organization of U.S. crop farming*. Economics Research report no. 152. Washington, DC: US Department of Agriculture.

Macedo, M. N., and E. A. Davidson. 2014. Climate and land use: Forgive us our carbon debts. *Nature Climate Change* 4 (7): 538–539.

Macedo, M. N., R. S. DeFries, D. C. Morton, C. M. Stickler, G. L. Galford, and Y. E. Shimabukuro. 2012. Decoupling of deforestation and soy production in the southern Amazon during the late 2000s. *Proceedings of the National Academy of Sciences of the United States of America* 109 (4): 1341–1346.

Malaysian Palm Oil Board. 2011. Malaysian palm oil industry. http://www.palmoilworld.org/about_malaysian-industry.html. (accessed August 12, 2015).

Mancini, A., E. Imperlini, E. Nigro, C. Montagnese, A. Daniele, S. Orrù, and P. Buono. 2015. Biological and nutritional properties of palm oil and palmitic acid: Effects on health. *Molecules* 20: 17339–17361.

Marchal, J. 2008. *Lord Leverhulme's ghosts: Colonial exploitation in the Congo*. London: Verso.

Marchetti, J. M., V. U. Miguel, and A. F. Errazu. 2007. Possible methods for biodiesel production. *Renewable and Sustainable Energy Reviews* 11 (6): 1300–1311.

Margono, B. A., P. V. Potapov, S. Turubanova, F. Stolle, and M. C. Hansen. 2014. Primary forest cover loss in Indonesia over 2000–2012. *Nature Climate Change* 4: 730–735.

Marin-Burgos, V. 2014. Access, power and justice in commodity frontiers: The political ecology of access to land and palm oil expansion in Colombia. PhD diss., University of Twente.

Marin-Burgos, V., J. S. Clancy, and J. C. Lovett. 2014. Contesting legitimacy of voluntary sustainability certification schemes: Valuation languages and power asymmetries in the Roundtable on Sustainable Palm Oil in Colombia. *Ecological Economics* 117: 303–313.

Marlier, M. E., R. S. DeFries, P. S. Kim, S. N. Koplitz, D. J. Jacob, L. J. Mickley, and S. S. Myers. 2015. Fire emissions and regional air quality impacts from fires in oil palm, timber, and logging concessions in Indonesia. *Environmental Research Letters* 10 (8): 1–9.

Marlier, M. E., R. S. DeFries, A. Voulgarakis, P. L. Kinney, J. T. Randerson, D. T. Shindell, Y. Chen, and G. Faluvegi. 2013. El Nino and health risks from landscape fire emissions in Southeast Asia. *Nature Climate Change* 3 (2): 131–136.

Martin, S M. 1988. *Palm oil and protest: An economic history of the Ngwa Region, South-eastern Nigeria, 1800–1980*. Cambridge: Cambridge University Press.

Martin, S. M. 2003. *The UP saga*. Copenhagen: NIAS Press.

Martinez Pelaez, G. 2013. Panorama de la agroindustria palmera: Retos y opportunidades. Paper presented at La Agroindustria de la Palma de Aceite: un Negocio Sostenible e Inclusive. Bogota, Colombia, October 2013.

MASDAR Ministry of Food and Agriculture. 2011. *Masterplan study on the oil palm industry in Ghana*. Eversley: MASDAR.

Masuda, T., and P. D. Goldsmith. 2009. World soybean production: Area harvested, yield, and long-term projections. *International Food and Agribusiness Management Review* 12 (4): 143–162.

Mattick, C., and B. Allenby. 2013. The future of meat. *Issues in Science and Technology* 30 (1): 64–70.

Mausch, K., D. Mithöfer, S. Asfaw, and H. Waibel. 2009. Export vegetable production in Kenya under the EurepGAP standard: Is large "more beautiful" than small? *Journal of Food Distribution Research* 40 (3): 115–129.

May, C. Y., and K. Nesaretnam. 2015. Research advancements in palm oil nutrition. *European Journal of Lipid Science and Technology* 116 (10): 1301–1315.

McCarthy, J. F. 2010. Processes of inclusion and adverse incorporation: Oil palm and agrarian change in Sumatra, Indonesia. *Journal of Peasant Studies* 37 (4): 821–850.

McCarthy, J. F. 2012. Certifying in contested spaces: Private regulation in Indonesian forestry and palm oil. *Third World Quarterly* 33 (10): 1871–1888.

McCarthy, J. F., and R. A. Cramb. 2009. Policy narratives, landholder engagement, and oil palm expansion on the Malaysian and Indonesian frontiers. *Geographical Journal* 175 (2): 112–123.

McCarthy, J. F., P. Gillespie, and Z. Zen. 2012. Swimming upstream: Local Indonesian production networks in "globalized" palm oil production. *World Development* 40 (3): 555–569.

McFarland, W., S. Whitley, and G. Kissinger. 2015. Subsidies to key commodities driving forest loss: Implications for private climate finance. Working paper. London: Overseas Development Institute.

Meena, R. K., S. S. Burark, and D. Goutam. 2013. Comparative analysis of soybean processing and employment generation by private and cooperative units: A study of Kota District of Rajasthan. *Indian Journal of Agricultural Research* 47 (6): 503–509.

Meher, L. C., D. V. Sagar, and S. N. Naik. 2006. Technical aspects of biodiesel production of transesterification: A review. *Renewable and Sustainable Energy Reviews* 10 (3): 248–268.

Meredith, D. 1984. Government and the decline of the Nigerian oil-palm export industry, 1919–1939. *Journal of African History* 25 (3): 311–329.

Micha, R., and D. Mozaffarian. 2010. Saturated fat and cardiometabolic risk factors, coronary heart disease, stroke and diabetes: A fresh look at the evidence. *Lipids* 45 (10): 893–905.

Mitchell, D. 2011. *Biofuels in Africa: Opportunities, prospects, and challenges.* Washington, DC: World Bank.

Morton, D. C., R. S. DeFries, Y. E. Shimabukuro, L. O. Anderson, E. Arai, F. del Bon Espirito-Santo, R. Freitas, and J. Morisette. 2006. Cropland expansion changes deforestation dynamics in the southern Brazilian Amazon. *Proceedings of the National Academy of Sciences of the United States of America* 103 (39): 14637–14641.

Moschini, G., J. Cui, and H. Lapan. 2012. Economics of biofuels: An overview of policies, impacts and prospects. *Bio-based and Applied Economics* 1 (3): 269–296.

Mosquera, M., K. Grogan, E. Evans, and T. Spreen. 2012. A dynamic model of pudricion del cogollo disease control in Colombian palm oil industry. Paper presented at the annual meeting of the Agricultural and Applied Economics Association, Seattle, Washington, August 2012.

Mpepereki, S., F. Javaheri, P. Davis, and K. E. Giller. 2000. Soyabeans and sustainable agriculture: Promiscuous soyabeans in southern Africa. *Field Crops Research* 65 (2): 137–149.

Msangi, S., and M. Batka. 2015. Aquaculture and global fish supply and demand: Implications for nutrition and the environment to 2030 and beyond. Paper presented at the 29th International Conference of Agricultural Economists, Milan, Italy, August 2015.

Mukti, A., B. Setiawan, H. Pramoedyo, and L. Fatah. 2014. The impact of palm oil plantation development towards income disparity. *Academic Research International* 5 (2): 96–106.

Murphy, D. J. 2014. The future of oil palm as a major global crop: Opportunities and challenges. *Journal of Oil Palm Research* 26 (1): 1–24.

Myers, R. J., S. R. Johnson, M. Helmar, and H. Baumes. 2014. Long-run and short-run comovements in energy prices and the prices of agricultural feedstocks for biofuel. *American Journal of Agricultural Economics* 96 (4): 991–1008.

Nagata, J., and S. W. Arai. 2013. Evolutionary change in the oil palm plantation sector in Riau Province, Sumatra. In: *The palm oil controversy in Southeast Asia: A transnational perspective,* ed. O. Pye and J. Bhattacharya, 76–96. Pasir Panjang: Institute of Southeast Asian Studies.

Nagiah, C., and R. Azmi. 2013. A review of smallholder oil palm production: Challenges and opportunities for enhancing sustainability: A Malaysian perspective. *Journal of Oil Palm, Environment and Health* 3: 114–120.

Nantha, H. S. 2014. The financial and political opportunity costs of orangutan conservation in the face of oil-palm expansion. Working Papers on Economics, Ecology and The Environment 195. Brisbane: School of Economics, University of Queensland.

Natella, S., V. Divan, and M. Giraldo. 2015. *Fat: The new health paradigm.* Zurich: Credit Suisse Ag.

National Renewable Energy Laboratory. 2013. International trade of biofuels. US Department of Energy, National Renewable Energy Laboratory. http://www.nrel.gov/docs/fy13osti/56792.pdf (accessed January 7, 2016).

Naylor, R. 2014. Biofuels, rural development, and the changing nature of agricultural demand. In: *Frontiers in food policy: Perspectives on sub-Saharan* Africa, ed. W. P. Falcon and R. L. Naylor, 343–376. Stanford, CA: Stanford University, Center on Food Security and the Environment.

Naylor, R. L., and W. P. Falcon. 2011. The global costs of American ethanol. *The American Interest* 7 (2): 66–76.

Naylor, R. L., R. W. Hardy, D. P. Bureau, A. Chiu, M. Elliott, A. P. Farrell, I. Forster, et al. 2009. Feeding aquaculture in an era of finite resources. *Proceedings of the National Academy of Sciences of the United States of America* 106 (36): 15103–15110.

Naylor, R., and M. Higgins. The political economy of biodiesel in an era of low energy prices. Preprint, submitted March 1, 2016.

Nchanji, Y. K., O. Tataw, R. N. Nkongho, and P. Levang. 2013. Artisanal milling of palm oil in Cameroon. Working paper 128. Bogor: Indonesia: Center for International Forestry Research.

Nepstad, D., D. McGrath, C. Stickler, A. Alencar, A. Azevedo, B. Swette, T. Bezerra, et al. 2014. Slowing Amazon deforestation through public policy and interventions in beef and soy supply chains. *Science* 344 (6188): 1118–1123.

Nepstad, D. C., C. M. Stickler, and O. T. Almeida. 2006. Globalization of the Amazon soy and beef industries: Opportunities for conservation. *Conservation Biology* 20 (6): 1595–1603.

Nesadurai, H. E. S. 2013. Food security, the palm oil–land conflict nexus, and sustainability: A governance role for a private multi-stakeholder regime like the RSPO? *Pacific Review* 26 (5): 505–529.

Neste Oil. 2014. *Hydrotreated vegetable oil (HVO): Premium renewable biofuel for diesel engines.* Espoo: Neste Oil.

Nkongho, R. N., L. Feintrenie, and P. Levang. 2014. The non-industrial palm oil sector in Cameroon. Working paper 139. Bogor, Indonesia: Center for International Forestry Research.

Nkongho, R. N., T. E. Ndjogui, and P. Levang. 2015. History of partnership between agro-industries and oil palm smallholders in Cameroon. *OCL* 22 (3): 1–15.

Nogués, J. J. 2014. Argentina. In: *Policy responses to high food prices in Latin America and the Caribbean*, ed. E. Krivonos and D. Dawe, 13–33. Rome, Italy: Food and Agriculture Organization of the United Nations.

Norfolk, S., and J. Hanlon. 2012. Confrontation between peasant producers and investors in northern Zambézia, Mozambique, in the context of profit pressures on European investors. Paper presented at the annual World Bank Conference on Land and Poverty, Washington. DC, April 2012.

Norwana, A. A. B. D., R. Kanjappan, M. Chin, G. Schoneveld, L. Potter, and R. Andriani. 2011. The local impacts of oil palm expansion in Malaysia: An assessment based on a case study in Sabah State. Working paper 78. Bogor Barat, Indonesia: Center for International Forestry Research.

Obidzinski, K. 2015. "Zero Deforestation commitments": Realizing the opportunities and overcoming challenges. Paper presented at the annual World Bank Conference on Land and Poverty, Washington, DC, March 2015.

Obidzinski, K., A. Dermawan, and A. Hadianto. 2014. Oil palm plantation investments in Indonesia's forest frontiers: Limited economic multipliers and uncertain benefits for local communities. *Environment, Development and Sustainability* 16 (6): 1177–1196.

Ofosu-Budu, K., and D. B. Sarpong. 2013. Oil palm industry growth in Africa: A value chain and smallholders' study for Ghana. In: *Rebuilding West Africa's food potential: Policies and market incentives for smallholder-inclusive food value chains*, ed. A. Elbehri, 349–389. Rome, Italy: Food and Agricultural Organization of the United Nations.

Oh, K., F. B. Hu, J. E. Manson, M. J. Stampfer, and W. C. Willett. 2005. Dietary fat intake and risk of coronary heart disease in women: 20 Years of follow-up of the Nurse's Health Study. *American Journal of Epidemiology* 161 (7): 672–679.

O'Kray, C. 2014. *Brazil: Oilseeds and products annual report*. GAIN report no BR0931. Washington, DC: US Department of Agriculture.

Oliveira, G. de L. T. 2015. Chinese and other foreign investments in the Brazilian soybean complex. Working paper 9, BRICS Initiative for Critical Agrarian Studies. http://www.iss.nl/fileadmin/ASSETS/iss/Research_and_projects/Research_networks/BICAS/BICAS_WP_9-Oliveira.pdf (accessed 31 January 2016).

O'Neil,. 2013. The European House Ambrosetti. Goldman Sachs, Asset Management. http://www.goldmansachs.com/s/GMeT_othermailings_attachments/634985868148530000841 51.pdf (accessed 31 January 2016).

Organisation for Economic Co-operation and Development. 2012a. Medium and long-term scenarios for global growth and imbalances. In: *OECD economic outlook*, 191–234. Paris: OECD Publishing.

Organisation for Economic Co-operation and Development. 2012b. Review of agricultural policies: Indonesia 2012. Paris: OECD Publishing.

Organisation for Economic Co-operation and Development. 2013. *Agricultural policy monitoring and evaluation 2013: OECD countries and emerging economies*. OECD Publishing. http://dx.doi.org/10.1787/agr_pol-2013-en (accessed 31 January 2016).

Organisation for Economic Co-operation and Development and Food and Agriculture Organization of the United Nations. 2014. Executive summary. In: *OECD-FAO agricultural outlook 2014*. Paris: OECD Publishing. *Agricultural Outlook 2014*. Paris: OECD Publishing. http://dx.doi.org/10.1787/agr_outlook-2014-3-en. Accessed January 8, 2016.

Organization of the Petroleum Exporting Countries. 2014. *2014 world oil outlook*. Vienna: Organization of the Petroleum Exporting Countries.

Osborn, D. Z. 2008. Soybean and soybean products in West Africa. In: *The world of soy* (*The food series*), ed. C. M. Du Bois, C.- B. Tan, and S. W. Mintz, 276–298. Urbana, IL: University of Illinois Press.

Osei-Amponsah, C., L. Visser, S. Adjei-Nsiah, P. C. Struik, O. Sakyi-Dawson, and T. J. Stomph. 2012. Processing practices of small-scale palm oil producers in the Kwaebibirem District, Ghana: A diagnostic study. *NJAS-Wageningen Journal of Life Sciences* 60–63: 49–56.

Overseas Development Institute. 2015. *The rising cost of a healthy diet: Changing relative prices of foods in high-income and emerging economies*. London: Overseas Development Institute.

Oxfam International. 2014. *Fair company–community partnerships in palm oil development*. Oxford: Oxfam International.

Oxford Economics. 2014. *Making FDI work for sub-Saharan Africa: Lessons from Liberia*. Oxford: Oxford Economics.

Pacheco, P. 2012. Soybean and oil palm expansion in South America: A review of main trends and implications. CIFOR working paper no. 90. Bogor, Indonesia: CIFOR.

Pakiam, R., and Y. Rusmana. 2014. Indonesia's "toothless" mandate for biofuels hurting palm. *Bloomberg Business*, November 28. http://www.bloomberg.com/news/articles/2014-11-28/indonesia-s-toothless-mandate-for-biofuel-seen-hurting-palm (accessed June 23, 2015).

Pan, S., S. Mohanty, and M. Welch. 2008. India edible oil consumption: A censored incomplete demand approach. *Journal of Agricultural and Applied Economics* 40 (3): 821–835.

Pankratz, H. 2014. Smart Balance dumps GMO oils from its line of 15 buttery spreads. *The Denver Post,* March 3. http://www.denverpost.com/business/ci_25267321/smart-balance-dumps-gmo-oils-from-its-line?source=infinite (accessed 31 Jan 2016).

Pardey, P. G., J. M. Alston, C. Chan-Kang, E. C. Magalhães, and S. A. Vosti. 2006. International and institutional R&D spillovers: Attribution of benefits among sources for Brazil's new crop varieties. *American Journal of Agricultural Economics* 88 (1): 104–123.

Parveez, G. K. A., O. A. Rasid, A. T. Hashim, Z. Ishak, S. K. Rosli, and R. Sambanthamurthi. 2012. Tissue culture and genetic engineering of oil palm. In: *Palm oil: Production, processing, characterization, and uses,* ed. O. M. Lai, C.- P. Tan, and C. C. Akoh, 87–136. Urbana, IL: AOCS Press.

Pauli, N., C. Donough, T. Oberthür, J. Cock, R. Verdooren, G. Rahmadsyah, K. Abdurrohim, et al. 2014. Changes in soil quality indicators under oil palm plantations following application of "best management practices" in a four-year field trial. *Agriculture, Ecosystems & Environment* 195: 98–111.

Payne, J. 2015. Nigerian firms in trouble as Central Bank measures backfire. September 25. http://www.reuters.com/article/nigeria-economy-idUSL5N11U3G520150925. (accessed 31 Jan 2016).

Perera, S. A. C. N. 2014. Oil palm and coconut. In: *Alien gene transfer in crop plants,* vol. 2, ed. A. Pratap and J. Kumar, 231–252. New York, NY: Springer-Verlag.

Persaud, S., and M. Landes. 2006. *The role of policy and industry structure in India's oilseed markets.* Economic Research report no 17. Washington, DC: US Department of Agriculture.

Pichler, M. 2013. "People, Planet & Profit": Consumer-oriented hegemony and power relations in palm oil and agrofuel certification. *Journal of Environment & Development* 22 (4): 370–390.

Pim, A. 1946. *Colonial agricultural production.* London: Oxford University Press.

Pinzon, L. 2012. *Colombia biofuels annual.* Bogotá: US Department of Agriculture Foreign Agricultural Service.

Pletcher, J. 1991. Regulation with growth: The political economy of palm oil in Malaysia. *World Development* 19 (6): 623–636.

Poku, K. 2002. Small-scale palm oil processing in Africa. FAO Agricultural Services bulletin 148. Rome, Italy: Food and Agriculture Organization of the United Nations.

Population Reference Bureau. 2014. 2013 World population data sheet. Washington, DC: Population Reference Bureau. http://www.prb.org/pdf13/2013-population-data-sheet_eng.pdf (accessed 31 January 2016).

Potts, J., M. Lynch, A. Wilkings, G. Huppe, M. Cunningham, and V. Voora. 2014. *The state of sustainability initiatives review 2014: Standards and the green economy.* Winnipeg: International Institute for Sustainable Development.

Pradhan, A., D. Shrestha, J. Van Gerpen, A. McAloon, W. Yee, M. Haas, and J. A. Duffield. 2012. Reassessment of life cycle greenhouse gas emissions for soybean biodiesel. *Transactions of the ASABE American Society of Agricultural and Biological Engineers* 55 (6): 2257–2264.

Prais, S. J. and H. S. Houthakker. 1955. *The analysis of family budgets.* London: Cambridge University Press.

Preechajarn, S. 2010. *Thailand: Oil palm situation field trip.* GAIN report no. TH0114. Washington, DC: US Department of Agriculture Foreign Agricultural Service.

Price Waterhouse and Coopers. 2015. *The world in 2050: Will the shift in global economic power continue?* London: Price Waterhouse and Coopers.

Pursell, G., A. Gulati, and K. Gupta. 2007. Distortions to agricultural incentives in India. Agricultural Distortions working paper 34. Washington, DC: World Bank.

Putz, F. E., and C. Romero. 2014. Futures of tropical forests (*sensu lato*). *Biotropica* 46 (4): 495–505.

Pye, O. 2015. Transnational space and workers' struggles. In: *The political ecology of agrofuels*, ed. K. Dietz, B. Engels, O. Pye, and A. Brunnengräber, 186–201. New York: Routledge.

Pye, O., R. Daud, Y. Harmono, and Tatat. 2012. Precarious lives: Transnational biographies of migrant oil palm workers. *Asia Pacific Viewpoint* 53 (3): 330–342.

Qaim, M. 2009. The economics of genetically modified crops. *Annual Review of Resource Economics* 1: 665–694.

Rabobank. 2012. *MATOPIBA: Brazil's last frontier*. Rabobank industry note no. 323. Amsterdam: Rabobank.

Radich, A. 2004. Biodiesel performance, costs, and use. http://www.eia.gov/oiaf/analysispaper/biodiesel/ (accessed June 23, 2015).

Rainforest Foundation UK. 2013. *Seeds of destruction: Expansion of industrial oil palm in the Congo Basin: Potential impacts on forests and people*. London: Rainforest Foundation UK.

Ramasamy, B., D. Ong, and M. C. H. Yeung. 2005. Firm size, ownership and performance in the Malaysian palm oil industry. *Asian Academy of Management Journal of Accounting and Finance* 1 (1): 81–104.

Ramdani, F., and M. Hino. 2013. Land-use changes and GHG emissions from tropical forest conversion by oil palm plantations in Riau Province, Indonesia. *PLoS One* 8 (7): e70323.

Ramteke, R., G. Gupta, P. Murlidharan, and S. K. Sharma. 2011. Genetic progress of soybean varieties released during 1969 to 2008 in India. *Indian Journal of Genetics and Plant Breeding* 71 (4): 330–340.

Rasiah, R. 2006. Explaining Malaysia's export expansion in oil palm and related products. In: *Technology, adaptation, and exports: How some developing countries got it right*, ed. V. Chandra, 163–192. Washington, DC: International Bank for Reconstruction and Development/World Bank.

Rasiah, R., and A. Shahrin. 2006. *Development of palm oil and related products in Malaysia and Indonesia*. Kuala Lumpur: University of Malaya.

Renewable Energy Policy Network for the 21st Century. 2014. Renewables 2014 global status report. Paris: Renewable Energy Policy Network for the 21st Century. http://www.ren21.net/Portals/0/documents/Resources/GSR/2014/GSR2014_full%20report_low%20res.pdf (accessed May 5, 2016).

Reydon, B. P., A. K. da Silva Bueno, and A. P. da Silva Bueno. Institutional innovation for the solution to Amazonia's land ownership problems: The Case of the Internal Affairs Department in Mato Grosso. Paper presented at the 2015 World Bank Conference on Land and Poverty, Washington, DC, March 2015.

Richards, P. D., R. J. Myers, S. M. Swinton, and R. T. Walker. 2012. Exchange rates, soybean supply response, and deforestation in South America. *Global Environmental Change* 22 (2): 454–462.

Richards, P., H. Pellegrina, L. VanWey, and S. Spera. 2015. Soybean development: The impact of a decade of agricultural change on urban and economic growth in Mato Grosso, Brazil. *PLoS One* 10 (4): e0122510.

Richards, P. D., R. T. Walker, and E. Y. Arima. 2014. Spatially complex land change: The indirect effect of Brazil's agricultural sector on land use in Amazonia. *Global Environmental Change* 29: 1–9.

Rifin, A. 2010. The effect of export tax on Indonesia's crude palm oil (CPO) export competitiveness. *Journal of Southeast Asian Economies* 27 (2): 173–184.

Rifin, A. 2011. The role of the palm oil industry in Indonesian economy and its export competitiveness. PhD diss., University of Tokyo.

Rist, L., L. Feintrenie, and P. Levang. 2010. The livelihood impacts of oil palm: Smallholders in Indonesia. *Biodiversity and Conservation* 19: 1009–1024.

Rival, A., and P. Levang. 2014. *Palms of controversies: Oil palm and development challenges.* Bogor: CIFOR.

Rodenburg, J. 2012. Iowa Soybean Association Asian Trade Mission: Recap. http://www.iasoybeans.com/chinablog2012/ (accessed 31 January 2016).

Rosegrant, M. W., M. S. Paisner, S. Meijer, and J. Witcover. 2001. *Global food projections to 2020: Emerging trends and alternative futures*. Washington, DC: IFPRI.

Rosegrant, M. W., S. Tokgoz, P. Bhandary, and S. Msangi. 2012. *Looking ahead: Scenarios for the future of food.* Global Food Policy report. Washington, DC: IFPRI.

Rosegrant, M. W., 2014 Personal communication to W.P. Falcon, "Vegetable oils". September 26.

Rosenbarger, A., B. Gingold, R. Prasodjo, A. Alisjahbana, A. Putraditama, and D. Tresya. 2013. *How to change legal land use classification to support more sustainable palm oil in Indonesia.* Washington, DC: World Resources Institute.

Rosillo-Calle, F., S. Teelucksingh, D. Thrän, and M. Seiffert. 2012. The potential and role of biofuels in commercial air transport: Biojetfuel. IAE Paris: IEA

Roundtable on Sustainable Palm Oil. 2013. *Principles and criteria for the production of sustainable palm oil.* Kuala Lumpur: Roundtable on Sustainable Palm Oil.

Roundtable on Sustainable Palm Oil. 2014. *An international multilevel stakeholder initiative transforming markets to make sustainable palm oil the norm.* Kuala Lumpur: Roundtable on Sustainable Palm Oil.

Rudorff, B. F. T., M. Adami, D. A. Aguiar, M. A. Moreira, M. P. Mello, L. Fabiani, D. F. Amaral, and B. M. Pires. 2011. The soy moratorium in the Amazon biome monitored by remote sensing images. *Remote Sensing* 3 (1): 185–202.

Rugeles, L. 2011. *La Cooperativa de Palmicultores de Colombia, COPALCOL: Un caso de empresa asociativa rural exitosa.* Bogotá: Universidad de Bogotá Jorge Tadeo Lozano, Facultad de Ciencias Económicas-Administrativas.

Rusike, J., C. Sukume, A. Dorward, S. Mpepereki, and K. Giller. 2000. *The economic potential for smallholder soybean production in Zimbabwe.* Soil Fertility Network special publication. University of Zimbabwe and University of London.

Saikkonen, L., M. Ollikainen, and J. Lankoski. 2014. Imported palm oil for biofuels in the EU: Profitability, greenhouse gas emissions and social welfare effects. *Biomass and Bioenergy* 68: 7–23.

Sanders, D. J., J. V. Balagtas, and G. Gruere. 2014. Revisiting the palm oil boom in South-East Asia: Fuel versus food demand drivers. *Applied Economics* 46 (2): 127–138.

Sandker, M., A. Suwarno, and B. M. Campbell. 2007. Will forests remain in the face of oil palm expansion? Simulating change in Malinau, Indonesia. *Ecology and Society: A Journal of Integrative Science for Resilience and Sustainability* 12 (2): 37.

Sands, R., C. Jones, and E. Marshall. 2014. *Global drivers of agricultural demand and supply*. Washington, DC: US Department of Agriculture.

Sanginga, P. C., A. A. Adesina, V. M. Manyong, O. Otite, and K. E. Dashiell. 1999. *Social impact of soybean in Nigeria's southern Guinea savanna*. Ibadan: International Institute of Tropical Agriculture.

Santana, C. A. M., D. A. P. Torres, R. do C. N. Guiducci, M. A. da Silva Alves, Fernando L. Garagorry, G. da Silva e Souza, E. D. Assad, et al. 2011. *Productive capacity of Brazilian agriculture: A long-term perspective*. Regional case study R5. London: Foresight, Government Office for Science.

Sapp, M. 2015b. Thailand slashes B7 mandate by half due to low palm oil supplies. *Biofuels Digest*, January 21. http://www.biofuelsdigest.com/bdigest/2015/01/21/thailand-slashes-b7-mandate-by-half-due-to-low-palm-oil-supplies/ (accessed June, 23 2015).

Saravanamuttu, J. 2013. The political economy of migration and flexible labour regimes: The case of the oil palm industry in Malaysia. In: *The palm oil controversy in Southeast Asia: A transnational perspective*, ed. O. Pye and J. Bhattacharya, 120–139. Singapore: ISEAS Publishing.

Sauer, S., and S. P. Leite. 2012. Agrarian structure, foreign investment in land, and land prices in Brazil. *Journal of Peasant Studies* 39 (3–4): 873–898.

Savilaakso, S., C. Garcia, J. Garcia-Ulloa, J. Ghazoul, M. Groom, M. R. Guariguata, Y. Laumonier, et al. 2014. Systematic review of effects on biodiversity from oil palm production. *Environmental Evidence* 3 (1): 1–21.

Sayer, J., J. Ghazoul, P. Nelson, and A. K. Boedhihartono. 2012. Oil palm expansion transforms tropical landscapes and livelihoods. *Global Food Security* 1 (2): 114–119.

Sayer, J., C. Margules, A. K. Boedhihartono, A. Dale, T. Sunderland, J. Supriatna, and R. Saryanthi. 2014. Landscape approaches: What are the pre-conditions for success? *Sustainability Science* 10 (2): 345–355.

Scherr, S. J., K. Mankad, S. M. Jaffee, and C. Negra. 2015. *Steps toward green: Policy responses to the environmental footprint of commodity agriculture in East and Southeast Asia*. Washington, DC: EcoAgriculture Partners and World Bank.

Schlesinger, S. 2004. A soja no Brasil: Brasil sustentável e democrático. Seminário do Cone Sul. http://uma.terra.free.fr/2Agrobusiness/Soja-Brasil.rtf (accessed 31 January 2016).

Schouten, G., and P. Glasbergen. 2012. Private multi-stakeholder governance in the agricultural market place: An analysis of legitimization processes of the Roundtables on Sustainable Palm Oil and Responsible Soy. *International Food and Agribusiness Management Review* 15: 63–88.

Schwarze, S., M. Euler, M. Gatto, J. Hein, E. Hettig, A. M. Holtkamp, L. Izhar, et al. 2015. *Rubber vs. oil palm: An analysis of factors influencing smallholders' crop choice in Jambi, Indonesia*. Goettingen: University of Goettingen.

Searchinger, T. D., L. Estes, P. K. Thornton, T. Beringer, A. Notenbaert, D. Rubenstein, R. Heimlich, R. Licker, and M. Herrero. 2015. High carbon and biodiversity costs from converting Africa's wet savannahs to cropland. *Nature Climate Change* 5: 481–486.

Selfer, T., C. Bain, and R. Moreno. 2013. Colombia. In: *Sustainable development of biofuels in Latin America and the Caribbean*, ed. B. D. Solomon and R. Bailis, 157–178. New York, NY: Springer.

Sentelhas, P. C., R. Battisti, G. M. S. Camara, J. R. B. Farias, A. C. Hampf, and C. Nendel. 2015. The soybean yield gap in Brazil: Magnitude, causes and possible solutions for sustainable production. *Journal of Agricultural Science* 153: 1394–1411.

Seymour, F., N. Birdsall, and W. Savedoff. 2015. *The Indonesia–Norway REDD+ agreement: A glass half-full*. Policy paper 56. Washington, DC: Center for Global Development.

Shah, D. 2014. Economic viability of soyabean crop cultivation in Maharashtra: Some emerging issues. *Indian Journal of Economics and Development* 10 (3): 256–267.

Sheil, D., A. Casson, E. Meijaard, M. van Noordwijk, J. Gaskell, J. Sunderland-Groves, K. Wertz, and M. Kanninen. 2009. *The impacts and opportunities of oil palm in Southeast Asia: What do we know and what do we need to know?* Bogor: Center for International Forestry Research.

Shibao, P. 2015. Big palm oil's financiers. *Mongabay*. July 27. http://news.mongabay.com/2015/07/big-palm-oils-financiers/ (accessed 31 January 2016).

Siakor, S. K. 2012. *Uncertain futures: The impacts of Sime Darby on communities in Liberia*. Montevideo: World Rainforest Movement.

Sills, E. O., S. S. Atmadja, C. de Sassi, A. E. Duchelle, D. L. Kweka, I. A. P. Resosudarmo, and W. D. Sunderlin, eds. 2013. *REDD+ on the ground: A case book of subnational initiatives across the globe*. Bogor, Indonesia: Center for International Forestry Research.

Sime Darby. 2009. *Palm oil industry in Malaysia: Skills and knowledge for sustained development in Africa*. Washington, DC: World Bank.

Sime Darby Berhad. 2013. *Global Reach local solutions annual report* 2013. Kuala Lumpur: Sime Darby Berhad.

Simeh, A. 2011. Oil palm smallholders in Malaysia. Paper presented at the Agribenchmark Conference, Middelfart, Denmark, June 2011.

Simeh, A., and T. M. A. T. Ahmad. The case study on the Malaysian palm oil. Paper presented at the Regional Workshop on Commodity Export Diversification and Poverty Reduction in South and Southeast Asia, Bangkok, Thailand, April 2001.

Skinner, E. B. 2013. Indonesia's palm oil industry rife with human-rights abuses. *Bloomberg Businessweek*. http://www.bloomberg.com/news/articles/2013-07-18/indonesias-palm-oil-industry-rife-with-human-rights-abuses. (Accessed 31 Jan 2016)

Small, R. 2014. *Review of smallholder models: Liberia and Sierra Leone*. Cambridge: Fauna and Flora International.

Smart, T., and J. Hanlon. 2014. Chickens and beer: Recipe for agricultural growth in Mozambique. http://www.open.ac.uk/technology/mozambique/sites/www.open.ac.uk.technology.mozambique/files/files/Chickens_and_beer-a_recipe_for_growth_in_Mozambique.pdf (accessed 31 January 2016).

Smeets, E., A. Tabeau, S. van Berkum, J. Moorad, H. van Meijl, and G. Woltjer. 2014. The impact of the rebound effect of the use of first generation biofuels in the EU on greenhouse gas emissions: A critical review. *Renewable and Sustainable Energy Reviews* 38: 393–403.

Smil, V. 2002. Nitrogen and food production: Proteins for human diets. *AMBIO: A Journal of the Human Environment* 31 (2): 126–131.

Smith, W. L., and R. L. Naylor. 2014. Land institutions and food security in sub-Saharan Africa. In: *The evolving sphere of food security*, ed. R. Naylor, 202–238. New York, NY: Oxford University Press.

SNV, World Wide Fund for Nature, and HONDUPALMA. 2009. *Biofuel value chain development in an African oil palm plantation case: HONDUPALMA*. Tegucigalpa: SNV.

Soares-Filho, B., R. Rajão, M. Macedo, A. Carneiro, W. Costa, M. Coe, H. Rodrigues, and A. Alencar. 2014. Cracking Brazil's forest code. *Science* 344 (6182): 363–364.

Soh, A. C. 2012. Breeding and genetics of the oil palm. In: *Palm oil: Production, processing, characterization, and uses*, ed. O.- M. Lai, C.- P. Tan, and C. C. Akoh, 31–59. Urbana, IL: AOCS Press.

Soh, A. C., C. K. Wong, Y. W. Ho, and C. W. Choong. 2010. Oil palm. In: *Oil crops: Handbook of plant breeding*, vol. 4, ed. J. Vollmann, and I. Rajcan, 333–367. New York, NY: Springer.

Spera, S. A., A. S. Cohn, L. K. VanWey, J. F. Mustard, B. F. Rudorff, J. Risso, and M. Adami. 2014. Recent cropping frequency, expansion, and abandonment in Mato Grosso, Brazil had selective land characteristics. *Environmental Research Letters* 9 (6): 64010–64021.

Spiertz, H. 2013. Challenges for crop production research in improving land use, productivity and sustainability. *Sustainability* 5 (4): 1632–1644.

Srinavas, S., K. C. Bell, K. Toha, A. Zaenal, and W. Collier. 2015. A review of Indonesian land based sectors with particular reference to land governance and political economy. Paper presented at the annual World Bank Conference on Land and Poverty, Washington, DC, March 2015.

Srivastava, S. C., B. S. Gupta, T. S. Singh, and H. P. Singh. 2015. Economics of production and resource use efficiency of soybean production in India. *Economic Affairs* 60 (2): 347–355.

Stads, G.-J., Haryono, and S. Nurjayanti. 2007. Key trends in agricultural R&D investments in Indonesia. Agricultural Science and Technology country brief, IFPRI. http://www.asti.cgiar.org/pdf/IndonesiaCRBr.pdf (accessed 31 January 2016).

Stevenson, J., D. Byerlee, N. Villoria, T. Kelley, and M. Maredia. 2011. *Agricultural technology, global land-use and deforestation: A review.* Rome, Italy: Independent Science and Partnership Council.

Stokes, S., M. Lowe, and S. Zoubak. 2014. *Deforestation and the Brazilian beef value chain.* Durham: Datu Research.

Strassburg, B. B. N., A. E. Latawiec, L. G. Barioni, C. A. Nobre, V. P. da Silva, J. F. Valentim, M. Vianna, and E. D. Assad. 2014. When enough should be enough: Improving the use of current agricultural lands could meet production demands and spare natural habitats in Brazil. *Global Environmental Change* 28: 84–97.

Straumann, L. 2014. *Money logging: On the trail of the Asian timber mafia.* Basel: Bergli Books.

Sun, Y., N. Neelakantan, Y. Wu, R. Lote-Oke, A. Pan, and R. M. van Dam. 2015. Palm oil consumption increases LDL cholesterol compared with vegetables low in saturated fat in a meta-analysis of clinical trials. *Journal of Nutrition* 145 (7): 1549–1558.

Susanti, A., and S. Budidarsono. 2014. Land governance and oil palm development: Examples from Riau Province, Indonesia. In: *The global land grab: Beyond the hype*, ed. M. Kaag and A. Zoomers, 119–134. London: Zed Books.

Susila, W. R. 2004. Contribution of oil palm industry to economic growth and poverty alleviation in Indonesia. *Jurnal Litbang Pertanian* 23 (3): 107–114.

Sutton, K. 1989. Malaysia's FELDA land settlement model in time and space. *Geoforum* 20 (3): 339–354.

Tate, D. J. M. 1996. *The RGA history of the plantation industry in the Malay Peninsula.* New York, NY: Oxford University Press.

Tavernise, S. 2013. F.D.A. ruling would all but eliminate trans fats. *The New York Times*, November 7. http://www.nytimes.com/2013/11/08/health/fda-trans-fats.html. (Accessed 31 Jan 2016)

Tawa, M., S. Amameishi, T. Noguchi, and S. Tamura. Inclusive development with special consideration to small-scale farmers: Addressing land rights issues in the Nacala Corridor, northern

Mozambique. Paper presented at the World Bank Conference on Land and Poverty, Washington, DC, March 2015.

Technoserve. 2011. Southern Africa soy roadmap: Zambia value chain analysis. http://www.technoserve.org/files/downloads/technoserve-bmgf-zambia.pdf (accessed 31 January 2016).

Teoh, C. H. 2002. *The palm oil industry in Malaysia: From seed to frying pan*. Selangor: World Wildlife Fund Malaysia.

Teoh, C. H. 2013. Malaysian corporations as strategic players in Southeast Asia's palm oil industry. In: *The palm oil controversy in Southeast Asia: A transnational perspective*, ed. O. Pye and J. Bhattacharya, 19–47. Pasir Panjang: Institute of Southeast Asian Studies.

Terauchi, D., N. Imang, M. Nanang, M. Kawai, M. A. Sardjono, F. Pambudhi, and M. Inoue. 2014. Implication for designing a REDD+ program in a frontier of oil palm plantation development: Evidence in East Kalimantan, Indonesia. *Open Journal of Forestry* 4 (3): 259–277.

The Economist. 2011. Global livestock counts: Counting chickens. *The Economist*, July 27. http://www.economist.com/blogs/dailychart/2011/07/global-livestock-counts (accessed 31 January 2016).

The Economist. 2012. Peace, land and bread: The hard bargaining starts. November 24. http://www.economist.com/news/americas/21567087-hard-bargaining-starts-peace-land-and-bread (accessed 31 January 2016).

The Economist. 2013. Special report: Brazil. September 28. http://www.economist.com/sites/default/files/20130928_brazil.pdf (accessed 31 January 2016).

The Economist. 2015. Indonesia's forest fire haze. November 6. http://www.economist.com/blogs/graphicdetail/2015/11/daily-chart-3 (accessed 31 January 2016).

Thongrak, S., and S. Kiatpahtomchai. 2012. *Impact study of the project on sustainable palm oil production for bio-energy in Thailand*. Songkla: Faculty of Economics, Prince of Songkla University.

Times of India. 2015. Indonesia, Malaysia form OPEC-like palm oil council. November 21. http://timesofindia.indiatimes.com/business/international-business/Indonesia-Malaysia-form-OPEC-like-palm-oil-council/articleshow/49873481.cms (accessed December 20, 2015).

Timmer, C. P. 1981. Is there "curvature" in the Slutsky Matrix? *Review of Economics and Statistics* 63 (3): 395–402.

Timmer, C. P., W. P. Falcon, and S. R. Pearson. 1983. *Food policy analysis*. Baltimore, MD: Johns Hopkins University Press.

Tree Crop Estate Statistics of Indonesia. 2013. *2012–2014 Kelapa sawit (palm oil)*. Jakarta: Directorate General of Estate Crops.

Udo, R. K. 1965. Sixty years of plantation agriculture in southern Nigeria: 1902–1962. *Economic Geography* 41 (4): 356–368

United Nations. 1999. *The world at six billion*. New York, NY: United Nations Population Division.

United Nations. 2009. *World investment report: Transnational corporations, agricultural production and development*. Geneva: United Nations Conference on Trade and Development.

United Nations. 2014. *World urbanization prospects: 2014 Revision*. New York, NY: United Nations, Department of Economic and Social Affairs.

Unruh, J. D. 2009. Land rights in postwar Liberia: The volatile part of the peace process. *Land Use Policy* 26 (2): 425–433.

US Department of Agriculture. 2011. Malaysia: Obstacles may reduce future palm oil production growth. http://www.pecad.fas.usda.gov/highlights/2011/06/malaysia/ (accessed 31 January 2016).

US Department of Agriculture. 2012a. *Argentina: Biodiesel tax increase and temporary soybean import policy*. Washington, DC: US Department of Agriculture Foreign Agriculture Service.

US Department of Agriculture. 2012b. Commodity and food elasticities. US Department of Agriculture, Economic Research Service. http://www.ers.usda.gov/data-products/commodity-and-food-elasticities/demand-elasticities-from-literature.aspx#.U724IF5X_1o (accessed 31 January 2016).

US Department of Agriculture. 2013. Indonesia: Palm oil expansion unaffected by forest moratorium. http://www.pecad.fas.usda.gov/highlights/2013/06/indonesia/ (accessed 31 January 2016).

US Department of Agriculture. 2014. *Nigeria provides export market for oilseeds and products*. Washington, DC: US Department of Agriculture Foreign Agricultural Service.

US Department of Agriculture. 2015a. New Indonesian palm oil export levy. GAIN report no. ID1520. Washington, DC: US Department of Agriculture Foreign Agricultural Service.

US Department of Agriculture. 2015b. Oil crops yearbook. http://www.ers.usda.gov/data-products/oil-crops-yearbook.aspx (accessed June 23, 2015).

US Department of Agriculture. 2015c. Published GAIN reports. http://gain.fas.usda.gov/Recent%20GAIN%20Publications/Forms/AllItems.aspx (accessed June 23, 2015).

US Department of Agriculture. 2015d. Trade. http://www.ers.usda.gov/topics/crops/soybean-oil-crops/trade.aspx (accessed May 5, 2015).

US Energy Information Administration. 2007. *Biofuels in the US transportation sector*. Washington, DC: US Energy Information Administration.

US Energy Information Administration. n.d. a. International energy statistics database. http://www.eia.gov/cfapps/ipdbproject/IEDIndex3.cfm (accessed June 23, 2015).

US Energy Information Administration. n.d. b. Petroleum and other liquids. http://www.eia.gov/petroleum/data.cfm (accessed June 23, 2015).

US Environmental Protection Agency. 2015. EPA proposes renewable fuel standards for 2014, 2015, and 2016, and the biomass-based diesel volume for 2017. Report no. EPA-420-F-15-028. US EPA, Office of Transportation and Air Quality. http://www.epa.gov/otaq/fuels/renewablefuels/documents/420f15028.pdf (accessed 31 January 2016).

US Food and Drug Administration. 2015. FDA cuts *trans* fat in processed foods. Washington DC: FDA Consumer Health Information.

US Congressional Senate. 2014. *To amend the Internal Revenue Code of 1986 to Modify the incentives for the production of biodiesel*. 113th Congress. 2nd sess. S. 2021. Washington, DC: Government Printing Office.

Valdes, C. 2011a. Brazil's ethanol industry: Looking forward. http://www.ers.usda.gov/media/126865/bio02.pdf (accessed 31 January 2016).

Valdes, C. 2011b. Can Brazil meet the world's growing need for ethanol? *Amber Waves* 9 (4): 36–45.

VanWey, L. K., S. Spera, R. de Sa, D. Mahr, and J. F. Mustard. 2013. Socioeconomic development and agricultural intensification in Mato Grosso. *Philosophical Transactions of the Royal Society B: Biological Sciences* 368 (1619): 20120168.

Varkkey, H. 2012a. Patronage politics as a driver of economic regionalisation: The Indonesian oil palm sector and transboundary haze. *Asia Pacific Viewpoint* 53 (3): 314–329.

Varkkey, H. 2012b. The growth and prospects of the oil palm industry in Indonesia. *Oil Palm Industry Economic Journal* 12 (2): 1–13.

Väth, S., and S. Gobien. 2014. Life satisfaction, contract farming and property rights: Evidence from Ghana. Joint Discussion Paper Series in Economics no. 15-2014. Marburg: Phillips-University Marburg.

Väth, S. J., and M. Kirk. 2013. Do land ownership and contract farming matter? Evidence from a large-scale investment in Ghana. Joint Discussion Paper Series in Economics no. 16-2014. Marburg: Philipps-University Marburg.

Veloo, R. 2013. *Managing oil palm plantation amidst low CPO prices and rising costs of production.* Kuala Lumpur: Incorporated Society of Planters.

Venter, O., E. Meijaard, H. Possingham, R. Dennis, D. Sheil, S. Wich, L. Hovani, and K. Wilson. 2009. Carbon payments as a safeguard for threatened tropical mammals. *Conservation Letters* 2 (3): 123–129.

Vera-Diaz, M. del C., R. K. Kaufmann, D. C. Nepstad, and P. Schlesinger. 2008. An interdisciplinary model of soybean yield in the Amazon Basin: The climatic, edaphic, and economic determinants. *Ecological Economics* 65 (2): 420–431.

Vermeulen, S., and L. Cotula. 2010. *Making the most of agricultural investment: A survey of business models that provide opportunities for smallholders.* London: FAO and IIED.

Vermeulen, S., and N. Goad. 2006. *Towards better practice in smallholder palm oil production.* London: International Institute for Environment and Development.

Victoria Transportation Policy Institute. 2014. Transport elasticities. http://www.vtpi.org/tdm/tdm11.htm (accessed October 6, 2014).

Villela, A. A., B. J. D'Alembert, L. P. Rosa, and M. V. Freitas. 2014. Status and prospects of oil palm in the Brazilian Amazon. *Biomass and Bioenergy* 67: 270–278.

Villoria, N. B., A. Golub, D. Byerlee, and J. Stevenson. 2013. Will yield improvements on the forest frontier reduce greenhouse gas emissions? A global analysis of oil palm. *American Journal of Agricultural Economics* 95 (5): 1301–1308.

von Braun, J., and E. T. Kennedy, eds. 1994. *Agricultural commercialization, economic development, and nutrition.* Washington, DC: International Food Policy Research Institute.

Wahid, M. B., S. N. A. Abdullah, and I. E. Henson. 2004. Oil palm: Achievements and potential. In: *New directions for a new planet: Proceedings of the 4th International Crop Science Congress, Brisbane, Australia, September–October 2004.* http://www.cropscience.org.au/icsc2004/symposia/2/4/187_wahidmb.htm. (accessed 31 Jan 2016)

Wakker, E. 2014. *Indonesia: Illegalities in forest clearance for large-scale commercial plantations.* Washington, DC: Forest Trends.

Walker, T. S., J. Alwang, A. Alene, J. Ndjuenga, R. Labarta, Y. Yigezu, A. Diagne, R. Andrade, R. M. Andriatsitohaina, H. De Groote, and K. Mausch. 2015. Varietal adoption, outcomes and impact. In: *Crop improvement, adoption and impact of improved varieties in food crops in sub-Saharan Africa*, ed. T. S. Walker and J. Alwang, 388–405. Wallingford, UK: CABI.

Walsh, B. 2014. Don't blame fat: Sources for further reading on the healthy-diet debate. *Time Magazine.* June 23. http://time.com/2912207/info-about-saturated-fat-and-sugar/ (accessed 31 January 2016).

Wang, D., J. Huang, and B. Lohmar. 2015. Feed conversion ratio, profitability and farm size in China's pig industry. Paper presented at the 29th International Conference of Agricultural Economists, Milan, Italy, August 2015.

Ward, K. 2012. *The world in 2050: From the top 30 to the top 100.* London: HSBC.

Weinhold, D., E. Killick, and E. J. Reis. 2013. Soybeans, poverty and inequality in the Brazilian Amazon. *World Development* 52: 132–143.

Wicke, B., R. Sikkema, V. Dornburg, and A. Faaij. 2011. Exploring land-use changes and the role of palm oil production in Indonesia and Malaysia. *Land Use Policy* 28 (1): 193–206.

Wiggins, S., and S. Keats. 2014. *Rural wages in Asia*. London: Overseas Development Institute.

Wiggins, S., G. Henley. and S. Keats. 2015. *Competitive or complementary? Industrial crops and food security in sub-Saharan Africa*. London: Overseas Development Institute.

Wilcock, D., and V. Kelly. 2014. *The economics of non-industrial oil palm production and processing in Ghana*. Utrecht: Solidaridad.

Wilkinson, J. M. 2011. Re-defining efficiency of feed use by livestock. *Animal* 5 (7): 1014–1022.

Wilmar International. 2013. *Gaining strength in adversity*. Singapore: Wilmar International.

Wilson, C. 1954. *The history of Unilever: A study in economic growth and social change*. London: Cassell.

Wirsenius, S., C. Azar, and G. Berndes. 2010. How much land is needed for global food production under scenarios of dietary changes and livestock productivity increases in 2030? *Agricultural Systems* 103 (9): 621–638.

Wisner, R. 2013. Biofuels mandates outside the US. AgMRC Renewable Energy and Climate Change Newsletter. February. http://www.agmrc.org/renewable_energy/biofuelsbiorefining_general/biofuels-mandates-outside-the-us/ (accessed June 23, 2015).

World Bank. 2008a. Cameroon agricultural value chain competitiveness study. Report no. AAA25-CM. Washington, DC: World Bank.

World Bank. 2008b. *Insecurity of land tenure, land law and land registration in Liberia*. Washington, DC: World Bank.

World Bank. 2009. *Awakening Africa's sleeping giant: Prospects for commercial agriculture in the Guinea savannah zone and beyond*. Washington, DC: World Bank.

World Bank. 2012. *Environmental, economic and social impacts of oil palm in Indonesia: A synthesis of opportunities and challenges*. Jakarta: World Bank.

World Bank. 2015a. Data. http://data.worldbank.org (accessed May 5, 2015).

World Bank. 2015b. GDP deflator (base year varies by country). http://data.worldbank.org/indicator/NY.GDP.DEFL.ZS (accessed June 23, 2015).

World Bank. 2015c. Indicators. http://data.worldbank.org/indicator?display=graph. (accessed June 25, 2015).

World Bank Group. 2008. *Reducing emissions from deforestation and forest degradation in Indonesia*. Washington, DC: World Bank.

World Food Programme. 2007. *PDPE market analysis tools: Price and income elasticities*. Rome: World Food Programme.

World Wildlife Fund—Indonesia. 2013. *Palming off a national park: Tracking illegal palm oil fruit in Riau, Sumatra*. Jakarta: World Wildlife Fund—Indonesia.

Xu, Z., D. A. Hennessy, K. Sardana, and G. Moschini. 2013. The realized yield effect of genetically engineered crops: US maize and soybean. *Crop Science* 53: 735–745.

Yilmaz, K. 1999. Optimal export taxes in a multicountry framework. *Journal of Development Economics* 60 (2): 439–465.

You, L., U. Wood-Sichra, S. Fritz, Z. Guo, L. See, and J. Koo. 2014. "MAPSPAM": Spatial Production Allocation Model (SPAM) 2005, version 2.0. http://mapspam.info (accessed February 2, 2016).

Yui, S., and S. Yeh. 2013. Land-use change emissions from oil palm expansion in Pará, Brazil depend on proper policy enforcement on deforested lands. *Environmental Research Letters* 8 (4): 044031.

Zen, Z., C. Barlow, and R. Gondowarsito. 2006. Oil palm in Indonesian socio-economic improvement: A review of options. *Oil Palm Industry Economic Journal* 6: 18–29.

Zhai, F., S. Du, Z. Wang, J. Zhang, W. Du, and B. Popkin. 2014. Dynamics of the Chinese diet and the role of urbanicity, 1991–2011. *Obesity Reviews* 15 (01): 16–26.

INDEX

Note: Page numbers followed by *b*, *f*, or *t* denote boxes, figures, or tables, respectively. Numbers followed by n indicate notes.